高等学校系列教材

水污染控制工程（双语）

王淑勤　苏金波　冯亚娜　编

中国建筑工业出版社

图书在版编目（CIP）数据

水污染控制工程：汉文，英文/王淑勤，苏金波，冯亚娜编. —北京：中国建筑工业出版社，2023.1
高等学校系列教材
ISBN 978-7-112-27924-1

Ⅰ.①水… Ⅱ.①王… ②苏… ③冯… Ⅲ.①水污染-污染控制-高等学校-教材-英、汉 Ⅳ.①X520.6

中国版本图书馆 CIP 数据核字（2022）第 177660 号

本教材按照污水处理厂的处理先后顺序、处理工艺的难易程度、由浅入深、系统讲解，既有生活污水处理，又有工业废水处理，还配备大量图表、实物照片、工程设计案例，采用中英文混合编写方式，常规的前沿性专业知识，水处理方法、原理、分类采用英文编写，处理工艺流程、设备组成、工艺对比分析、设计计算案例、例题采用中文编写，便于基础差的学生深刻理解，课后复习；每章后面附有思考题和练习题，便于学生提前预习和课后复习，也有助于提高英文水平的同时，掌握专业知识和进行后续的课程设计。

本书不仅适合环境工程、环境科学、给排水科学与工程等专业的本科生作为教材，还可作为资源与环境、环境科学与工程等相近专业研究生的教材和参考书，同时也适用于相关科研人员作为参考用书。

为便于教学，作者特制作了与教材配套的电子课件，如有需求，可发邮件（标注书名、作者名）至 jckj@cabp.com.cn 索取，或到 http://edu.cabplink.com 下载，电话（010）58337285。

责任编辑：王美玲
文字编辑：勾淑婷
责任校对：姜小莲

高等学校系列教材
水污染控制工程（双语）
王淑勤　苏金波　冯亚娜　编

*

中国建筑工业出版社出版、发行（北京海淀三里河路 9 号）
各地新华书店、建筑书店经销
霸州市顺浩图文科技发展有限公司制版
北京圣夫亚美印刷有限公司印刷

*

开本：787 毫米×1092 毫米　1/16　印张：19　字数：461 千字
2023 年 9 月第一版　2023 年 9 月第一次印刷
定价：**48.00** 元（赠教师课件）
ISBN 978-7-112-27924-1
（40060）

前　　言

本教材共包括10章，第1章概述水质特点、水质指标、水污染处理方法的分类及选择原则；第2章介绍预处理和初级处理方法，主要是物理方法，包括沉砂、沉淀、混凝、过滤、气浮原理、流程、设备特点、设计计算；第3章介绍二级生物处理方法，主要包括好氧、厌氧方法，生物膜法、活性污泥法原理、流程、设备特点；第4章介绍好氧生物膜法流程、设备特点、设计计算；第5章介绍好氧活性污泥法流程及发展、设备特点、设计计算；第6章介绍厌氧法流程及发展、设备特点、设计计算；第7章介绍脱氮除磷方法，主要包括好氧-厌氧联合方法原理、流程、设备特点、设计计算；第8章介绍三级或高级处理，主要包括吸附、化学氧化、膜析法等工艺特点；第9章介绍污泥处理和处置方法、流程、设备特点、设计计算；第10章介绍水污染控制技术案例，包含流程选择、总体布置、设备特点，尤其增加了电力工业、煤化工业废水的处理案例，还包括膜生物反应器、三级废水处理工艺仿真模拟的运行结果。

教材按照污水处理厂的处理先后顺序、处理工艺的难易程度，由浅入深、系统讲解，既有生活污水处理，又有工业废水处理，还配备大量图表、实物照片、工程设计案例，采用中英文混合编写方式，常规的前沿性专业知识，水处理方法、原理、分类采用英文编写，处理工艺流程、设备组成、工艺对比分析、设计计算案例、例题采用中文编写，便于学生深刻理解，课后复习；每章后面附有思考题和练习题，便于学生提前预习和课后复习，也有助于在提高英文水平的同时，掌握专业知识和进行后续的课程设计。

本书不仅适合环境工程、环境科学、给排水科学与工程等专业的本科生作为教材，还可作为资源与环境、环境科学与工程等相近专业研究生的教材和参考书，同时也适用于相关科研人员作为参考用书。

本书是河北省教改项目（2019GJJG416）、教育部教改项目（E-HGZY20202003）的研究成果之一。

本书由王淑勤教授、苏金波讲师、冯亚娜讲师编，张玉玲、张敬红、郭天祥、陈岚副教授参编了部分章节和案例，陈司晗、周巧云、李晓雪、董剑鑫参加了资料收集和排版工作，肖惠宁教授主审，一并表示感谢。

目　　录

Chapter 1 Introduction

Water pollution is an imprecise term that reveals nothing about either the type of polluting material or its source. The way we deal with the waste depends upon whether the contaminants are oxygen demanding, algae promoting, infectious, toxic, or simply unsightly. Pollution of water resources can come directly from sewer outfalls or industrial discharges (point sources) or indirectly from air pollution or agricultural or urban runoff (nonpoint sources).

In considering effects on water quality, it is useful to distinguish *point sources* and *nonpoint sources* of pollution. Point sources are factories and other industrial and commercial installations that release hazardous substances into the water. In recent years, hazardous releases from point sources have been substantially reduced, especially in developed countries. Nonpoint sources have become a much harder problem. They include emissions from transport vehicles, agricultural runoff, which can carry excess nutrients, pesticides, and silt into streams and groundwaters, and urban runoff, which can carry toxic metals and organics through storm drains into sewage treatment plants or directly into rivers and lakes. Advances in controlling point sources of pollution have led to a greater focus on nonpoint sources, which account for an increasing fraction of the total pollutant load.

Throughout the vast land of China, many regions have been developing rapidly for years however large amounts of pollutants are dumped into water bodies from municipal, industrial, and agricultural sources either untreated or inadequately treated. Water pollution is the main cause of the increasingly serious ecological and environmental pollution in the industrial and mining areas, and the situation is "*terrible and shocking*". A report on municipal wastewater treatment by Professor Sheng Guangfan of the China Association of Environmental Protection Industry, in the late 1990s, noted that water pollution is still serious for historical and other reasons. Traditional pollutants (COD and BOD) have not been controlled, while pollution caused by toxic compounds and eutrophication increases. It is worth noting that more efforts were needed in the construction of municipal wastewater treatment plants while much attention had been paid to industrial wastewater with encouraging results.

A consultancy program carried out by the Chinese Academy of Engineering (CAE) was reported in early 2004. Heavily polluted areas included the mainstreams of the Yellow River, Weihe River and Shule River, and some sections of the Yili River where 21 cities and prefectures are located. The report stated that water quality in these areas stood at Grade V or lower, which can't meet the requirements of agricultural irrigation. Water in some sections had even turned black and stinking. The Weihe River valley had become one of the most heavily polluted areas in the country. The main pollutants were found to be COD (organic pollutants), BOD (biological pollutants), and ammonia nitrogen. Heavy metal contents were found to be excessive in the mainstream of the Yellow River flowing through Gansu Province and the Inner Mongolia Autonomous Region.

In China's coastal waters, a major concern is the occurrence of red tide (named from water discoloration caused by toxins) resulted from untreated industrial waste and domestic sewage that discharged into the sea. Coastal provinces have invested heavily in aquatic farming for the export market. Red tide can destroy fishing grounds and lose millions of revenue. Red tide is a harmful ecological phenomenon caused by the explosive growth or high concentration of microorganisms, and occurs naturally all over the world. Not all red tides are harmful. However, some sing-cell organisms produce a toxin that paralyzes and kills fishes, and they can also suffocate fishes by consuming nearly all the oxygen in the water. In April and May of each year, the frequent occurrence of red tides in the East China Sea is due to the increasing discharges of agricultural pollutants (containing nitrogen and phosphorus), domestic waste and industrial waste. Other attacks have occurred in the northern part of the Bohai Sea further and in the southern waters near Guangdong and Hainan.

Municipalities treat water supplies for both domestic and commercial uses to ensure that people are freed from disease and to eliminate odors and turbidity; they treat sewage to reduce water pollution and eutrophication.

Sewage treatment relies to a large extent on settling and screening to remove solids; this physical separation step is called primary treatment. Most municipalities also carry out secondary treatment that uses bacteria to metabolize organic compounds and converts them to CO_2.

This chapter is an introduction to wastewater engineering. The purpose of this chapter in the study of wastewater treatment is to understand the effluent quality standards required for both domestic and industrial wastewater.

The goal of wastewater treatment is to protect the quality of the receiving waters and this is achieved by wastewater plants designed to reduce the BOD_5, the TSS, N and P, and fecal coliforms. There are other goals regarding effluent quality that depend on the type of water body into which the effluent is discharged.

1.1 Sources and Characteristics of Wastewater

1.1.1 Undesirable Wastewater Characteristics

It is not possible to enumerate all types of wastes since many wastes are specific to particular industries. Eckenfelder (2001) cites the following among undesirable waste constituents:

(1) Soluble organics that cause depletion of dissolved oxygen.

(2) Suspended solids.

(3) Priority pollutants such as phenol and other organics discharged into industrial wastes which will cause tastes and odors in water and have carcinogen in some cases.

(4) Heavy metals, cyanide and toxic organics.

(5) Color and turbidity.

(6) Nitrogen and phosphorus.

(7) Refractory substances resistant to biodegradation.

(8) Oils and floating materials.

(9) Volatile materials.

(10) Aquatic toxicity.

Sewage sources are generally divided into domestic sewage, industrial wastewater and initial polluted rainwater. The water quality of sewage from different sources varies greatly (Table 1-1 and Table 1-2). Table 1-1 shows concentrations of typical major pollutants such as BOD_5 and suspended solids in industrial wastewater.

Concentrations of BOD_5 and suspended solids in industrial wastewater **Table 1-1**

Industry	BOD_5(mg/L)	Suspended solids(mg/L)
Ammunition	50～300	70～1700
Fermentation	4500	10000
Slaughterhouse(cattle)	400～2500	400～1000
Pulp and paper(kraft)	100～350	75～300
Tannery	700～7000	4000～20000

Typical compositions of major pollutants in industrial wastewater **Table 1-2**

Factory category	Main harmful substances in wastewater
Coking plant	Phenols, benzene, chloride, sulfide, tar, pyridine, ammonia, etc
Fertilizer plant	Phenols, benzene, cyanide, fluoride, copper, mercury, alkali, ammonia, etc
Synthetic ammonia plant	Volatile phenol, cyanide, ammonia nitrogen, sulfide, petroleum, etc
Chemical plant	Acid, alkali, cyanide, sulfide, mercury, lead, arsenic, benzene, naphthalene, nitro compounds, etc
Petrochemical Plant	Oil, acid, alkali, cyanide, sulfide, phenol, aromatics, pyridine, arsenic, etc
Synthetic rubber factory	Chloroprene, butadiene, benzene, xylene, styrene, etc
Shipyard	Chromium, zinc, copper, cadmium, nickel, cyanide, benzene, toluene, xylene, etc
Aerospace propellant plant	Formaldehyde, cyanide, aniline, methylhydrazine, dimethylhydrazine, triethylamine, diethylenetriamine, etc
Resin plant	Cresol, formaldehyde, styrene, vinyl chloride, mercury, etc
Chemical fiber plant	Carbon disulfide, amines, ketones, acrylonitrile, ethylene glycol, etc
Textile mill	Sulfide, cellulose, detergent, etc

Continued

Factory category	Main harmful substances in wastewater
Leather factory	Sulfide, alkali, chromium, formic acid, aldehyde, detergent, etc
Paper mill	Lignin, sulfide, alkali, cyanide, mercury, phenols, etc
Insecticide factory	Various pesticides, benzene, chloral, chlorobenzene, phosphorus, arsenic, fluorine, lead, acid, alkali, etc
Electroplating plant	Oxide, chromium, zinc, copper, cadmium, nickel, etc
Paint factory	Phenol, benzene, formaldehyde, lead, manganese, chromium, cobalt, etc
Steel works	Phenol, cyanide, pyridine, acid, etc
Nonferrous metal treatment plant	Cyanide, fluoride, boron, manganese, zinc, copper, cadmium, lead, germanium, other rare metals, etc
Coking plant	Phenols, benzene, chloride, sulfide, tar, pyridine, ammonia, etc

1.1.2 Domestic Wastewater Characteristics

Domestic wastewater is sewage that does not include rainwater runoff. Urban (or municipal——the traditional term) wastewater is defined as the combination of domestic wastewater and industrial wastewater with or without rainwater runoff. The words "urban" and "municipal" are synonymous in this text.

The main characteristics pollutants of wastewater are physical, chemical and microbiological pollutants, as shown in Table 1-3. Domestic wastewater is usually not as complex as industrial wastewater, in which specific toxic and hazardous compounds may exist, such as phenols and toxic organics.

Typical water quality of domestic wastewater **Table 1-3**

Water quality indexes	High	Medium	Low
Total soilds(mg/L)	1230	720	390
Suspended solids(mg/L)	400	210	120
BOD_5 (mg/L)	350	190	110
COD(mg/L)	800	430	250
Total nitrogen(Calculated as N)(mg/L)	70	40	20
Ammonia nitrogen(Calculated as N)(mg/L)	45	25	12
Total phosphorus(Calculated as P)(mg/L)	12	7	4
Chloride(mg/L)	90	50	30
Alkalinity(Calculated as $CaCO_3$, mg/L)	200	100	50
Grease(mg/L)	100	90	50
Volatile organic compounds(VOCs)(mg/L)	>400	100~400	<100
Total coliforms[Pcs. $(100mL)^{-1}$]	10^7~10^{10}	10^7~10^9	10^6~10^8
Cryptosporidium oocysts[Pcs. $(100mL)^{-1}$]	10^{-1}~10^2	10^{-1}~10^1	10^{-1}~10^0

1.2 Catalogue of Wastewater Treatment Processes

The main objective of conventional wastewater treatment processes is reduction of biochemical oxygen demand, suspended solids, and pathogenic organisms. In addition, it may be necessary to remove nutrients, toxic components, non-biodegradable compounds, and dissolved solids. Since most contaminants are present in low concentrations, the treatment processes must be able to work effectively with dilute streams.

1.2.1 Pretreatment and Primary Treatment

Pretreatment processes are used to screen out coarse solids, to reduce the size of solids, to separate floating oils and to equalize fluctuations in flow or concentration through short-term storage. Primary treatment usually refers to the removal of suspended solids by settling or floating.

Sedimentation is currently the most widely used in the primary treatment process. In the sedimentation unit, solid particles are allowed to settle to the bottom of a tank under quiescent conditions. Chemicals may be added during primary treatment to neutralize the stream or to improve the removal of small suspended solid particles. Primary reduction of solids reduces oxygen requirements in a subsequent biological step and also reduces the solids loading to the secondary sedimentation tank.

1.2.2 Secondary Treatment

Secondary treatment generally involves a biological process to remove organic matter through biochemical oxidation. The particular biological process selected depends on factors such as quantity of wastewater, biodegradability of waste, and availability of land. Activated sludge reactors and trickling filters are the most commonly used biological processes.

In the activated sludge or suspended growth process, wastewater is fed to an aerated tank where microorganisms consume organic wastes for maintenance and for generation of new cells. The resulting microbial floc (activated sludge) is settled in a sedimentation vessel called a clarifier or thickener. A portion of the thickened biomass (activated sludge) is usually recycled to the reactor (aerated tank) to improve performance through higher cell concentrations. The trickling filter is a kind of bed packed with rocks, plastic structures, or other media. Microbial films grow on the surface of the packing and remove soluble organics from the wastewater flowing over the packing. Excess biological growth can wash off the packing and is shown in the clarifier.

1.2.3 Tertiary Treatment

Many effluent standards require tertiary treatment or advanced wastewater treatment to remove particular contaminants or to prepare the water for reuse. Some common tertiary operations include: the removal of phosphorus compounds by coagulation with chemicals, the removal of nitrogen compounds by ammonia stripping with air or by nitrification-denitrification in biological reactors, the removal of residual organic and color compounds by adsorption on activated carbon, and the removal of dissolved solids by membrane processes (reverse osmosis and electrodialysis). The effluent water is often treated with chlorine or ozone to destroy pathogenic organisms before being discharged into the receiving waters.

Many treatment processes are used to purify water before it is discharged into the environment. A partial listing of these treatment processes and their purposes are given in Table 1-4. These treatment processes will be discussed briefly to show where they fit into an overall treatment plant.

Wastewater treatment processes and major purposes **Table 1-4**

Treatment processes	Purpose of operation
Bar screens and racks	Coarse solids removal
Comminutor	Grinding up of screenings
Grit chamber	Grit and sand removal
Skimmer and grease trap	Floating liquid and solid removal
Equalization tank	Smoothing out flow and concentration
Neutralization	Neutralizing acids and bases
Sedimentation and flotation	Suspended solids removal
Activated sludge reactor, trickling filter, aerated lagoon	Biological removal of soluble organics
Activated carbon adsorber	Soluble nonbiodegradable organics removal
Chemical coagulation	Precipitation of phosphates
Nitrification-denitrification	Biological removal of nitrates
Air stripping	Ammonia removal
Ion exchange	Charged species removal
Bed filtration	Fine solids removal
Reverse osmosis and electrodialysis	Dissolved solids removal
Chlorination and ozonation	Pathogenic organism destruction

Since many different combinations of these treatment processes operations are possible. Each situation must be evaluated to select the best combination.

The application of many wastewater treatment processes is related to both the characteristics of the waste and the degree of treatment required. Pretreatment or primary treatment is used for removal of floating and suspended solids and oils, neutralization and equalization, and to prepare the wastewater for subsequent treatment or discharge into the receiving water.

The selection of combination of wastewater treatment processes depends upon: the characteristics of the wastewater, required effluent quality, cost and availability of land

and future upgrading of water quality standards.

For any given wastewater treatment problem, several treatment combinations can produce the desired effluent. Only one of these alternatives, however, is the most cost-effective. Each process and its relevant design and criteria will be discussed in detail in subsequent chapters.

One example is the addition of powdered activated carbon (PAC) to the biological treatment process to adsorb organics that the microorganisms cannot degrade or slowly degrade; this is marketed as the PACT process. Another example is adding coagulants at the end of the biological treatment basin or tank to remove phosphorus and residual suspended solids.

1.3 Water Quality Indicators

Water quality refers to the comprehensive characteristics of water and its impurities, including physical, chemical and microbial properties. The quality of water quality usually needs to be characterized by three corresponding water quality indicators or indexes. In other words, the water quality indicators of sewage are usually divided into physical, chemical and biological water quality indicators.

Physical water quality indicators mainly include temperature, chroma, turbidity, transparency, odor and taste, solid content and conductivity, among which temperature, chroma, turbidity, odor and taste are called sensory physical indicators; Total solids (TS) refer to the total amount of all residues left after sewage evaporation at 103~105℃, including suspended solids (SS) and dissolved solids (DS). The content of soluble salts in water can be known according to the measured conductivity.

Sewage often contains chemical substances such as aerobic organic matter, plant nutrients, heavy metals, inorganic non-metallic compounds and toxic and harmful organic pollutants. The total content of chemical substances is characterized by general organic comprehensive indicators or indexes (including BOD, COD, TOC and TOD), and the indicators or indexes of plant nutrients are expressed by ammonia nitrogen, Kjeldahl nitrogen, nitrite, nitrate, total nitrogen and total phosphorus; Heavy metal indicators or indexes are expressed by the amount of mercury, cadmium, lead, nickel and chromium; Inorganic non-metallic compound indicators or indexes such as total arsenic, selenium, sulfide, cyanide and fluoride; and toxic and harmful organic pollutants indicators or indexes such as phenols, organophosphorus pesticides, organochlorine pesticides, organic dyes, organometallic compounds, benzene and polycyclic aromatic hydrocarbons are also used.

The biological water quality indicators or indexes of sewage mainly include the total number of bacteria, the number of coliforms, various pathogens and viruses, etc.

Although the interpretation of most waste characteristics is straightforward and definitive, special consideration must be given to the organic content. The organic content of the waste can be estimated by BOD (Biochemical Oxygen Demand), COD (Chemical Oxygen Demand), TOC (Total Organic Carbon), or TOD (Total Oxygen Demand).

1.3.1 Biochemical Oxygen Demand

By definition, BOD is the quantity of oxygen required for the stabilization of the oxidizable organic matter present in the water over 5d of incubation at 20℃.

BOD_5 test can be used to measure biodegradable organic carbon and, under certain conditions, the oxidizable nitrogen present in the waste. The current practice is to suppress nitrification, so only carbonaceous oxidation is recorded as $CBOD_5$.

1.3.2 Chemical Oxygen Demand

Among the many drawbacks of the BOD test, the most important is that it takes five days to run. If the organics are oxidized chemically instead of biologically, the test can be shortened considerably. Such oxidation is accomplished with the chemical oxygen demand (COD) test. Because nearly all organics are oxidized in the COD test and only some are decomposed during the BOD test, COD values are always higher than BOD values. One example is wood pulping wastes in which compounds such as cellulose are easily chemically oxidized (high COD), but are very slow to decompose biologically (low BOD).

Potassium dichromate is generally used as an oxidizing agent. It is an inexpensive compound which is available in a very pure form. A known amount of this compound is added to a measured amount of the sample and the mixture is then boiled. The reaction in unbalanced form is

$$\underset{\text{(organic)}}{C_xH_yO_z} + \underset{\text{(dichromate)}}{Cr_2O_7^{\ 2-}} + H^+ \xrightarrow{\triangle} CO_2 + H_2O + Cr^{3+}$$

After boiling with acid, the excess dichromate (not used for oxidizing) is measured by adding a reducing agent (usually ferrous ammonium sulfate). The difference between the chromate originally added and the chromate remaining is the chromate used to oxidize the organics. The more chromate used, the more organics in the sample, and hence the higher the COD.

COD test measures the total organic carbon with the exception of certain aromatics, such as benzene, which is not completely oxidized in the reaction. COD test is oxidation-reduction reaction, so other reduced substances, such as sulfides, sulfites, and ferrous iron, will also be oxidized and reported as COD. NH_3-N will not be oxidized in the COD test.

1.3.3 Total Organic Carbon

Since the ultimate oxidation product of organic carbon is CO_2, the total combustion of a sample will yield some significant information about the amount of organic carbon present in the wastewater sample. Without elaboration, this is done by allowing a little of the sample to be burned in a combustion tube and measuring the amount of CO_2 emitted. This test is not widely used at present, mainly because of the expensive instrumentation required. It has significant advantages over the BOD and COD tests, however, it will undoubtedly be more widely used in the future.

1.3.4 Turbidity

Turbidity is caused by many substances. In the treatment of water for drinking purposes, turbidity is of great importance firstly because of the aesthetic considerations, and secondly because pathogenic organisms can be hidden on (or in) the tiny colloidal particles.

In recent years, electronic devices that measure light scatter or transmittance have been developed.

1.3.5 Color and Odor

Color and odor are both important measurements in water treatment. Along with turbidity, they are called the physical parameters of drinking water quality. From an aesthetic point of view, color and odor are very important. If the water looks colored or smells bad, people will instinctively avoid using it, even though it might be perfectly safe from the public health aspect. Both color and odor can be caused by organic substances such as algae or humic compounds.

Color is measured by comparison with standards. Colored water made with potassium chloroplatinate when tinted with cobalt chloride closely resembles the color of many natural waters. Where multicolored industrial wastes are involved, such color measurement is meaningless.

Odor is measured by successive dilutions of sample with odor-free water until the odor is no longer detectable. This test is obviously subjective and depends entirely on the olfactory senses of the tester.

1.3.6 pH

pH of a solution is a measure of concentration of hydrogen ions. The abundance of hydrogen ions makes the solution acidic, while the dearth of H^+ ions makes it basic or alkaline. A basic solution has, instead, an abundance of hydroxide ions OH^-.

The measurement of pH is now almost universally by electronic means. Electrodes that are sensitive to hydrogen ion concentration (strictly speaking, hydrogen ion activity) convert the signal to electrical current.

pH is important in almost all phases of water and wastewater treatment. Aquatic organisms are sensitive to pH changes, and biological treatment requires either pH control or pH monitoring. In water treatment, pH is important to ensure proper chemical treatment as well as in disinfection and corrosion control. Mine drainage often involves the formation of sulfuric acid (high H^+ concentration) that is extremely detrimental to aquatic life.

1.4 污水排放标准

污水排放标准根据控制形式可分为浓度标准和总量控制标准，根据地域管理权限可分为国家排放标准、行业排放标准、地方排放标准。

浓度标准规定了排出口排放污染物的浓度限值，其单位一般为“mg/L”。我国现有的国家标准和地方标准基本上都是浓度标准。浓度标准对每个污染指标都有一个明确的执行标准，这样管理方便，但其并未考虑排放量的大小以及接受水体的环境容量、性状和要求等。当污染总量超过接纳水体的环境容量时，水体的环境质量就不能得到保证。此外，排污企业可以通过稀释的方法降低污水排放浓度，使之达到浓度标准，但其反而造成水资源浪费和水环境污染。

总量控制标准是以通过水质模型法计算出的水体环境容量为依据而设定的，其可保证水体的质量，但对管理技术要求高，需要与排污许可证制度结合进行总量控制。我国已经开始推行总量控制标准，《污水排入城镇下水道水质标准》GB/T 31962—2015 提到在有条件的城市可根据标准采用总量控制。

国家排放标准按照污水排放去向，规定了水污染物最高允许排放浓度，适用于排污单位水污染物的排放管理，以及建设项目的环境影响评价，建设项目环境保护设施设计、竣工验收及其投产后的排放管理。我国现行的国家排放标准主要有《污水综合排放标准》GB 8978—1996，《城镇污水处理厂污染物排放标准》GB 18918—2002，《污水排入城镇下水道水质标准》GB/T 31962—2015 等。根据国家排放标准，工业污水第一类及第二类污染物最高允许排放浓度分别见表 1-5 和表 1-6，污水排入城镇下水道水质控制项目限制见表 1-7。

同时针对部分行业排放废水的特点和治理技术发展水平，我国制定了部分行业水污染物排放标准。根据地方社会经济发展水平和管辖地水体污染控制需要，省、自治区、直辖市等可以制定地方污水排放标准，但应注意地方污水排放标准不能减少污染物控制指标数，相反地方污水排放标准可以增加污染物控制指标数，不能降低污染物排放标准的要求，可以提高污染物排放标准的要求。

此外，近年来根据水质回用目的不同，我国还颁发了一系列回用水水质标准，主要有《城市污水再生利用分类》GB/T 18919—2002、《城市污水再生利用 城市杂用水水质》GB/T 18920—2020、《城市污水再生利用 景观环境用水水质》GB/T 18921—2019、《农田灌溉水质标准》GB 5084—2021、《城市污水再生利用 工业用水水质》GB/T 19923—2005 等。

工业污水第一类污染物最高允许排放浓度（单位：mg/L） **表 1-5**

序号	污 染 物	最高允许排放浓度
1	总汞	0.05
2	烷基汞	不得检出
3	总镉	0.1
4	总铬	1.5

续表

序号	污染物	最高允许排放浓度
5	六价铬	0.5
6	总砷	0.5
7	总铅	1.0
8	总镍	1.0
9	苯并(a)芘	0.00003
10	总铍(按 Be 计)	0.005
11	总银(按 Ag 计)	0.5
12	总 α 放射性	1Bq/L
13	总 β 放射性	10Bq/L

注：摘自《污水综合排放标准》GB 8978—1996。

工业污水第二类污染物最高允许排放浓度（部分）（单位：mg/L）　　表 1-6

序号	污染物	适用范围	一级标准	二级标准	三级标准
1	pH	一切排污单位	6～9	6～9	6～9
2	色度(稀释倍数)	一切排污单位	50	80	—
3	悬浮物(SS)	采矿、选矿选煤工业	70	300	—
		脉金选矿	70	400	—
		边远地区砂金选矿	70	800	—
		城镇二级污水处理厂	20	30	—
		其他排污单位	70	150	400
4	5d 生化需氧量(BOD_5)	甘蔗制糖、苎麻脱胶、湿法纤维板、染料、洗毛工业	20	60	600
		甜菜制糖、酒精、味精、皮革化纤浆粕工业	20	100	600
		城镇二级污水处理厂	20	30	—
		其他排污单位	20	30	300
5	化学需氧量(COD)	甜菜制糖、合成脂肪酸、湿法纤维板、染料、洗毛、有机磷农药工业	100	200	1000
		味精、酒精、医药原料药、生物制药、苎麻脱胶、皮革、化纤浆粕工业	100	300	1000
		石油化工工业(包括石油炼制)	60	120	500
		城镇二级污水处理厂	60	120	—
		其他排污单位	100	150	500
6	石油类	一切排污单位	5	10	20
7	动植物油	一切排污单位	10	15	100
8	挥发酚	一切排污单位	0.5	0.5	2.0
9	总氰化合物	一切排污单位	0.5	0.5	1.0

续表

序号	污染物	适用范围	一级标准	二级标准	三级标准
10	硫化物	一切排污单位	1.0	1.0	1.0
11	氨氮	医药原料药、染料、石油化工工业	15	50	—
		其他排污单位	15	25	—
12	氟化物	黄磷工业	10	15	20
		低氟地区(水体氟含量<0.5mg/L)	10	20	30
		其他排污单位	10	10	20
13	磷酸盐(以P计)	一切排污单位	0.5	1.0	—
14	甲醛	一切排污单位	1.0	2.0	5.0
15	苯胺类	一切排污单位	1.0	2.0	5.0
16	硝基苯类	一切排污单位	2.0	3.0	5.0
17	阴离子表面活性剂(LAS)	一切排污单位	5.0	10	20
18	总铜	一切排污单位	0.5	1.0	2.0
19	总锌	一切排污单位	2.0	5.0	5.0
20	总锰	合成脂肪酸工业	2.0	5.0	5.0
		其他排污单位	2.0	2.0	5.0
21	彩色显影剂	电影洗片	1.0	2.0	3.0
22	显影剂及氧化物总量	电影洗片	3.0	3.0	6.0
23	元素磷	一切排污单位	0.1	0.1	0.3
24	有机磷农药(以P计)	一切排污单位	不得检出	0.5	0.5

注：适用于1998年1月1日后建设的单位，摘自《污水综合排放标准》GB 8978—1996。

污水排入城镇下水道水质控制项目限制（部分） **表 1-7**

序号	控制项目名称	单位	A 等级	B 等级	C 等级
1	水温	℃	40	40	40
2	色度	倍	64	64	64
3	pH	—	6.5~9.5	6.5~9.5	6.5~9.5
4	悬浮物	mg/L	400	400	250
5	溶解性总固体	mg/L	1500	2000	2000
6	石油类	mg/L	15	15	10
7	5d生化需氧量(BOD_5)	mg/L	350	350	150
8	化学需氧量(COD)	mg/L	500	500	300
9	氨氮(以N计)	mg/L	45	45	25
10	总氮(以N计)	mg/L	70	70	45
11	总磷(以P计)	mg/L	8	8	5

续表

序号	控制项目名称	单位	A 等级	B 等级	C 等级
12	总氰化物	mg/L	0.5	0.5	0.5
13	总余氯(以 Cl_2 计)	mg/L	8	8	8
14	硫化物	mg/L	1	1	1
15	氟化物	mg/L	20	20	20
16	氯化物	mg/L	500	800	800
17	硫酸盐	mg/L	400	600	600
18	苯胺类	mg/L	5	5	2
19	硝基苯类	mg/L	5	5	3
20	挥发酚	mg/L	1	1	0.5
21	阴离子表面活性剂(LAS)	mg/L	20	20	10
22	总汞	mg/L	0.005	0.005	0.005
23	总镉	mg/L	0.05	0.05	0.05
24	总铬	mg/L	1.5	1.5	1.5
25	六价铬	mg/L	0.5	0.5	0.5
26	总砷	mg/L	0.3	0.3	0.3
27	总铅	mg/L	0.5	0.5	0.5
28	总镍	mg/L	1	1	1
29	总硒	mg/L	0.5	0.5	0.5
30	总锌	mg/L	5	5	5
31	总锰	mg/L	2	5	5
32	总铁	mg/L	5	10	10
33	甲醛	mg/L	5	5	2
34	三氯甲烷	mg/L	1	1	0.6
35	五氯酚	mg/L	5	5	5

注：摘自《污水排入城镇下水道水质标准》GB/T 31962—2015。

1.5 水污染处理设备及分类

在国家环境保护局颁布的标准——《环境保护设备分类与命名》HJ/T 11—1996 中，按环保设备的原理和用途划分，水污染治理设备可分为物理法、化学法、物理化学法、生物法处理设备及组合式水处理设备等亚类别；按环保设备的功能原理划分，每个亚类别又可分为不同组别，例如物理法处理设备的亚类可进一步分为沉淀装置、澄清装置、气浮分离装置、离心分离装置、过滤装置、压滤和吸滤装置等；按环保设备的结构特征和工作方式划分，每个组别又可分为不同型别，例如气浮分离装置可划分为溶气气浮装置、真空气浮装置、分散空气气浮装置及电解气浮装置等型别。常用的水污染处理设备分类见表 1-8。

水污染处理设备分类简表 **表 1-8**

亚类别	组别
物理法处理设备	沉淀装置 澄清装置 上浮分离装置 气浮分离装置 离心分离装置 磁分离装置 筛滤装置 过滤装置 微孔过滤装置 压滤和吸滤装置 蒸发装置
化学法处理设备	酸碱中和装置 氧化还原和消毒装置 混凝装置
物理化学法处理设备	萃取装置 汽提和吹脱装置 吸附装置 离子交换装置 膜分离装置
生物法处理设备	好氧处理装置 供氧曝气装置 厌氧装置 厌氧-好氧处理装置

本章关键词（Keywords）

来源	Source
特点	Characteristics
水质指标	Water quality indicators
色度	Chroma
浊度	Turbidity
总固体量	Total solid volume（TS）
悬浮物	Suspended solids（SS）
溶解性固体	Dissolved solids（DS）
氨氮	Ammonia nitrogen
凯氏氮	Kjeldahl nitrogen
亚硝酸盐	Nitrite
硝酸盐	Nitrate
总氮	Total nitrogen
总磷	Total phosphorus
大肠菌群	Coliforms
处理工艺	Treatment process
调节池	Equalization tank

粉碎机	Comminutor
沉砂池	Grit chamber
中和	Neutralization
沉降	Sedimentation
曝气塘	Aerated lagoon
混凝	Coagulation
硝化	Nitrification
反硝化	Denitrification
离子交换	Ion exchange
反渗透	Reverse osmosis
氯化	Chlorination
臭氧化	Ozonation
三级处理	Tertiary treatment

思考题

1. 简述水质污染指标在水体污染控制中的作用，举例说明“一个水质指标可能包括几种污染物，而一种污染物可能对应几种水质指标”。
2. 分析物理性指标 TS、DS、SS、VSS 互相之间的关系，画出关系图。
3. 表征污水有机污染特性的指标有哪些？分析它们之间的区别和联系。
4. 简述排污标准、水环境质量标准和环境容量的关系。
5. 分析我国工业污水排放标准里第一类污染物和第二类污染物的划分理由。
6. 分析导致水体黑臭的原因和具体成分。
7. 为何要控制排水的 pH 和温度？
8. 污水的处理程度有哪几个？分别适用于什么场合？

Chapter2　Pretreatment and Primary Treatment

The objective of pretreatment and primary treatment is to make wastewater suitable for discharge to a POTW (publicly owned treatment work) or a subsequent biological treatment plant. Common pretreatment technologies are as follows.

By definition, pretreatment refers to the process or processes that preparing wastewater to be further processed in conventional secondary treatment biological processes. In municipal wastewater, it means removing floating debris and oily scums. These pollutants can inhibit biological processes and possibly damage mechanical equipments. There may be occasions when municipal wastewater (if also taking industrial effluents) may have a pH either too acidic or too alkaline to optimum biological degradation and may thus require pH correction. This can be achieved by adding of sulphuric acid (H_2SO_4) or lime. When the flow rate is inconsistent (e. g. five days a week of industrial effluent), it may also be desirable to provide flow balance in a storage tank. The balance or equalization tank may also be used to balance the organic loading if that varies substantially. If the wastewater is deficient in nutrients, essential for biological treatment, nutrients may be added during the pretreatment stage. Pretreatment of municipal wastewater is normally only physical, i. e. flow balancing, screenings removal, and grit or oily scum removal. Industrial wastewater may additionally require chemical pretreatment in the form of air stripping (ammonia removal), oxidation, reduction (heavy metal precipitation), and air flotation (oil removal).

2.1　Pretreatment Technologies

2.1.1　Equalization

The objective of equalization is to minimize or control fluctuations in wastewater characteristics in order to provide optimal conditions for subsequent treatment processes. The size and type of the equalization tank or basin vary with the amount of waste and the variability of the wastewater stream. The tank or basin should be of a sufficient size to adequately absorb waste fluctuations caused by variations in plant's production scheduling and to inhibit concentrated batches periodically dumped or spilled into the sewer.

The specific purposes of equalization for industrial treatment facilities are:

(1) To provide adequate suppression of organic fluctuations in order to prevent shock loading of biological systems. The effluent concentration from a biological treatment plant will be proportional to the influent concentration. If the wastewater is readily degradable,

an increase in the influent will result in a less increase in the effluent due to an increase in bio-metabolism. By contrast, if the influent contains bio-inhibitors, it will result in an increase in the effluent concentration.

(2) To provide adequate pH control or to minimize the chemical requirements required for neutralization.

(3) To minimize flow fluctuations to physical-chemical treatment systems and permit chemical feed rates to be compatible with feeding equipment.

(4) To provide continuous feed to biological systems over periods when the manufacturing plant is not operating.

(5) To provide capacity for controlling discharge of wastes to municipal systems in order to distribute waste loads more evenly.

(6) To prevent high concentrations of toxic substances from entering biological treatment plants.

Mixing is usually provided to ensure adequate equalization and to prevent settleable solids from depositing in the basin. In addition, oxidation of reduced compounds in the waste stream or reduction of BOD by air stripping can be achieved by mixing and aeration. Methods that have used for mixing include: distribution of inlet flow and baffling; turbine mixing; diffused air aeration; mechanical aeration and submerged mixers.

The application of flow equalization in wastewater treatment is illustrated by the two flow diagrams shown in Figure 2-1. In the in-line arrangement (Figure 2-1a), all flow passes through the equalization tank or basin. This arrangement can be used to achieve a considerable amount of constituent concentration and flowrate damping. In the off-line arrangement (Figure 2-1b), only the flow above some predetermined flow limit is diverted into the equalization tank or basin. Although the pumping requirements are minimized in this arrangement, the amount of constituent concentration damping is considerably reduced. Off-line equalization is sometimes used to capture the "first-flush" from combined collection systems. Flow equalization can be applied after grit removal, after primary sedimentation, and after secondary treatment where advanced treatment is used.

2.1.2 Neutralization

Many industrial wastes contain acidic or alkaline substances that require neutralization prior to discharge to receiving waters or prior to chemical or biological treatment. For biological treatment, the pH of the biological system should be maintained between 6.5 and 8.5 to ensure optimal biological activity. Therefore, the biological process itself provides neutralization and a buffering capacity as a result of pre-neutralization required which depends on the ratio of BOD removed and the causticity or acidity present in the waste. These requirements are discussed in Chap. 3.

1. Types of Processes

(1) Mixing Acidic and Alkaline Waste-stream

This process requires sufficient equalization capacity to achieve the desired neutralization.

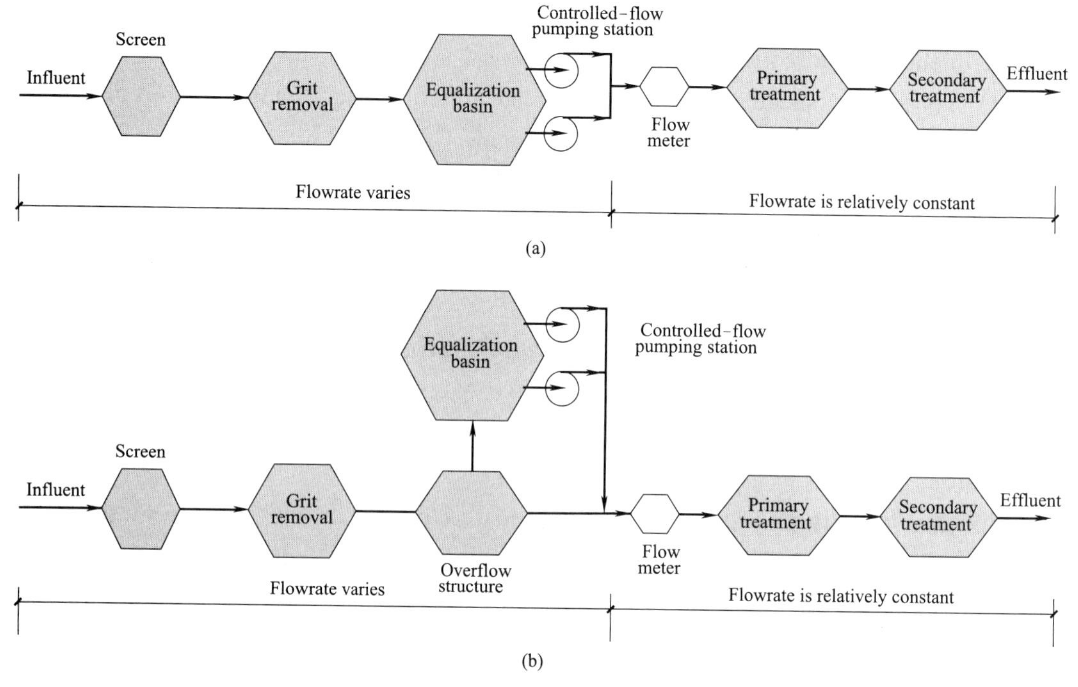

Figure 2-1 Typical flow diagram incorporating flow equalization in a wastewater treatment plant

(a) in-line equalization; (b) off-line equalization

(2) Acid Waste Neutralization through Limestone Beds

The schematic of neutralization equipment structure is shown in Figure 2-2.

(3) Mixing Acid Waste with Lime Slurry

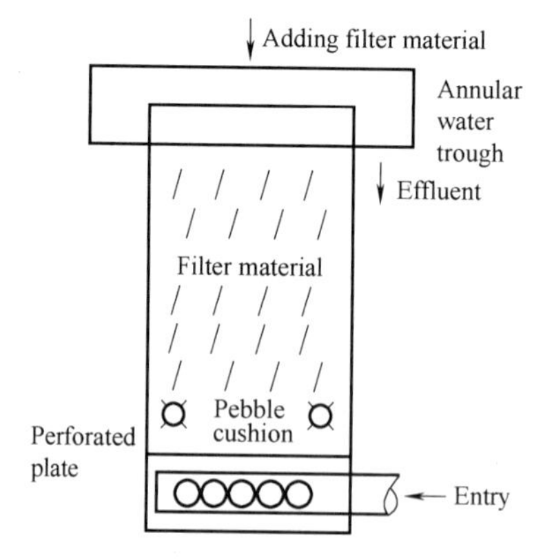

Figure 2-2 Schematic of neutralization equipment structure

2. Control of Processes

Automatic control of the pH of waste-stream is one of the most troublesome problems for the following reasons:

(1) The relation between pH and neutralization, particularly when close to neutral

(pH 7.0). The nature of the titration curve favors multi-staging in order to ensure close control of the pH.

(2) The influent pH can vary at a rate as fast as 1 pH unit per minute.

(3) The waste-stream flow rates can double in a few minutes.

(4) A relatively small amount of reagent must be thoroughly mixed with a large volume of liquid in a short time interval.

2.1.3 Screens or Screening

二维码 1 粗格栅运行视频

The objective of screens is to remove large floating materials (e.g. rags, plastic bottles, etc.) to protect downstream mechanical equipments (pumps). There are four types of screens or screenings that are normally used:

(1) Coarse screens with openings greater than 6mm that remove large materials.

(2) Fine screens with openings in the range 1.5 to 6mm, which are sometimes used as a substitute for primary clarification (e.g. when activated sludge is used). The structure of fine screen is shown in Figure 2-3.

(3) Very fine screens with openings in the range 0.2 to 1.5mm, which reduce the SS to primary clarification levels.

(4) Micro screens with openings in the range 0.001 to 0.3mm, which are used for effluent polishing as a final treatment step. These processes are not used in pre-treatment except as a single one step treatment process for predominantly inorganic wastewater, e.g. quarry washings.

Figure 2-3 Structure of fine screen

2.1.4 Grit Channels or Chambers

Grit is inorganic sand or gravel particles of approximately 1mm in size which are washed into sewer collection systems from roads and pavements. Grit is usually not present in industrial process wastewater, but is part of municipal systems where the collection systems combine foul water and stormwater. The grit is removed (after screenings) because it can abrade mechanical equipment and also because it can settle out in the biological treatment plant to reduce its space efficiency.

Two common types of grit collection devices are: helical flow aerated grit chambers and horizontal flow grit channels. The more traditional grit chamber is the horizontal flow type. The design principle is based on a minimum through-flow velocity of about 0.3m/s and a detention time of about 1min. They are typically designed to remove grit particles larger than 0.15mm with associated settling velocities of about 0.01m/s.

Aerated grit chambers have proven to be more efficient than the horizontal flow type and the grit tends to be "cleaner". The aerated grit chamber is shown in Figure 2-4.

Figure 2-4 Aerated grit chamber

2.1.5 Heavy Metals Removal Principle

There are a number of technologies available for the removal of heavy metals from a wastewater. Chemical precipitation is most commonly employed for most of the metals. Common precipitants include OH^-, CO_3^{2-} and S^{2-}. Metals are precipitated as the hydroxide through the addition of lime or caustic to a pH of minimum solubility. However, several of these compounds are amphoteric and exhibit a point of minimum solubility. The pH of minimum solubility varies with the metal in question. Metals can also be precipitated as the sulfide or in some cases as the carbonate.

In treating industrial wastewater containing metals, it is frequently necessary to pretreat the wastewater to remove substances that will interfere with the precipitation of the metals. Cyanide and ammonia can form complexes with many metals that limit the removal that can be achieved by precipitation for the case of ammonia. Cyanide can be removed by alkaline chlorination or other processes such as catalytic oxidation of carbon. Cyanide wastewater containing nickel or silver is difficult to treat by alkaline chlorination because of the slow reaction rate of these metal complexes. Ferrocyanide [$Fe(CN)_6^{4-}$] is oxidized to ferricyanide [$Fe(CN)_6^{3-}$], which resists further oxidation. Ammonia can be removed by stripping, break-point chlorination, or other suitable methods prior to the removal of metals.

For many metals, such as arsenic and cadmium, co-precipitation with iron or aluminum is highly effective for removal to low residual levels. In this case, the metal adsorbs to the alum or iron floc. In order to meet low effluent requirements, it may be necessary in some cases to provide filtration to remove floc carried over from the precipitation process. With precipitation and clarification alone, effluent metals concentrations may be as high as 1 to 2mg/L. Filtration can reduce these concentrations to 0.5mg/L or less (Table 2-1). Carbonate salts can be employed for enhanced precipitation. Because of the chemical cost, this type of precipitation is usually employed as a polishing step following conventional precipitation.

Summary of trivalent chromium treatment results **Table 2-1**

Method	pH	Chromlum(mg/L)	
		Initial	Final
Precipitation	7～8	140	1.0
	7.8～8.2	16.0	0.06～0.15
	8.5	47～52	0.3～1.5
	8.8	650	18
	8.5～10.5	26.0	0.44～0.86
	8.8～10.1	—	0.6～30
	12.2	650	0.3
Precipitation with sand filtration	8.5	7400	1.3～4.6
	8.5	7400	0.3～1.3
	9.8～10.0	49.4	0.17
	9.8～10.0	49.4	0.05

Metals can be removed by adsorption on activated carbon, aluminum oxides, silica, clays, and synthetic material such as zeolites and resins. In the case of adsorption, higher pH favors the adsorption of cations while a lower pH favors the adsorption of anions. Complexing agents will interfere with cationic species. There will be competition from major background ions such as calcium or sodium. For chromium waste treatment, hexavalent chromium must first be reduced to the trivalent state, Cr^{3+}, and then precipitated with lime. This is referred to as the process of reduction and precipitation.

1. Arsenic

Arsenic and arsenical compounds are present in wastewater from the metallurgical industry, glassware and ceramic production, tannery operation, dyestuff manufacture, pesticide manufacture, some organic and inorganic chemicals manufacture, petroleum refining, and the rare-earth industry. Arsenic is removed from wastewater by chemical precipitation. Enhanced performance is achieved as arsenate (AsO_4^{3-}, As^{5+}) rather than arsenite (AsO^+, As^{3+}). Arsenite is therefore usually oxidized to arsenate prior to precipitation. Effluent arsenic levels of 0.05mg/g are obtainable by precipitation of the arsenic as the sulfide by the addition of sodium or hydrogen sulfide at pH of 6.0 to 7.0. In order to meet reported effluent levels, polishing of the effluent by filtration would usually be required.

Arsenic present in low concentrations can also be reduced by filtration through activated carbon. Effluent concentrations of 0.06mg/L arsenic have been reported from an initial concentration of 0.2mg/L. Arsenic is removed by co-precipitation with a ferric hydroxide floc that ties up the arsenic and removes it from solution. Effluent concentrations of less than 0.005mg/L have been reported from this process.

2. Cadmium

Cadmium is present in wastewater from metallurgical alloying, ceramics, electroplating, photography, pigment works, textile printing, chemical industries, and lead mine drainage. Cadmium is removed from wastewater by precipitation or ion exchange. In some cases, electrolytic and evaporative recovery processes can be employed, provided the wastewater is in a concentrated form. Cadmium forms an insoluble and highly stable hydroxide at an alkaline pH. Cadmium in solution is approximately 1mg/L at pH 8.0 and 0.05mg/L at pH 10.0 to 11.0. Co-precipitation with iron hydroxide at pH 6.5 will reduce cadmium to 0.008mg/L; iron hydroxide at pH 8.5 reduces cadmium to 0.05mg/L. Sulfide and lime precipitation with filtration will yield 0.002 to 0.03mg/L at pH 8.5 to 10.0. Cadmium is not precipitated in the presence of complexing ions, such as cyanide. In these cases, it is necessary to pretreat the wastewater to destroy the complexing agent. In the case of cyanide, cyanide destruction is necessary prior to cadmium precipitation. A hydrogen peroxide oxidation precipitation system has been developed that simultaneously oxidizes cyanides and forms the oxide of cadmium, thereby yielding cadmium, whose recovery is feasible.

3. Chromium

The reducing agents commonly used for chromium waste are ferrous sulfate, sodium meta-bisulfite, or sulfur dioxide (SO_2). Ferrous sulfate and sodium meta-bisulfite may be dry-or solution-fed; SO_2 is diffused into the system directly from gas cylinders. Since the reduction of chromium is most effective at acidic pH values, a reducing 2 agent with acidic properties is desirable. When ferrous sulfate is used as the reducing agent, the Fe^{2+} is oxidized to Fe^{3+}; if meta-bisulfite or SO_2 is used, the negative radical SO^{3-} is converted to SO_4^{2-}. The general reactions are:

$$Cr^{6+} + Fe^{2+} \text{ or } SO_2 \text{ or } Na_2S_2O_5 + H^+ \rightarrow Cr^{3+} + Fe^{3+} \text{ or } SO_4^{2-} \quad (2\text{-}1)$$

$$Cr^{3+} + 3OH^- \rightarrow Cr(OH)_3 \quad (2\text{-}2)$$

Ferrous ion reacts with hexavalent chromium in an oxidation-reduction reaction, reducing the chromium to a trivalent state and oxidizing the ferrous ion to the ferric state. This reaction occurs rapidly at pH levels or values below 3.0. The acidic properties of ferrous sulfate are low at high dilution; acid must therefore be added for pH adjustment. The use of ferrous sulfate as a reducing agent has the disadvantage that a contaminating sludge of $Fe(OH)_3$ is formed when an alkali is added. In order to obtain a complete reaction, an excess dosage of 2.5 times than the theoretical addition of ferrous sulfate must be used.

Reduction of chromium can also be accomplished by using either sodium meta-bisulfite or SO_2. In either case, reduction occurs by reaction with the H_2SO_3 produced in the reaction.

Above pH 4.0, only 1% of sulfite is present as H_2SO_3 and the reaction is very slow. During this reaction, acid is required to neutralize the NaOH formed. The reaction is highly dependent on both pH and temperature. At pH levels below 2.0, the reaction is practically instantaneous and close to theoretical requirements.

At pH levels above 3.0, when a basic chrome sulfate is produced, the quantities of lime required for subsequent neutralization are reduced. At pH 8.0 to 9.9, $Cr(OH)_3$ is virtually insoluble. Experimental investigations have shown that the sludge produced will compact to 1% to 2% by weight.

The acid requirements for the reduction of Cr^{6+} depend on the acidity of the original waste, the pH of the reduction reaction, and the type of reducing agent used. Since it is difficult, if not impossible, to predict these requirements, it is usually necessary to titrate a sample to the desired pH endpoint with standardized acid.

In cases where the chrome content of the rinse-water varies markedly, equalization should be provided before the reduction tank to minimize fluctuations in the chemical feed system. The fluctuation in chrome content can be minimized by provision of a drain station before the rinse tanks.

Successful operation of a continuous chrome reduction process requires instrumentation and automatic control. pH and redox control are provided for the reduction tank. The addition of lime should be modulated by a second pH control system.

4. Lead

Lead is present in wastewater from storage-battery manufacture. Lead is generally removed from wastewater by precipitation as the carbonate, $PbCO_3$, or the hydroxide, $Pb(OH)_2$. Lead is effectively precipitated as the carbonate by the addition of soda ash, resulting in effluent-dissolved lead concentrations of 0.01 to 0.03mg/L at a pH of 9.0 to 9.5. Precipitation with lime at pH 11.5 resulted in effluent concentrations of 0.02 to 0.2mg/L. Precipitation as the sulfide to 0.01mg/L can be accomplished with sodium sulfide at a pH of 7.5 to 8.5.

5. Mercury

Mercury is used as a catalyst in the chemical and petrochemical industry. Mercury is also found in most laboratory wastewater. Power generation is a large source of mercury release into the environment through the combustion of fossil fuel. When scrubber devices are installed on thermal power plant stacks for sulfur dioxide removal, accumulation of mercury is possible if extensive recycle is practiced. Mercury can be removed from wastewater by precipitation, ion exchange, and adsorption. Mercury ions can be reduced upon contact with other metals such as copper, zinc, or aluminum. In most cases, mercury recovery can be achieved by distillation. For precipitation, mercury compounds must be oxidized to the mercuric ions. Table 2-2 shows effluent levels or concentrations achievable by candidate technology.

6. Nickel

Wastewater containing nickel originates from metal-processing industries, steel foundries, motor vehicle and aircraft industries, printing; and in some cases, chemicals

Mercury removal, effluent levels　　Table 2-2

Technology	Effluent(μg/L)	Technology	Effluent(μg/L)
Sulfide precipitation	10～20	Influent	—
Alum Co-precipitation	1～10	High	20
Iron Co-precipitation	0.5～5	Moderate	2
Ion exchange	1～5	Low	0.25
Carbon adsorption			

industries. In the presence of complexing agents such as cyanide, nickel may exist in a soluble complex form. The presence of nickel cyanide complexes interferes with both cyanide and nickel treatment. Nickel forms insoluble nickel hydroxide upon the addition of lime, resulting in a minimum solubility of 0.12mg/L at pH 10.0 to 11.0. Nickel hydroxide precipitates have poor settling properties. Nickel can also be precipitated as the carbonate or the sulfate associated with recovery systems. In practice, lime addition (pH 11.5) may be expected to yield residual nickel concentrations in the order of 0.15mg/L after sedimentation and filtration. Recovery of nickel can be accomplished by ion exchange or evaporative recovery, provided the nickel concentrations in the wastewater are at a sufficiently high level.

2.2 Primary Treatment Technologies

2.2.1 Sedimentation

Sedimentation, the oldest and most widely used from water and wastewater treatment, uses gravity settling to remove particles from water. It is relatively simple and inexpensive and can be implemented in basins that are round, square, or rectangular. As noted earlier, sedimentation may follow coagulation and flocculation (for highly turbid water) or be omitted entirely (with moderately turbid water). Particulates suspended in surface water can range in size from 10^{-7} to 10^{-1}mm in diameter, the size of fine sand and small clay particles, respectively. Turbidity or cloudiness in water is caused by those particles larger than 10^{-4}mm, while particles smaller than 10^{-4}mm contribute to the water's color and taste.

Sedimentation is employed for the removal of suspended solids from wastewater. The process can be considered in three basic classification, depending on the nature of the solids present in the suspension: discrete, flocculent, and zone settling.

Discrete Settling: A particle will settle when the impelling force of gravity exceeds the inertia and viscous forces.

Flocculent Settling: Flocculent settling occurs when the settling velocity of the particle increases as it settles through the tank depth, because of coalescence with other particles. This increases the settling rate, yielding a curvilinear settling path.

Zone Settling: Zone settling is characterized by activated sludge and flocculated chemical suspensions when the concentration of solids exceeds approximately 500mg/L. The floc particles adhere together and the mass settle as a blanket, forming a distinct interface between the floc and the supernatant.

2.2.2 Coagulation

Coagulation is used to remove waste materials in suspended or colloidal form. Colloids are particles within the size range of 1nm (10^{-7}cm) to 0.1nm (10^{-8}cm). These particles do not settle out on standing and cannot be removed by conventional physical treatment processes.

Colloids present in wastewater can be either hydrophobic or hydrophilic. Hydrophobic colloids (clays, etc.) have no affinity for liquid medium and lack stability in the presence of electrolytes. They are readily susceptible to coagulation. Hydrophilic colloids, such as proteins, have a marked affinity for water. The absorbed water retards flocculation and frequently requires special treatment to achieve effective coagulation.

1. Zeta Potential

Since the stability of the colloid is primarily due to electrostatic forces, neutralization of this charge is necessary to induce flocculation and precipitation. Although it is not possible to measure the psi potential, the zeta potential can be determined, and hence the magnitude of the charge and the resulting degree of stability can be determined as well. The zeta potential is defined as:

$$\xi=\frac{4\pi\eta v}{\varepsilon \chi}=\frac{4\pi EM}{\varepsilon} \tag{2-3}$$

where v——particle velocity;

ε——dielectric constant of the medium;

η——viscosity of the medium;

χ——applied potential per unit length of cell;

EM——electrophoretic mobility.

The zeta potential is lowered by: change in the concentration of the potential determining ions; addition of ions of opposite charge or contraction of the diffuse part of the double layer by increasing in the ion concentration in solution.

2. Mechanism of Coagulation

Colloids have electrical properties that generate repelling forces and prevent agglomeration and settling. Stabilizing ions are strongly absorbed into an inner fixed layer that provides a particle charge that varies with the valence and number of adsorbed ions. Ions of all opposite charge form a diffuse outer layer that is held near the surface by electrostatic forces. The psi potential (ψ) is defined as the potential drop between the interface of the colloid and the body of solution. The zeta potential (ζ) is the potential drop between the slipping plane and the body of solution and is related to the particle charge and the thickness of the double layer. The thickness of the double layer (χ) is inversely proportional to the concentration and valence of nonspecific electrolytes. Van der Waals attractive force is effective in close proximity to the colloidal particle.

The addition of high-valence cations depresses the particle charge and the effective distance of the double layer, thereby reducing the zeta potential. As the coagulant dissolves, the cations serve to neutralize the negative charge on the colloids. This occurs before visible floc formation, and rapid mixing which "coats" the colloid is effective in this phase. Microflocs are then formed which retain a positive charge in the acid range because of the adsorption of H^+. These microflocs also serve to neutralize and coat the colloidal particle. Flocculation agglomerates the colloids with hydrous oxide floc. In this phase, surface adsorption is also active. The initially unadsorbed colloids are removed by enmeshment in the flocs.

If necessary, alkalinity should first be added. Alum or ferric salts should be also added next; they coat the colloid with Al^{3+} or Fe^{3+} and positively charged microflocs. Coagulant aids, such as activated silica and/or polyelectrolyte for floc buildup and zeta potential control, are finally added. After adding the alkali and coagulant, a rapid mixing of 1 to 3min is recommended, and then flocculate with addition of coagulant aids for 20 to 30min. Destabilization can also be accomplished by the addition of cationic polymers, which can bring the system to the isoelectric point without changing the pH. Although the effectiveness of polymers is 10 to 15 times that of alum, they are considerably more expensive as coagulants.

3. Coagulant

The most popular coagulant in waste-treatment application is aluminum sulfate, or alum $[Al_2(SO_4)_3 \cdot 18H_2O]$, which can be obtained in either solid or liquid form. When alum is added to water in the presence of alkalinity, the reaction is:

$$Al_2(SO_4)_3 \cdot 18H_2O + 3Ca(OH)_2 \rightarrow 3CaSO_4 + 2Al(OH)_3 + 18H_2O \tag{2-4}$$

The aluminum hydroxide is actually of the chemical form $Al_2O_3 \cdot xH_2O$ and is amphoteric, it can act as either an acid or a base.

The alum floc has the least solubility at a pH of approximately 7.0. The floc charge is positive below pH 7.6 and negative above pH 8.2. Between these limits the floc charge is mixed. Ferric salts are also commonly used as coagulants but have the disadvantage of being more difficult to handle. An insoluble hydrous ferric oxide is produced over a pH range of 3.0 to 13.0:

$$Fe^{3+} + 3OH^- \rightarrow Fe(OH)_3 \tag{2-5}$$

$$[Fe^{3+}][OH^-]^3 = 10^{-36} \tag{2-6}$$

The floc charge is positive in the acid range and negative in the alkaline range, with mixed charges ranging from 6.5 to 8.0 in the pH range.

4. Coagulant Aids

The addition of some chemicals will enhance coagulation by promoting the growth of large, rapid-settling flocs. Activated silica is a short-chain polymer used to bond microfine aluminum hydrate particles together. At high dosage, silica inhibits floc formation due to its electronegative properties. The usual dosage is 5 to 10mg/L.

Polyelectrolytes are high-molecular-weight polymers that contain adsorbable groups and form bridges between particles or charged flocs. Large flocs (0.3 to 1mm) are thus created when small dosage of polyelectrolyte (1 to 5mg/L) is added in conjunction with alum or ferric chloride. The polyelectrolyte is substantially unaffected by pH and can serve as a coagulant itself by reducing the effective charge on the colloid. There are three types of polyelectrolytes: a cation, which adsorbs on a negative colloid or floc particle; an anion, that replaces anionic groups on the colloidal particle and permits hydrogen bonding between the colloid and the polymer; and a nonion, which adsorbs and flocculates by hydrogen bonding between the solid surfaces and the polar groups in the polymer.

5. Coagulation Equipment

There are two types of basin equipment adaptable to the flocculation and coagulation of industrial wastes. The conventional system uses a rapid-mixing tank, followed by a flocculation tank containing longitudinal paddles which provide slow mixing. The flocculated mixture is then settled in conventional settling tanks. The structure is shown in Figure 2-5.

A sludge-blanket unit combines mixing, flocculating, and settling in one unit. Although colloidal destabilization might be less effective than in the conventional system, there are distinct advantages in recycling preformed floc. With lime and a few other coagulants, the time required to form a settleable floc is a function of the time necessary for calcium carbonate or other calcium precipitates to form nuclei on which other calcium materials can deposit and grow large enough to settle. It is possible to reduce both coagulant dosage and floc formation time by seeding the influent wastewater with previously formed nuclei or by recycling portion of the precipitated sludge. Recycling preformed floc usually reduces chemical dosage, the blanket itself is a filter for improved effluent clarity, and denser sludges are frequently attainable. The structure of paddle mixing tank is shown in Figure 2-5 and hydraulic circulating clarifier is shown in Figure 2-6.

2.2.3 Coagulation of Industrial Wastes

Coagulation may be used for the clarification of industrial wastes containing colloidal and suspended solids. Paperboard waste can be effectively coagulated with low dosage of alum. Silica or polyelectrolyte will aid in the formation of a rapid settling floc.

Waste containing emulsified oil can be clarified by coagulation. An emulsion can consist of droplets of oil in water. The oil droplets are of approximately 10^{-5}cm and are stabilized by adsorbed ions. Emulsifying agents include soaps, an anion-active agent. The emulsion can be broken by "salting out" with the addition of salts, such as $CaCl_2$. Flocculation will then effect charge neutralization and entrainment, resulting in clarification. An emulsion can also frequently be broken by lowering of the pH of the waste solution.

The presence of anionic surface agents in a waste will increase the coagulant dosage.

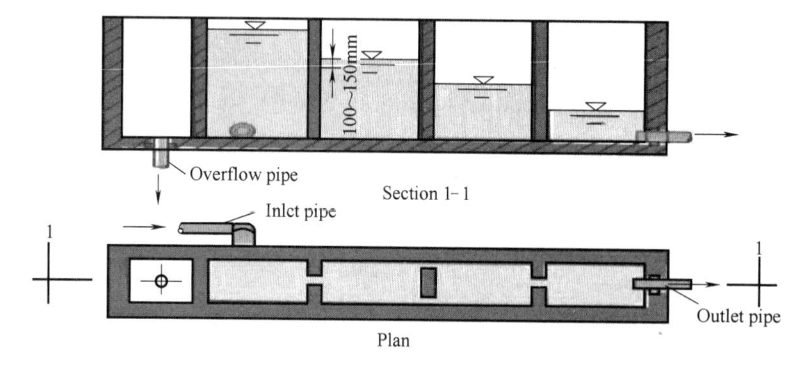

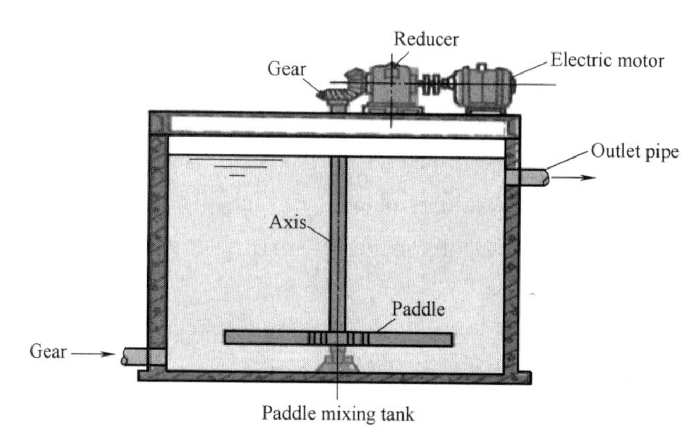

Figure 2-5 Structure of paddle mixing tank

Figure 2-6 Hydraulic circulating clarifier

The polar head of the surfactant molecule enters the double layer and stabilizes the negative colloids. Industrial laundry wastes have been treated with H_2SO_4 followed by lime and alum; this has resulted in a reduction of COD and a reduction of suspended solids.

The coagulation results obtained for the clarification of industrial wastes are summarized in Table 2-3.

2.2.4 Filtration

Filtration is widely used for removing particles from water. Filtration can be defined as any process for the removal of solid particles from a suspension (a two-phase system containing particles in a fluid) by passage of the suspension through a porous medium.

Coagulation of textile wastewater **Table 2-3 (a)**

Plant	Coagulant	Dosage (mg/L)	pH	Color*		COD	
				Influent	Removal (%)	Influent (mg/L)	Removal (%)
1	$Fe_2(SO_4)_3$	250	7.5~11.0	0.25	90	584	33
	Alum	300	5.0~9.0		86		39
	Lime	1200			68		30
2	$Fe_2(SO_4)_3$	500	3.0~4.0, 9.0~11.0	0.74	89	840	49
	Alum	500	8.5~10.0		89		40
	Lime	2000			65		40
3	$Fe_2(SO_4)_3$	250	9.5~11.0	1.84	95	825	38
	Alum	250	6.0~9.0		95		31
	Lime	600			78		50
4	$Fe_2(SO_4)_3$	1000	9.0~11.0	4.60	87	1570	3
	Alum	750	5.0~6.0		89		44
	Lime	2500			87		44

* Color sum of absorbances at wavelengths of 450, 550, and 650nm.

Color removal from pulp and paper mill effluent **Table 2-3 (b)**

Plant	Coagulant	Dosage (mg/L)	pH	Color		COD	
				Influent	Removal (%)	Influent (mg/L)	Removal (%)
1	$Fe_2(SO_4)_3$	500	3.5~4.5	2250	92	776	60
	Alum	400	4.0~5.0		92		53
	Lime	1500	—		92		38
2	$Fe_2(SO_4)_3$	275	3.5~4.5	1470	91	480	53
	Alum	250	4.0~5.5		93		48
	Lime	1000	—		85		45
3	$Fe_2(SO_4)_3$	250	4.5~5.5	940	85	468	53
	Alum	250	5.0~6.5		91		44
	Lime	1000	—		85		40

Coagulation of tannery wastewater **Table 2-3 (c)**

Parameter	Influent(mg/L)	Effluent(mg/L)	Removal(%)
COD	7800	2900	63
BOD	3500	1450	58
SO_4	1800	1200	33
Chromium	100	3	97

Granular media filtration is employed for the removal of suspended solids as a pretreatment for low suspended solids wastewater, following coagulation in physical-chemical treatment or as tertiary treatment following a biological wastewater treatment process.

Suspended solids are removed on the surface of a filter by straining, and through the depth of a filter by both straining and adsorption. Adsorption is related to the zeta potential on the suspended solids and the filter media. Particles normally encountered in a wastewater vary in size and particle charge and some will pass the filter continuously. The efficiency of the filtration process is therefore a function of:

(1) The concentration and characteristics of the solids in suspension;

(2) The characteristics of the filter medium and other filtration aids;

(3) The method of filter operation.

Granular media filters may be either gravity or pressure. Gravity filters may be operated at a constant rate with influent flow control and flow splitting, or at a declining rate with four or more units fed through a common header. To achieve constant flow, an artificial head loss (flow regulator) is used. As suspended solids are removed and head loss increases, the artificial head loss is reduced so the total head loss remains constant. In a declining-rate filter design, the decrease in flow rate through one filter as the head loss increases raises the head and rate through other filters. A maximum filtration rate of 0.24 $m^3/(min \cdot m^2)$ is used when one unit is out of service. The filter run terminates when the total head loss reaches the available driving force or when excess suspended solids or turbidity appear in the effluent.

Medium size is an important consideration in filter design. The sand size is chosen on the basis that it provides slightly better removal than is required. In dual-media filters, the coal size is selected to provide 75% to 90% suspended solids removal across 0.46 to 0.6m of media. For example, if 90% suspended solids removal is desired across a filter bed, 68% to 80% should be removed through the coal layer and the remaining 10% to 25% through the sand layer. If the feed suspended solids particle size is large than 5% of the granular medium particles, mechanical straining will occur.

A 2μm particle will be mechanically strained by a 0.5mm filter media. If the feed solids particles have a density of 2 to 3 times that of the suspending medium, then particles as small as 0.5% of the filter media particle size can be effectively by in-depth granular medium filtration.

The fine media are usually found in propriety-type filters such as the automatic backwash filter or the pulse-bed-type filter and rely on a straining mechanism for removal. Frequent backwash (or pulsing) is required, especially during plant upset and/or high influent turbidity conditions. The coarse media are usually much deeper than large media requiring scour for cost-effective backwash. Operation of coarse medium filters is characterized by longer filter runs and the ability to respond to plant upset conditions. Dual media and multimedia have traditionally been used in potable water applications and their use has carried over into the tertiary filtration application wastewater treatment.

Filtration rate will affect the buildup of head loss and the effluent quality attainable. The optimum filtration rate is defined as the filtration rate that results in the maximum volume of filtrate per unit filter area while achieving an acceptable effluent quality.

Too high filtration rate will permit solids to penetrate the coarse media and accumulate on the fine media. Too low filtration rate is insufficient to achieve good solids penetration of the coarse media, resulting in head loss buildup at the top of the coarse media. Filtration rate will also effect influent quality, depending on the nature of the particles to be removed.

Improved suspended solids removals can be achieved by the additional of coagulant to the wastewater prior to filtration. The use of alum also results in the precipitation and removal of phosphors through the filter. Flocculation is not needed since the filter serves as a flocculator. Since the suspended solids are removed by filtration rather than by sedimentation, 25% to 50% less chemicals are required in many cases. For most applications, a maximum of 100mg/L suspended solids removal is used in order to avoid excessive backwash volumes.

In granular filtration, the porous medium is a thick bed of granular material such as sand. The most common granular filtration technology in water treatment is *rapid filtration*. The term is used to distinguish it from *slow sand filtration*, an older filtration technology with a filtration rate 50 to 100 times lower than rapid filtration. Rapid filtration has several features that allow it to operate at rates up to 100 times greater than slow sand filtration. Key features of rapid filtration include granular media sieved for greater uniformity, coagulation pretreatment, back washing to remove accumulated particles, and a reliance on *depth filtration* as the primary particle removal mechanism. The first feature is that the filter material in rapid filters is processed to a fairly uniform size. Media uniformity allows the filters to operate at a higher hydraulic loading rate with lower head loss but results in a filter bed with void spaces significantly larger than the particles being filtered. As a result, straining is not the dominant removal mechanism. Instead, particles are removed when they adhere to the filter grains or previously deposited particles. Particles are removed throughout the entire depth of the filter bed by a process called *depth filtration*, which gives the filter a high capacity for solids retention without clogging rapidly. In depth filtration, particles accumulate throughout the depth of the filter bed by colliding with and adhering to the media. Captured particles can be many times smaller than the pore spaces in the bed. Rapid filtration uses mechanical and hydraulic systems to efficiently remove collected solids from the bed.

Rapid filters are also classified by the number of layers of filter materials. The common filter materials are sand, anthracite, granular activated carbon (GAC), garnet, and ilmenite. Some are used alone, and others are used only in combination with other media. The structure of filter is shown in Figure 2-7.

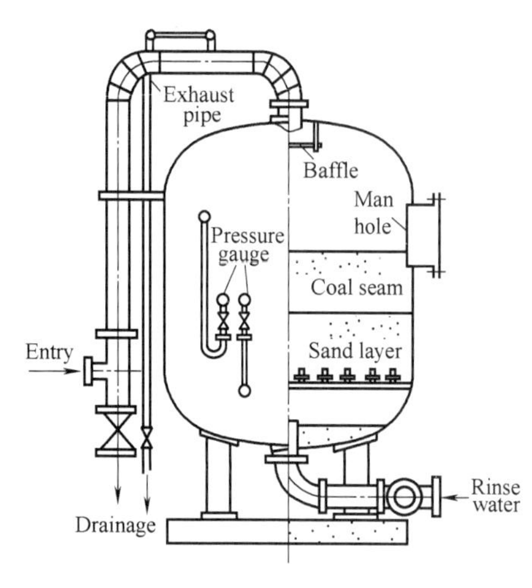

Figure 2-7 Structure of filter

By using a filtration process similar to that used in water treatment plants, it is possible to remove the residual suspended solids including the unsettled microorganisms. Removing the microorganisms also reduces the residual BOD_5. Conventional sand filters that are identical to those used in water treatment can be used, but they often clog quickly, thus requiring frequent backwashing. To lengthen filter runs and reduce backwashing, it is desirable to have the larger filter grain sizes at the top of the filter. This arrangement allows some of the larger particles of biological floc to be trapped at the surface without plugging the filter. Multimedia filters accomplish this by using low-density coal for the large grain sizes, medium-density sand for intermediate sizes, and high-density garnet for the smallest size filter grains. Thus, during backwashing, the greater density offsets the smaller diameter so that the coal remains on top, the sand remains in the middle, and the garnet remains on the bottom.

Typically, plain filtration can reduce activated sludge effluent suspended solids from 25 to 10mg/L. Plain filtration is not as effective on trickling filter effluents because these effluents contain more dispersed growth. However, the use of coagulation and sedimentation followed by filtration can yield suspended solids concentrations that are virtually zero. Typically, filtration can achieve 80% suspended solids reduction for activated sludge effluent and 70% reduction for trickling filter effluents.

2.2.5 Flotation

1. Oil Separtion Principle

In an oil separator, free oil is floated to the surface of a tank and then skimmed off. The same conditions holding for the subsidence of particles apply, except that the lighter-than-water oil globules rise through the liquid. The design of gravity separators as specified by the American Petroleum Institute is based on the removal of all free oil globules larger than 0.015cm. The Reynolds number is less than 0.5, so Stokes' law applies. A design procedure considering short circuiting and turbulence has been developed.

Several types of filtration devices have proved effective in removing free and emulsified oils from refinery-petrochemical wastewater. These vary from filters with sand media to those containing special media which exhibit a specific affinity for oil. One type is an up flow unit using a graded silica medium as the filtering and coalescing section. Even the small particles and globules are separated and retained on the medium. The oil particles, which flow upward by gravity differential and fluid flow, rise through the coalescent medium and through the water phase from which the oil is separated and collected near the top of the separator. The bed is regenerated by introducing wash water at a rapid rate and evacuating the solids and remaining oil. This filtration and coalescing process is often enhanced by using of polymer resin media. The primary application of these units is for selected in-plant streams which are dirt-free. Another application would be ballast water treatment following phase separation.

Emulsified oily materials require special treatment to break the emulsion so that the oily materials will be free and can be separated by gravity, coagulation, or air flotation. The breaking of emulsion is a complex art and may require laboratory or pilot-scale investigations prior to developing a final process design.

Emulsified can be broken by a variety of techniques. Quick-breaking detergents form unstable emulsion which break in 5 to 60min to 95% to 98% completion. Emulsion can be broken by acidification, the addition of alum or iron salts, or the use of emulsion-breaking polymers. The disadvantage of alum or iron is the large quantities of sludge generated.

2. Flotation Principle

Flotation is used for the removal of suspended solids and oil and grease from wastewater and for the separation and concentration of sludge. The waste flow or a portion of clarified effluent is pressurized 3. 4 to 4. 8atm in the presence of sufficient air to approach saturation. When this pressurized air-liquid mixture is released to atmospheric pressure in the flotation unit, minute air bubbles are released from solution. The sludge flocs, suspended solids, or oil globules are floated by these minute air bubbles, which attach themselves to and become enmeshed in the floc particles. The air-solids mixture rises to the surface, where it is skimmed off. The clarified liquid is removed from the bottom of the flotation unit; at this time a portion of the effluent may be recycled back to the pressure chamber. When flocculent sludge is to be clarified, pressurized recycle will usually yield a superior effluent quality since the flocs are not subjected to shearing stresses through the pumps and pressurizing system.

The performance of a flotation system depends on having sufficient air bubbles present to float substantially all of the suspended solids. An insufficient quantity of air will result in only partial flotation of the solids, and excessive air will yield no improvement.

The flow diagram is illustrated in Figure 2-8. Table 2-4 shows the types and characteristics of the air flotation method.

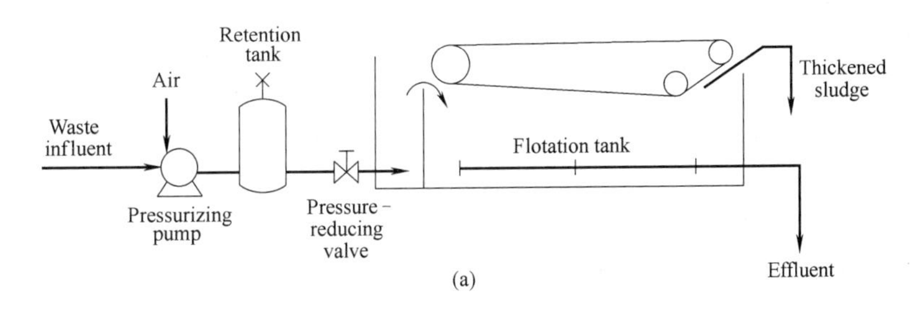

Figure 2-8 Schematic representation of flotation systems

(a) Flotation system without recirculation

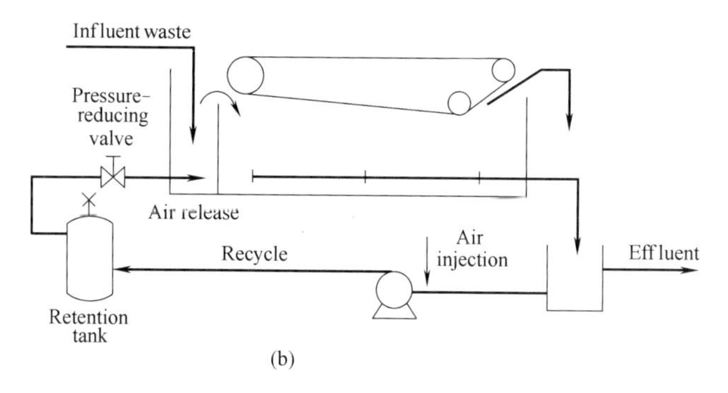

Figure 2-8 Schematic representation of flotation systems

(b) Flotation system with recirculation

Air flotation treatment of oily wastewater **Table 2-4**

Wastewater	Coagulant(mg/L)	Oil concentration(mg/L)		
		Influent	Effluent	Removal(%)
Refinery	0	125	35	72
	100 alum	100	10	90
	130 alum	580	68	88
	0	170	52	70
Oil tanker ballast water	100 alum+1mg/L polymer	133	15	89
Paint manufacture	150 alum+1mg/L polymer	1900	0	100
Aircraft maintenance	30 alum+10mg/L activated	250～700	20～50	>90
Meat packing	silica	3830	270	93
		4360	170	96

2.3 初级处理设备及运行管理

2.3.1 沉砂池

除了传统的平流式沉砂池和曝气沉砂池（图 2-9）之外，近年还出现了旋流沉砂池（也称涡流沉砂池）。其专指污水由池下部呈旋转方向流入，从池上部四周溢流而出，利用机械力控制水流流态与流速、加速砂粒的沉淀，并使有机物随水流带走的沉砂装置。这类沉砂池结构紧凑，占地面积小，土建费用低，维护管理较方便，对中小型污水处理厂具有较好的适用性。目前应用较多的有英国 Jones & Attwod 公司的钟式（Jeta）沉砂池（图 2-10）和美国 Smith & Loveless 公司的佩斯塔（Pista）沉砂池等，可根据设计流量直接选型。

污水由流入口切线方向流入沉砂区，旋转的涡轮叶片使砂粒呈螺旋形流动，促进有机物和砂粒的分离，由于所受离心力不同，相对密度较大的砂粒被甩向池壁，在重力作用下沉入砂斗，有机物随出水旋流带出池外，通过调整流速，可达到最佳沉砂效果（图 2-11）。砂斗内沉砂可用空气提升、排砂泵排砂等方式排出，经砂水分离达到清洁排砂标准。

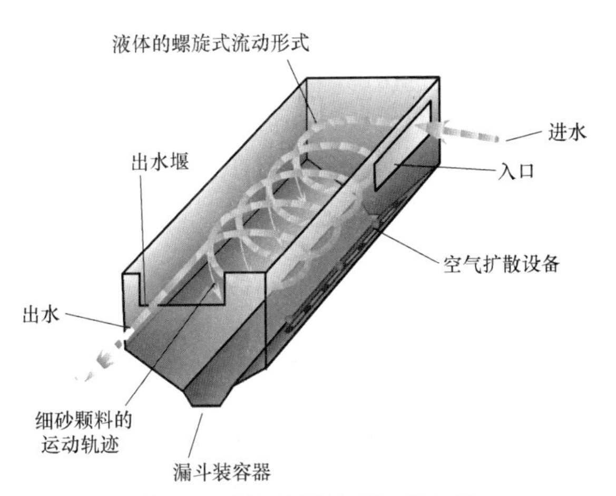

图 2-9　曝气沉砂池螺旋状水流

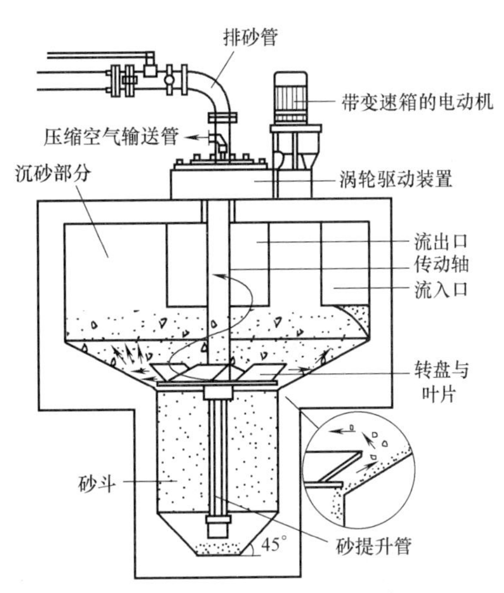

图 2-10　钟式沉砂池构造示意图

旋流沉砂池进水管最大流速为 0.3m/s；池内最大流速为 0.1m/s，最小流速为 0.02m/s；最大流量时停留时间不小于 20s，一般采用 30～60s；设计表面水力负荷为 150～200m^3/(m^2·h)，有效水深为 1.0～2.0m；池径与池深比以 2.0～2.5 为宜。

沉砂池日常运行时，操作人员需根据水量变化，调节沉砂池进水闸阀，以保持沉砂池污水设计流速和停留时间。

曝气沉砂池多采用穿孔管曝气，设有调节阀门，以根据水量的变化及时调节进入曝气沉砂池的空气量，气水比不大于 0.2 时大部分砂粒恰好呈悬浮状态。

各种类型的沉砂池均应定时排砂或连续排砂，以避免砂粒密度造成管道内的沉积和堵塞，另一方面，沉砂在池内堆积会减少池内有效容积的利用，使流速增大，降低沉砂效率。

沉砂池里除去的砂粒宜每年化验一次，对沉砂颗粒进行筛分，分析砂粒中有机物含量、含水率、砂粒的粒径、沉砂量等，根据情况调整气水比，确定排砂间隔时间。

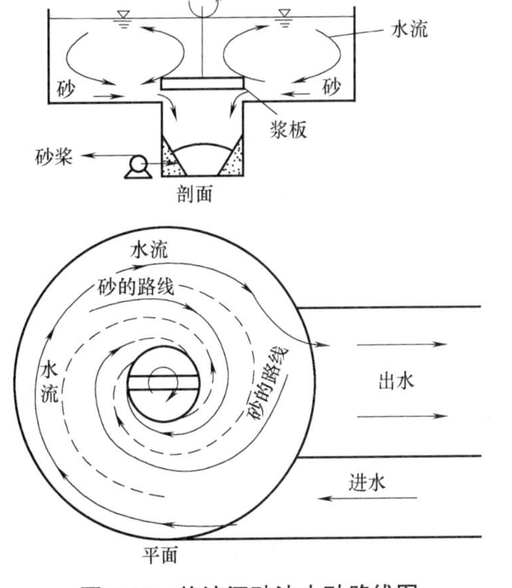

图 2-11　旋流沉砂池水砂路线图

砂粒中的有机物含量宜小于 35%，否则长期放置后容易腐败发臭。运行中需依靠调整空气量、控制流速等方法，尽量降低砂粒中的有机物含量。

沉砂池每运行两年，应放空一次，彻底清理池内所有污物，避免形成死角，同时检修除砂设备和设施。

2.3.2　沉淀池

沉淀池，是一种用于去除液体中的固体颗粒或悬浮固体的设备，其结构及实物如图 2-12 所示。

(a)

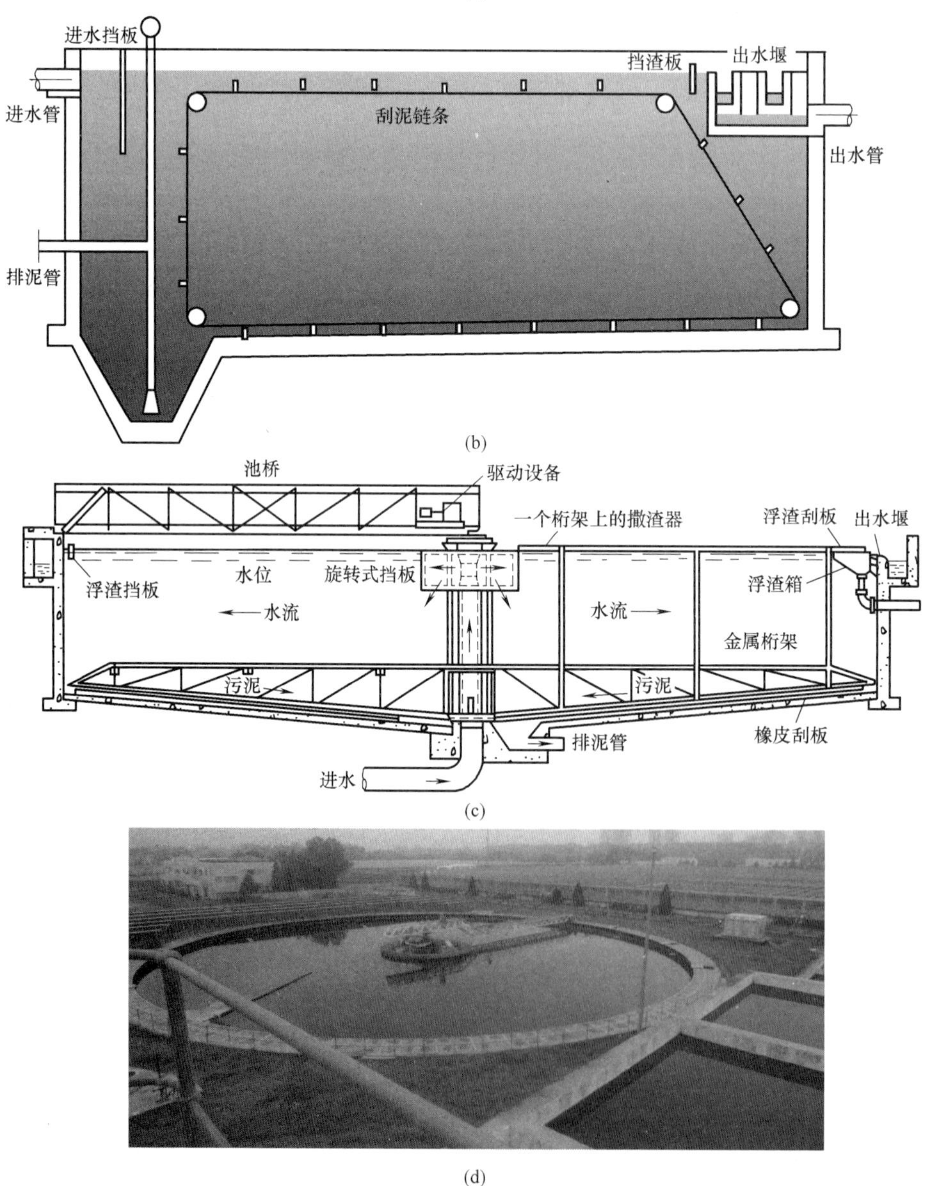

(d)

图 2-12　沉淀池结构及实物图

(a) 平流式沉淀池实物图；(b) 平流式沉淀池结构；(c) 辐流式沉淀池结构；(d) 辐流式沉淀池实物图

传统的沉淀池包括平流式、竖流式和辐流式三种类型，均存在以下缺点：①去除效率不高。污水在初沉池沉淀 1.5h，悬浮物去除率为 40%～60%，BOD_5 去除率为 20%～30%；②占地面积大。因此，实际运行时往往需要对传统沉淀池进行强化与改进。

二维码 2 沉淀池的运行视频

强化与改进的措施包括两个方向：一是从原水水质和工艺方面着手，采取措施如混凝或预曝气，改变水中悬浮物质的状态，使其易于与水分离沉淀；二是从沉淀池结构方面着手，创造更适宜于颗粒沉淀分离的条件，提高沉淀池的容积利用率。目前以改变传统沉淀池结构来提高沉淀效率的沉淀池有向心辐流式沉淀池和斜板（管）沉淀池。

预曝气就是在污水进入沉淀池之前，先进行短时间（10～20min）的曝气，以改善悬浮物的沉淀性能，使相对密度接近于 1 的微小颗粒絮凝后被沉淀去除。预曝气的类型分两种：①单纯曝气，即只进行曝气不投加任何物质，利用气泡的搅动促使污水中的悬浮颗粒相互作用，产生自然絮凝。采用此法可使沉淀效率提高 5%～80%，每立方米污水曝气量约为 $0.5m^3$；②生物絮凝，就是在曝气的同时，投加生物处理单元排出的剩余生物污泥，利用这些污泥具有的活性产生絮凝作用。该法已在国内外得到广泛应用，可使沉淀效率提高 10%～15%，BOD_5 的去除率也能增加 15%以上，活性污泥的投加量一般为 100～400mg/L。预曝气一般在专设的构筑物（预曝气池或生物絮凝池）中进行。预曝气池与沉淀池可以合建，称为预曝气沉淀池，即将预曝气与沉淀两种功能置于同一池中的不同部位。预曝气所使用的曝气装置与生物处理曝气池使用的相同。

向心辐流式沉淀池特指周边进水的辐流式沉淀池，其进水渠布置在沉淀池四周，上清液经过设在沉淀池四周或中间的出水堰溢流而出。相对于普通辐流式沉淀池的中心进水方式，周边进水的结构能使布水比较均匀，增大过水面积，使向下的流速变小，避免冲击池底沉泥，因此向心辐流式沉淀池的去除效率和容积利用率比普通辐流式沉淀池有明显提高。

斜板（管）沉淀池是根据浅层沉淀原理设计的新型沉淀池，目前污水处理中通常是在原有的普通沉淀池中加设斜板（管）构成的。各类型沉淀池的优缺点和适用条件见表 2-5。

各类型沉淀池的优缺点和适用条件　　表 2-5

池型	优点	缺点	适用条件
平流式沉淀池	1. 沉淀效果较好； 2. 耐冲击负荷及温度变化； 3. 施工简单，造价低	1. 配水不易均匀； 2. 多斗排泥时，每个泥斗需单独设排泥管各自操作； 3. 机械排泥时，设备机件易腐蚀，难维修	1. 适用于大、中、小型污水处理厂； 2. 地下水位较高及地质较差的地区
竖流式沉淀池	1. 排泥方便，管理简单； 2. 占地面积较小	1. 池子深度大，施工困难； 2. 对冲击负荷和温度变化适应性差； 3. 池径大，布水不匀	小型污水处理厂

续表

池型	优点	缺点	适用条件
辐流式沉淀池	1. 机械排泥，运行可靠； 2. 排泥设备已定型化	1. 水流速度不稳定； 2. 易出现异重流现象，影响沉淀效果； 3. 机械排泥设备复杂，对施工质量、运行管理水平均要求高	1. 适用于大型（大于 $5\times10^4m^3/d$）、中型污水处理厂； 2. 适用于地下水位较高的地区
斜板（管）沉淀池	1. 去除效率高； 2. 停留时间短； 3. 占地面积小	1. 排泥困难，易堵，不耐冲击负荷； 2. 基于自由沉淀机理而设计，不宜用作二沉池	1. 占地面积受限时，适用作初沉池； 2. 已有的污水处理厂需要扩大处理能力时

初沉池在设计和运行管理时的主要控制参数有 3 个，除在设计计算中必须用的水力表面负荷 q、水力停留时间 h 之外，还有一个参数是堰板溢流负荷 [$m^3/(m^2\cdot h)$]。堰板溢流负荷指单位堰板长度在单位时间内所溢流的污水量，初沉池一般控制在小于 $10m^3/(m^2\cdot h)$，在这个控制参数内能使污水在沉淀池的出水端保持一个均匀而稳定的流态，防止污泥及浮渣的流失。

二维码 3 初沉池出水视频

沉淀池的运行巡视主要包括观察沉淀池的运行状况，刮浮渣板是否把浮渣准确刮进浮渣斗里，链条刮渣机的齿轮链条是否有缠绕物，刮泥板在水下行走是否平衡；要注意出水三角堰板的堰口是否被浮渣堵住；进出水闸门是否需要维修。

初沉池运行时出现的异常情况包括污泥上浮、污泥短路流出及排泥浓度降低。

1. 污泥上浮

（1）若是经常性的污泥上浮，应从控制参数上核算表面水力负荷、水力停留时间、堰板溢流负荷的数据是否都在控制参数内，否则需加以调整。

（2）若进水是腐败严重的污水，也能造成污泥上浮，此时应加强去除浮渣的工作。

（3）采用生物絮凝法时二沉池回流污泥能进入初沉池一部分，由于其硝酸盐含量较高，进入初沉池后缺氧可使硝酸盐反硝化，还原成氮气附着于污泥中，使之上浮。这时可控制后面生化处理系统，使污泥龄减小，降低消化程度，也可加大回流污泥量，使之停留时间减少。

2. 污泥短路流出

（1）若是由于堰板溢流负荷超标或堰板不平整造成，应减少堰板的负荷或调整堰板出水高度一致。

（2）可能是由于刮泥机故障造成污泥上浮。

（3）可能是辐流式沉淀池池面受大风影响出现偏流。

3. 排泥浓度降低

（1）排泥时间过长导致含固率下降，污泥浓度降低。

（2）刮泥与排泥步调不一致，各单体池排泥不均匀。

（3）积泥斗严重积砂，有效容积减小。

2.3.3 中水回用工艺中的快滤池

普通快滤池一般为单层细砂级配滤料或煤、砂双层滤料，单水冲洗，搭配大阻力配水系统，总高度约3.2～3.6m，设计过滤周期为12～24h。目前在我国使用较多的是V型滤池和变孔隙滤池。

V型滤池（又名均粒滤料滤池）是法国德格雷蒙（DEGREMONT）公司设计的一种气水反冲洗快滤池，因两侧（或一侧也可）进水槽设计成V字形而得名。V型滤池的滤速可达7～20m/h，一般为12.5～15.0m/h；滤料采用单层加厚均粒滤料，粒径一般为0.95～1.35mm，允许扩大到0.7～2.0mm，不均匀系数在1.2～1.6（或1.8）之间；过滤周期一般采用24～48h。

V型滤池的主要工艺结构一般由5部分组成：进水系统、过滤系统、反冲洗系统、反冲洗扫洗系统和排水系统。滤池采用V型槽进水，布水均匀；一般采用较粗、较厚的均匀颗粒的石英砂滤料，滤层的纳污能力得到增强；V型滤池在反冲洗过程中引入气洗和横向表面扫洗，可以快速地将杂质排入污水槽中，从而减少冲洗时间，节约冲洗水量；另外由于采用自动化控制，使滤层保持微膨胀状态，可减少滤池深度。总之，V型滤池能明显提升滤池的反冲洗效果，改善V型滤池过滤能力的再生状况，从而增大滤池的截污能力，降低滤池的反冲洗频率。该滤池的优点是出水水质好、滤速高、运行周期长、反冲洗效果好、节能和便于自动化管理等；缺点是需要依靠相应的自控设备和滤池专用仪表。

20世纪90年代以来，我国新建的大、中型给水处理厂多数采用V型滤池这种滤水工艺。近几年来，V型滤池作为工艺处理核心单元出现在我国很多钢铁企业总排口的废水处理及回用工程中。只要符合进水SS小于10～15mg/L的条件，即可推广用于城市污水处理厂的深度处理及中水回用、工业废水处理回用工艺。

变孔隙滤池的结构和普通快滤池相似，其主要不同点在于滤料层的组成。它采用一种比常规滤料粒径更大的滤料和另一种细颗粒滤料，按一定比例混合而组成滤床，粗滤料的粒径和所占比例与细滤料相距较大。因为主要使用粗滤料，依靠整个滤层进行整体过滤，不会形成细滤料过滤时表面过滤的情况，所以滤层阻力小，滤速较高，且不易堵塞。而每次反冲洗后用压缩空气将滤料混合均匀，使较细的滤料均匀地填充在较粗滤料的孔隙之中，也避免了因水力筛分作用造成的细滤料集中在滤层表面的现象，使滤层继续保持较高的纳污能力。同时，由于加入的有限量细滤料均匀分散在整个滤层，降低粗滤料的局部孔隙率，实现对水中细小悬浮颗粒的"同向凝聚"效果，提高对细小悬浮颗粒的去除率。

变孔隙滤池的优点是滤速高、截污能力强、运行周期长、出水水质好、反冲洗水耗量小等，缺点是对滤料选择要求高；运行、反冲洗操作要求严格；专门需要一套冲洗设备。反冲洗分三个阶段进行。第一阶段冲洗强度较大，为15.8L/(m^2·s)，清洗部分较松散的截留杂质；第二阶段用较小冲洗强度11.9L/(m^2·s)水冲洗，并辅以压缩空气吹洗，清洗吸附在滤层表面的杂质，并使滤料混合均匀；第三阶段以冲洗强度11.9L/(m^2·s)的低流速漂洗，不使滤层膨胀，漂洗出残余杂质和滤层中的空气。

当城市中水用于火力发电厂的循环冷却水补充水时，需要将污水处理厂二级处理后的出水进一步去除碱度、硬度、重金属等。目前，石灰澄清过滤工艺被广泛地应用于中水深度处理，而变孔隙滤池则成为该工艺中的一个重要单元。国内部分火力发电厂中水处理工程的实例表明：在石灰处理工艺中，变孔隙滤池不仅能发挥滤速高、出水水质好的优点，且反冲洗彻底，滤料不易板结，与常规的石英砂过滤器和纤维球过滤器相比，运行更稳定，检修更方便。

2.3.4 气浮池

1. 气浮法的类型和特点

表 2-6 汇总了气浮法的类型和特点。

气浮法的类型和特点 **表 2-6**

类型		特点	适用条件
电解气浮法		1. 气泡微小均匀； 2. 表面负荷率低于 $4m^3/(m^2 \cdot h)$； 3. 有电耗高，管理复杂，电极结垢等问题	仅适用于 $10 \sim 20m^3/h$ 水量的工业废水处理
分散空气气浮法	微孔曝气气浮法	1. 简单易行； 2. 气泡大，效果不高； 3. 微孔容易堵塞	适用于矿物浮选或含油脂、大量表面活性剂等废水的初级处理
	剪切气泡气浮法	1. 气泡比微孔产生得小，但能量消耗大； 2. 管理复杂	适用于水量不大、污染物浓度高的废水
溶解空气气浮法	真空气浮法	1. 空气的溶解度低，气泡释放量有限； 2. 处理设备需密闭，真空条件的运行维护困难	特殊场合
	加压溶气气浮法	1. 产生的气泡量充足，气泡细微； 2. 部分回流加压溶气流程比全加压溶气流程节省电耗； 3. 回流的方式避免了释放阀堵塞	目前最常用的气浮处理方法

2. 气浮法的运行管理

压力溶气气浮法包括压力溶气系统、空气释放系统和气浮池三个部分。溶气方式目前主要采用两种：水泵出水管射流溶气式和空压机供气式。

水泵出水管射流溶气式的优点是不需另设空压机，没有空压机产生的油污染和噪声，但本身的能量损失高达 30%。相对使用更广泛的是空压机供气式，因其能耗较低。

压力溶气系统的核心设备是压力溶气罐，其中以罐内填充填料的溶气罐效率最高。影响填料溶气罐效率的主要因素为：填料特性、填料层高度、罐内液位高、布水方式和温度等。其主要工艺技术参数为：

过流密度：$2500 \sim 5000m^3/(m^2 \cdot d)$；

填料层高度：填料多为阶梯环、拉西环或波纹片卷，填料高度 $0.8 \sim 1.3m$；

压力罐压力：大于 0.6MPa；

液位控制高度：$0.6 \sim 1m$；

空气量：按25%的过量空气考虑，实际空气用量取处理水量的1%～5%（体积比）或应去除细小颗粒物量的0.5%～1%（质量比）；

使用的回流水量：进水的25%～50%；

水在溶气罐内的停留时间：3～5min。

空气释放系统是由溶气释放装置和溶气水管路组成的。常用的溶气释放装置有减压阀、溶气释放器等。

溶气释放装置的作用是将压力溶气水减压，使溶气水中的气体以直径在20～100μm之间的微气泡形式均匀释放出来，并与水中的细小颗粒物黏附，形成浮渣。由于减压阀安装在气浮池外，而释放器安装在气浮池接触室的池底，二者有一段距离，如果此段管道较长，就会使气泡减压合并现象严重，从而影响气浮效果。

国内同济大学开发了TS型、TJ型和TV型等节约能耗的气浮专用释放器。这些释放器在0.2MPa以上的低压下工作，能取得良好的气浮效果，瞬时释放99%溶气量。释放出的微气泡平均直径为20～40μm，气泡密集、附着力好。

应用广泛的气浮池有平流式和竖流式两种。平流式气浮池是目前最常用的一种形式，其反应池与气浮池合建。废水进入反应池完全混合后，经挡板底部进入气浮接触室以延长絮体与气泡的接触时间，然后由接触室上部进入分离室进行固-液分离。池面浮渣由刮渣机刮入集渣槽，清水由底部集水槽排出，平流式气浮池的优点是池身浅、造价低、构造简单、运行方便，缺点是分离部分的容积利用率不高等。气浮池的有效水深通常为2.0～2.5m，一般以单格宽度不超过10m，长度不超过15m为宜。

废水在反应池中的停留时间与混凝剂种类、投加量、反应形式等因素有关，一般为5～15min。为避免打碎絮体，废水经挡板底部进入气浮接触室时的流速应小于0.11m/s，废水在接触室中的上升流速一般为10～20mm/s，停留时间应大于60s。

废水在气浮分离室的停留时间一般为10～20min，其表面负荷率为6～8$m^3/(m^2 \cdot h)$，最大不超过10$m^3/(m^2 \cdot h)$。

竖流式气浮池的基本工艺参数与平流式气浮池相同。其优点是接触室在池中央，水流向四周扩散，水力条件较好。缺点是与反应池较难衔接，容积利用率较低。有经验表明，当处理水量大于150～200m^3/h、废水中悬浮物固体浓度较高时，宜采用竖流式气浮池。

3. 投加化学药剂提高气浮效果

疏水性很强的物质（如植物纤维、油滴及炭粉末等），不投加化学药剂即可获得满意的固-液和液-液分离效果。一般的疏水性或亲水性的悬浮物质，均需投加化学药剂，以改变颗粒的表面性质，增加气泡与颗粒的吸附。这些化学药剂包括混凝剂、助凝剂、浮选剂与破乳剂。

（1）混凝剂和助凝剂

各种无机或有机高分子混凝剂不仅可以使废水中的胶体污染物脱稳，改变其亲水性能，而且还能使废水中的微小颗粒絮凝成较大的絮状体以吸附、截获气泡，加速颗粒上浮。

助凝剂的作用是提高悬浮颗粒表面的水密性，以提高颗粒的可浮性。常用的有聚丙烯酰胺等。

（2）浮选剂

浮选剂大多数由极性-非极性分子组成。极性-非极性分子的结构一般用符号○—表示，圆头表示极性基，易溶于水（因为水是强极性分子），尾端表示非极性基，难溶于水，表现出疏水性。

投加浮选剂之后能否使亲水性物质转化为疏水性物质，主要取决于浮选剂的极性基能否附着在亲水性悬浮颗粒的表面，而与气泡相黏附的强弱则取决于非极性基中碳链的长短。当浮选剂的极性基被吸附在亲水性悬浮颗粒表面后，非极性基则朝向水中，这样就可以使亲水性物质转化为疏水性物质，从而能使其与微细气泡相黏附。

浮选剂的种类很多，如松香油、石油、表面活性剂、硬脂酸盐等。

（3）破乳剂

当一种或几种难溶于水的微粒大量存在于水溶液中，在水力或者外在动力的搅动下，这些固体可以以乳化的状态存在于水中，形成乳浊液。理论上讲这种体系是不稳定的，但如果存在一些乳化剂，就使得乳化状态很严重，甚至两相难于分离，最典型的例子是在油水分离中的乳化油以及在污水处理中的水油混合物。

在此情况下，往稳定的乳化体系里投入一些药剂，以破坏乳化状的液体结构，如破坏油滴界面的稳定薄膜，从而达到两相分离的目的。这些为了达到破坏乳化作用的药剂统称为破乳剂。

污水处理现场破乳的途径有下述几种：

1）投加换型乳化剂，例如，氯化钙可以使以钠皂为乳化剂的水包油乳状液转换为以钙皂为乳化剂的油包水乳状液。在转型过程中存在着一个由钠皂占优势转化为钙皂占优势的转化点，这时的乳状液非常不稳定，可借此进行油水分离。因此控制“换型剂”的用量，即可达到破乳的目的，这一转化点用量应由试验确定。

2）投加盐类、酸类物质可使乳化剂失去乳化作用。

3）投加某种本身不能成为乳化剂的表面活性剂，例如，异戊醇可从两相界面上挤掉乳化剂而使其失去乳化作用。

（4）投加抑制剂。抑制剂的作用是暂时或永久性地抑制某些物质的气浮性能，而又不妨碍需要去除的悬浮颗粒的上浮过程，常用的抑制剂有石灰、硫化钠等。

（5）投加调节剂。调节剂主要用于调节废水的 pH，改进和提高气泡在水中的分散度以及提高悬浮颗粒与气泡的黏附能力，如各种酸、碱等。

2.4 设计计算

2.4.1 调节池的设计计算

1. 以调节水量为主的调节池

以调节水量为主的调节池，可采用公式计算确定调节池容积。

生产周期 T 内废水总量为：

$$Q_T=\sum_{t_i=0}^{T}q_it_i \tag{2-7}$$

式中 Q_T——生产周期 T 内废水总量，m^3；

q_i——t_i 时间段内废水的平均流量，m^3/h；

t_i——对应废水排放量 q_i 时的排放时间，h。

在生产周期 T 内废水平均流量为：

$$\overline{Q}=\frac{Q_T}{T}=\frac{\sum_{t_i=0}^{T}q_it_i}{T} \tag{2-8}$$

式中 $\overline{Q}$——生产周期 T 内废水平均流量，m^3/h；

T——废水产生时间，h。

调节池容积为：

$$V=\overline{Q}t \tag{2-9}$$

式中 V——调节池容积，m^3；

t——污水在调节池的水力停留时间，h。

【例 2-1】 某工厂逐时废水流量变化见表 2-7，求水力停留时间为 6h 的调节池容积。

某工厂废水流量监测表 **表 2-7**

时间(h)	流量(m^3/h)	时间(h)	流量(m^3/h)	时间(h)	流量(m^3/h)	时间(h)	流量(m^3/h)
0～1	20	6～7	24	12～13	40	18～19	55
1～2	22	7～8	30	13～14	35	19～20	50
2～3	18	8～9	50	14～15	50	20～21	45
3～4	16	9～10	60	15～16	60	21～22	30
4～5	20	10～11	65	16～17	65	22～23	20
5～6	22	11～12	60	17～18	60	23～24	20

【解】（1）计算废水总量

生产周期 T 内废水总量为：

$$\begin{aligned}Q_T=\sum_{t_i=0}^{T}q_it_i&=(20+22+18+16+20+22+24+30+50+60+65+60\\&+40+35+50+60+65+60+55+50+45+30+20+20)\\&=937m^3\end{aligned}$$

（2）计算平均流量

生产周期 T 内废水平均流量为：

$$\overline{Q}=\frac{Q_T}{T}=\frac{937}{24}=39m^3/h$$

（3）计算调节池有效容积

已知，污水停留时间 $t=6h$，则调节池有效容积为：

$$V=\overline{Q}t=39\times6=234m^3$$

（4）计算调节池几何尺寸

调节池有效水深 h 取 4.5m，则调节池表面积为：$A=\frac{V}{h}=\frac{234}{4.5}=52m^2$

调节池采用方形池体，池长 L 与池宽 B 相等，则

$$L=B=\sqrt{A}=\sqrt{52}=7.21\text{m} \qquad 取\ L=B=7.5\text{m}$$

设调节池超高 $h_1=0.5$m，调节池总高度为：

$$H=h+h_1=0.5+4.5=5.0\text{m}$$

则调节池的几何尺寸为：$L\times B\times H=7.5\text{m}\times7.5\text{m}\times5.0\text{m}$

2. 以调节水质为主的调节池

对调节池可写出物料平衡方程：

$$C_1Qt_i+C_0V=C_2Qt_i+C_2V \tag{2-10}$$

式中 Q——取样间隔时间内进入调节池的平均流量，m^3/h；

C_1——取样间隔时间内进入调节池的污染物浓度，mg/m^3；

t_i——取样间隔时间，h；

C_0——取样开始时调节池内的污染物浓度，mg/m^3；

V——调节池容积，m^3；

C_2——调节池出水污染物浓度，mg/m^3。

当一个取样间隔时间内出水浓度 C_2 不变时，则式（2-10）为：

$$C_2=\frac{C_1t_i+C_0V/Q}{t_i+V/Q} \tag{2-11}$$

根据式（2-11）可以计算出各时间间隔后出水浓度，由此计算出调节池出水的 P 值：

$$P=\frac{调节池出水最大浓度}{调节池出水平均浓度} \tag{2-12}$$

根据调节池出水的 P 值不大于 1.2 的要求，确定污水在调节池的水力停留时间，计算调节池容积为：

$$V=\overline{Q}t \tag{2-13}$$

式中 t——污水在调节池的停留时间，h。

【例 2-2】 某化工厂生产周期为 8h，进入调节池中的废水水量和 BOD_5 变化见表 2-8，开始时调节池内废水的 BOD_5 为 360mg/L，计算停留时间为 6h 时的调节池出水 BOD_5。

调节池进水水量和 BOD_5 变化 **表 2-8**

时间	流量(m^3/h)	进水浓度(mg/L)	出水浓度(mg/L)
0～1	3.5	200	344
1～2	4.5	320	341
2～3	5.5	400	350
3～4	5.0	350	350
4～5	4.5	310	345
5～6	5.0	350	344
6～7	6.5	450	361
7～8	7.5	500	387

【解】（1）计算平均流量

$$\overline{Q}=\frac{Q_T}{T}=\frac{\sum_{t_i=0}^{T}q_it_i}{T}=\frac{3.5+4.5+5.5+5.0+4.5+5.0+6.5+7.5}{8}=5.25\text{m}^3/\text{h}$$

（2）计算调节池容积

当停留时间为 $t=6$h，调节池容积为：$V=\overline{Q}t=5.25\times6=31.5\text{m}^3$

（3）计算不同时间调节池出水污染物浓度

时间为 0～1h 时，调节池出水 BOD_5 为：

$$C_2=\frac{C_1t_i+C_0V/Q}{t_i+V/Q}=\frac{200+360\times\frac{31.5}{3.5}}{1+\frac{31.5}{3.5}}=344\text{mg/L}$$

其余时间间隔出水 BOD_5，见表 2-8。

2.4.2 格栅的设计计算

1. 格栅设计规定

格栅应设置在污水处理系统或水泵之前，根据现场要求选择适宜的栅条间隙。粗格栅栅渣宜采用带式输送机输送，细格栅栅渣宜采用螺旋输送机输送。格栅所截的栅渣必须定时清除，并妥善处理和处置。格栅除污机、输送机和压榨脱水机的进出料口宜采用密闭形式，根据周围环境情况，可设置除臭处理装置。格栅间应设置通风设施和有毒有害气体的检测与报警装置。格栅的设计包括格栅形式选择、尺寸计算、水力计算、栅渣量计算及清渣机械的选用，图 2-13 为格栅的计算简图。

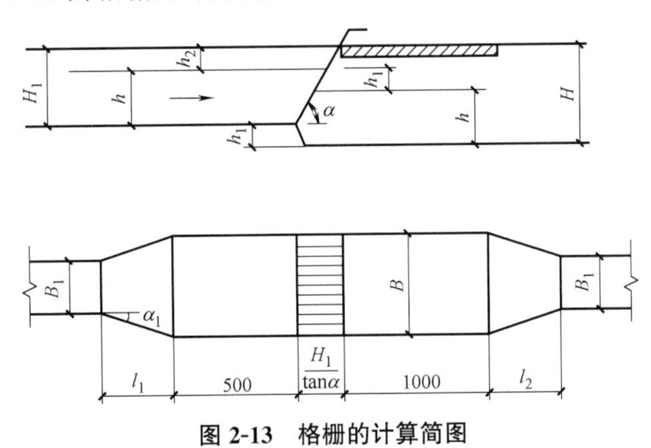

图 2-13 格栅的计算简图

为了防止格栅发生堵塞，污水通过栅条间距的流速一般采用 0.6～1.0m/s。除转鼓式格栅除污机外，机械清理的格栅安装倾角一般为 60°～90°，人工清除的格栅安装角度一般为 30°～60°。机械清渣格栅的过水面积，一般应不小于进水渠道有效面积的 1.2 倍。格栅工作平台过道宽度，采用人工清理栅渣时应大于 1.2m，机械清理时应大于 1.5m。为防止格栅前渠道出现阻流回水现象，一般在格栅的渠道与栅前渠道的连接部位，设置一展开角 $\alpha_1=20°$的渐扩部位。格栅基本参数应符合表 2-9 的规定。

当格栅截留的栅渣量大于 0.2m³/d 时，一般采用机械清渣；栅渣量小于 0.2m³/d 时，一般采用人工清渣。粗格栅栅渣宜采用带式输送机输送，细格栅栅渣宜采用螺旋输送机输送。常见格栅除污机性能参数见表 2-10。

格栅基本参数（单位：mm） **表 2-9**

格栅类型	基本参数	参数值
平面格栅	格栅宽度	600、800、1000、1200、1400、1600、1800、2000、2200、2400、2600 2800、3000、3200、3400、3600、4000、4500、5000、6000
	格栅长度	600、800、1000、1200……以 200 为一级增大，其上限值由水深等参数确定
	栅条间隙	细格栅 5～10，以 1 递增；粗格栅 15～100，30 以下以 5 递增，30 以上以 10 递增
弧形格栅	格栅宽度	300、400、500、600、800、1000、1200、1400、1600、1800、2000
	回转半径	1000、1200、1600、2000
	栅条间隙	10、15、20、25、30、40、50、60、80

格栅除污机性能参数 **表 2-10**

格栅除污机类型	宽度（mm）	回转半径（mm）	安装角度（°）	运行速度（m/min）	栅条间隙（mm）	网孔净尺寸（mm）
钢丝绳牵引式格栅除污机	500～4000		60～85	1.0～3.5	10～100	
回转式链条传动格栅除污机	300～3000		60～85	1.5～3.5	10～100	
回转式齿耙链条格栅除污机	300～3000		60～85	1.5～3.5	2～100	
高链式格栅除污机	300～2000		60～80	≤4.5	8～60	
阶梯式格栅除污机	300～2000		45	1.5～3.5（r/min）	2～10	
弧形格栅除污机	300～2000	300～2000		5～6	5～80	
转鼓式格栅除污机		500～3000（栅筒直径）	35	5～15	0.2～12	
移动式格栅除污机	500～1500（齿耙或抓斗宽度）		60～85	≤4.5	10～100	
回转滤网式格栅除污机	500～4000		90	1.5～4.5	5～80	(0.5×0.5)～(50×50)

注：300mm≤格栅除污机宽度≤1000mm 时，格栅除污机宽度系列的间隔为 50mm；1000mm≤格栅除污机宽度≤4000mm 时，格栅除污机宽度系列的间隔为 100mm。

2. 格栅设计计算

（1）栅条间隙数量

$$n=\frac{Q_{max}\sqrt{\sin\alpha}}{b\cdot h\cdot v} \quad (2\text{-}14)$$

式中 n——栅条间隙数量，个；

Q_{max}——最大设计流量，m^3/s；

b——格栅栅条间隙，m；

h——格栅前水深，m；

v——污水流经格栅的速度，一般取 0.6～1.0m/s；

α——格栅安装倾角，°。

（2）格栅槽总宽度

$$B=b\cdot n+S(n-1) \tag{2-15}$$

式中 B——格栅槽总宽度，m；

S——栅条宽度，m。

（3）污水经过格栅的水头损失

$$h_2=k\cdot h_0 \tag{2-16}$$

$$h_0=\xi\cdot\frac{v^2}{2g}\cdot\sin\alpha \tag{2-17}$$

式中 h_2——污水经过格栅的水头损失，m；

k——考虑由于格栅受污染物堵塞后，格栅阻力系数增大，一般 $k=3$；

h_0——计算水头损失，m；

v——污水流经格栅的速度，m/s；

ξ——阻力系数，其值与格栅栅条的断面形状有关，可按表 2-11 计算；

g——重力加速度，m/s^2。

格栅阻力系数计算公式 **表 2-11**

栅条断面形状	计算公式	说明
正方形	$\xi=\left(\frac{b+S}{\varepsilon b}-1\right)^2$ ε 为收缩系数	$\varepsilon=0.64$
圆形	$\xi=\beta\left(\frac{S}{b}\right)^{\frac{4}{3}}$ β 为形状系数	$\beta=1.79$
迎水面为半圆形的矩形		$\beta=1.83$
迎水面、背水面均为半圆形的矩形		$\beta=1.67$
锐边矩形		$\beta=2.42$

（4）格栅后槽的总高度

$$H=h+h_1+h_2 \tag{2-18}$$

式中 H——格栅后槽的总高度，m；

h——栅前水深，m；

h_1——格栅前渠道超高，m，一般 $h_1=0.3$m。

h_2——格栅的水头损失，m。

（5）格栅槽的总长度

$$L=L_1+L_2+1.0+0.5+\frac{H_1}{\tan\alpha} \tag{2-19}$$

式中 L——格栅槽的总长度，m；

L_1——进水渠道渐宽部位的长度，m；

L_2——格栅槽与出水渠道连接处的渐窄部位长度，一般 $L_2=0.5L_1$；

H_1——格栅前槽的总高度，m。

$$L_1=\frac{B-B_1}{2\tan\alpha_1} \tag{2-20}$$

式中　B_1——进水渠道宽度，m；

α_1——进水渠道渐宽部位的展开角度，一般取 $\alpha_1=20°$。

（6）每日栅渣量

$$W=\frac{86400Q_{max}\cdot W_1}{1000K_z} \tag{2-21}$$

式中　W——每日栅渣量，m^3/d；

W_1——单位污水栅渣量，$m^3/(10^3m^3$ 污水)；格栅间隙为 1.5～10mm 时，W_1 为 0.15～0.12；格栅间隙为 10～25mm 时，W_1 为 0.12～0.05；格栅间隙为 25～100mm 时，W_1 为 0.05～0.004。

K_z——污水流量总变化系数，可按当地实际污水变化量资料选用，没有测量资料时，按表 2-12 和表 2-13 选用。

城镇综合生活污水量总变化系数　　**表 2-12**

平均日流量(L/s)	5	15	40	70	100	200	500	≥1000
K_z	2.7	2.4	2.1	2.0	1.9	1.8	1.6	1.5

注：当污水平均日流量为中间数值时，总变化系数可用内插法求得。

村镇综合生活污水量总变化系数　　**表 2-13**

平均日流量(L/s)	<2	5	15	40	70	100
K_z	4	2.5	2.2	1.9	1.8	1.6

注：1. 当污水平均日流量为中间数值时，总变化系数可用内插法求得；

2. 当污水日平均流量大于 100L/s，总变化系数按表 2-12 选用。

3. 设计计算

【例 2-3】 某城镇污水处理厂日平均处理流量为 $5.0\times10^4m^3/d$，污水流量总变化系数 $K_z=1.35$，求格栅各部分尺寸，同时选择格栅除污设备。

【解】（1）格栅间隙数量

$$Q_{max}=\frac{50000}{24\times3600}\times1.35=0.78m^3/s$$

取格栅栅前水深为 $h=0.5m$，格栅栅条间隙 $b=20mm$，过栅流速 $v=0.9m/s$，格栅安装倾角 $\alpha=60°$，设置 3 台机械格栅，2 用 1 备，则每台格栅间隙数为：

$$n=\frac{Q_{max}\sqrt{\sin\alpha}}{2b\cdot h\cdot v}=\frac{0.78\times\sqrt{\sin60°}}{2\times0.02\times0.5\times0.9}=40$$

（2）格栅槽总宽度

取栅条宽度 $S=0.01m$，则格栅槽总宽度为：

$$B=S(n-1)+bn=0.01\times(40-1)+0.02\times40=1.19m$$

（3）过栅水头损失

设格栅断面形状为圆形，阻力系数 $\beta=1.79$，重力加速度 $g=9.8m/s^2$，格栅受污染物堵塞后的水头损失增大倍数 $k=3$，则：

$$\xi=\beta\left(\frac{S}{b}\right)^{\frac{4}{3}}=1.79\times\left(\frac{0.01}{0.02}\right)^{\frac{4}{3}}=0.71$$

$$h_0=\xi \cdot \frac{v^2}{2g} \cdot \sin\alpha=0.71\times\frac{0.9^2}{2\times9.8}\times\sin60°=0.025\text{m}$$

$$h_1=k \cdot h_0=3\times0.025=0.075\text{m}$$

(4) 进水渠道渐宽部分长度

取进水渠道宽度 $B_1=0.80$m，渐宽部分展开角度 $\alpha_1=20°$；

$$L_1=\frac{B-B_1}{2\tan\alpha_1}=\frac{1.19-0.80}{2\tan20°}=0.54\text{m}$$

(5) 出水渠道渐窄部分长度

$$L_2=\frac{1}{2} \cdot L_1=\frac{1}{2}\times0.54=0.27\text{m}$$

(6) 格栅后槽总高度

设栅前渠道超高 $h_2=0.3$m，栅前水深 $h=0.5$m，则：

$H=h+h_1+h_2=0.5+0.075+0.3=0.875$m，取 0.9m。

(7) 格栅前槽总高度

$$H_1=h+h_2=0.50+0.3=0.8\text{m}$$

(8) 格栅槽总长度

$$\begin{aligned}L&=L_1+L_2+0.5+1.0+\frac{H_1}{\tan\alpha}\\&=0.54+0.27+0.5+1.0+\frac{0.80}{\tan60°}\\&=2.77\text{m}\end{aligned}$$

(9) 每日产生的栅渣量

格栅间隙为 20mm，取污水栅渣量 $W_1=0.05\text{m}^3/(10^3\text{m}^3$ 污水)，则：

$$W=\frac{86400Q_{\max} \cdot W_1}{1000K_z}=\frac{86400\times0.78\times0.05}{1.35\times1000}=2.496\text{m}^3/\text{d}$$

每台格栅每日栅渣量 $W'=\frac{W}{2}=1.248\text{m}^3/\text{d}>0.2\text{m}^3/\text{d}$，宜采用机械清渣。

(10) 格栅除污机的选用

根据某环境工程有限公司提供的有关技术资料，选用 3 台回转格栅机除污机，2 用 1 备，所选设备技术参数为：①过水流量，3600m^3/h；②安装角度，60°～75°；③有效宽度，1200mm；④电机功率，1.5kW；⑤水槽宽度，1550mm；⑥耙齿栅隙，20mm；⑦耙齿节距，100mm；⑧设备标准沟深为 1535mm，可根据用户需要及实际使用情况加长。

2.4.3 沉砂池的设计计算

1. 平流式沉砂池设计规定

(1) 一般按去除相对密度 2.65，粒径大于 0.2mm 的砂粒确定。

(2) 沉砂池的座数或分格数不应少于两个，宜按并联系列设计。污水量较小时，可考虑一备一用。

(3) 设计流量应按最大设计流量计算，合流制处理系统中，按合流流量计算。

(4) 最大设计流量时，污水在池内的最大流速为 0.3m/s，最小流速为 0.15m/s。

（5）污水在池内停留时间不应小于45s。

（6）设计有效水深应不大于1.5m，一般采用0.25～1.0m，每格池宽不宜小于0.6m，超高不宜小于0.3m。

（7）城镇污水的沉砂量在分流制中一般按$3m^3/(10^5m^3$污水）计算，在合流制中适当放大；沉砂含水率约为60%，容重为$1500kg/m^3$。

（8）池底坡度一般为0.01～0.08，并可根据除砂设备要求，考虑池底的形状。

（9）一般采用机械方法除砂，当采用人工方法除砂时，排砂管直径应大于0.2m。采用重力排砂时，沉砂池和贮砂池应尽量靠近，以缩短排砂管长度，便于维护管理。

2. 平流式沉砂池设计计算公式

（1）沉砂池长度

$$L=vt \tag{2-22}$$

式中 L——沉砂池部分长度，m；

v——最大设计流量时的水平速度，m/s；

t——最大设计流量时的停留时间，s。

（2）水流断面面积

$$A=\frac{Q_{max}}{v} \tag{2-23}$$

式中 A——水流断面面积，m^2；

Q_{max}——最大设计流量，m^3/s。

（3）沉砂池总宽度

$$B=\frac{A}{h_2} \tag{2-24}$$

式中 B——沉砂池总宽度，m；

h_2——沉砂池设计有效水深，m。

（4）沉砂斗所需容积

$$V=\frac{86400Q_{max}\cdot T\cdot X}{1000} \tag{2-25}$$

式中 V——沉砂斗所需容积，m^3；

X——污水沉砂量，$L/(m^3$污水），一般$X=0.03L/(m^3$污水）；

T——排砂时间间隔，d。

也可根据污水日平均流量计算沉砂斗所需容积，即

$$V=\frac{Q\cdot T\cdot X\cdot K_z}{1000} \tag{2-26}$$

式中 Q——日平均设计流量，m^3/d。

K_z——污水流量总变化系数。

（5）沉砂池总高度

$$H=h_1+h_2+h_3 \tag{2-27}$$

式中 H——沉砂池总高度，m；

h_1——沉砂池超高，m，不宜小于0.3m；

h_3——沉砂室高度，m。

(6) 核算最小流速

$$v_{\min}=\frac{Q_{\min}}{n_1 \cdot A_{\min}} \tag{2-28}$$

式中 $Q_{\min}$——设计最小流量，m^3/s；

n_1——最小流量时工作的沉砂池数目，个；

$A_{\min}$——最小流量时沉砂池的过水断面面积，m^2。

【例 2-4】 已知某城市污水处理厂最大设计流量 $Q_{\max}=0.5m^3/s$，最小设计流量 $Q_{\min}=0.25m^3/s$，污水流量变化系数 $K_z=1.35$，设计平流式沉砂池。

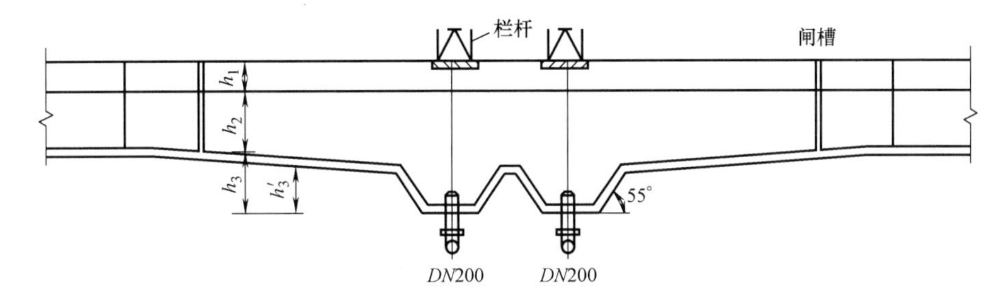

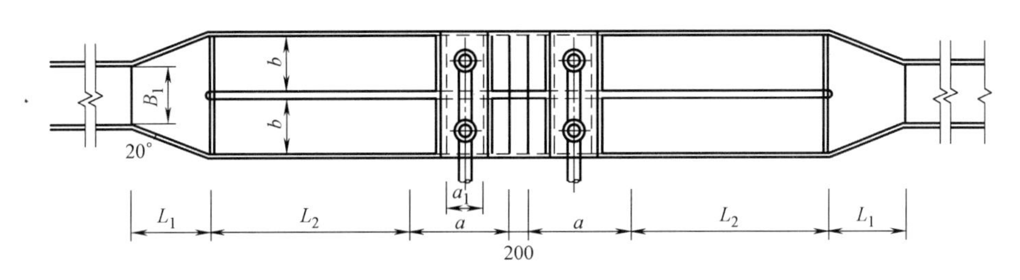

图 2-14 平流式沉砂池计算图

【解】 (1) 沉砂池长度

设水平流速 $v=0.20m/s$，水力停留时间 $t=50s$，则沉砂池长度为：

$$L=vt=0.20\times50=10m$$

(2) 水流断面面积

$$A=\frac{Q_{\max}}{v}=\frac{0.5}{0.20}=2.5m^2$$

(3) 沉砂池总宽度

设沉砂池有效水深 $h_2=0.5m$（图 2-14），则沉砂池总宽度为：

$$B=A/h_2=\frac{2.5}{0.5}=5m$$

沉砂池共分为 3 格，每格宽度为 $b=\frac{B}{3}=\frac{5}{3}=1.67m$。

(4) 沉砂斗所需容积

设排砂周期 $T=2d$，污水沉砂量 $X=0.03L/(m^3$ 污水)，沉砂斗所需容积为：

$$V=\frac{86400Q_{\max}\cdot X\cdot T}{1000}=\frac{86400\times0.5\times0.03\times2}{1000}=2.6\text{m}^3$$

设每一分格有 2 个沉砂斗，共 6 个沉砂斗，每个沉砂斗的容积为：

$$V_1=\frac{2.6}{3\times2}=0.43\text{m}^3$$

（5）沉砂斗各部分尺寸计算

设沉砂池沉砂斗底宽 $b_1=0.7$m；斗壁与水平面的倾角为 60°，沉砂斗高度 h'_3 为 0.5m，则上口宽 b_2 为：

$$b_2=\frac{2h'_3}{\tan60^\circ}+b_1=\frac{2\times0.7}{\tan60^\circ}+0.5=1.28\text{m}，取 b_2 为 1.3\text{m}。$$

沉砂斗的容积为：

$$V_1=\frac{1}{3}h'_3(S_1+S_2+\sqrt{S_1\cdot S_2})=\frac{1}{3}\times0.5\times(0.7^2+1.3^2+\sqrt{0.7^2\times1.3^2})=0.52\text{m}^3>0.43\text{m}^3,$$

符合要求。

（6）沉砂斗的高度

设排砂方式采用重力排砂，池底坡度 $i=6\%$，坡向砂斗，则

$$h_3=h'_3+\frac{0.06(L-2b_2-b')}{2}=0.5+0.06\times\frac{10-2\times1.3-0.2}{2}\approx0.7\text{m}$$

（b' 为两沉砂斗间隔壁厚，一般取 0.2m）

（7）沉砂池总高度

取超高 $h_1=0.3$m，则　$H=h_1+h_2+h_3=0.3+0.5+0.7=1.5\text{m}$

（8）核算最小流速

最小流量时，只有一格工作（$n_1=1$），则：

$$v_{\min}=\frac{Q_{\min}}{n_1\cdot A_{\min}}=\frac{0.25}{1\times1.67\times0.5}=0.3\text{m/s}\geqslant0.15\text{m/s}，符合要求。$$

（9）选择砂水分离器

平流式沉砂池排出的沉砂，实际上是砂水混合物，含水率 60%，密度 1500kg/m³，为进一步将砂和水分离，需要配套砂水分离器。

沉砂池排砂时间间隔为 2d，根据计算的排砂量，选择 1 套螺旋式砂水分离器。该砂水分离器主要技术参数为：①处理量，5～12L/s；②螺旋槽宽度，260mm；③螺旋公称直径，220mm；④安装角度，20°～30°可选；⑤配套电机功率，0.37kW；⑥进水口直径，100mm；⑦溢水口直径，150mm。⑧分离效率，96%～98%。

3. 曝气沉砂池的设计参数

（1）水平流速不宜大于 0.1m/s；旋流速度应保持 0.25～0.3m/s。

（2）污水在池内停留时间宜大于 5min。

（3）沉砂池的有效水深宜为 2.0～3.0m。池宽与池深比为 1.0～1.5，当池长宽比达到 5 时，可考虑设置横向挡板。

（4）曝气沉砂池多采取多孔管曝气，布置在池的一侧，穿孔孔径为 2.5～6.0mm，距池底约 0.6～0.9m，每组穿孔曝气管应有调节阀门。

（5）污水所需曝气量宜为 5.0～12.0L 空气/（m・s）。

（6）曝气沉沙池形状应尽可能不产生死角和偏流，集砂槽附近可安装纵向挡板。进水方向应与池中旋流方向一致，出水方向与进水方向垂直，并宜设置挡板，防止产生短流。

（7）曝气沉沙池内应设置浮渣收集及排除装置，根据环保要求采取封闭及除臭措施。

4. 曝气沉砂池设计计算

（1）曝气沉砂池有效容积

$$V=60Q_{max}t \tag{2-29}$$

式中 V——曝气沉砂池有效容积，m^3；

Q_{max}——最大设计流量，m^3/s；

t——最大设计流量时停留时间，min。

（2）曝气沉砂池断面面积

$$A=\frac{Q_{max}}{v} \tag{2-30}$$

式中 A——曝气沉砂池的断面面积，m^2；

v——最大流量时的水平速度，m/s。

（3）曝气沉砂池总宽度

$$B=\frac{A}{h_2} \tag{2-31}$$

式中 B——曝气沉砂池总宽度，m；

h_2——曝气沉砂池有效水深，m。

（4）曝气沉砂池长度

$$L=60vt \text{ 或 } L=\frac{V}{A} \tag{2-32}$$

式中 L——曝气沉砂池长度，m。

（5）曝气沉砂池所需空气量

$$Q=0.06qL \tag{2-33}$$

式中 Q——曝气沉砂池所需空气量，m^3/min；

q——污水所需曝气量，L/(m·s)。

（6）曝气沉砂池沉砂斗容积

$$V_0=\frac{a+a_1}{2}\cdot h_4\cdot L \tag{2-34}$$

式中 a_1——沉砂槽下底宽，m；

a——沉砂槽上底宽，m；

h_4——沉砂槽高度，m。

曝气沉砂池其余尺寸计算方法同平流式沉砂池。

【例 2-5】 已知某城市污水处理厂的最大设计流量 $Q_{max}=1.2m^3/s$，污水流量总变化系数 $K_z=1.2$，设计曝气沉砂池。

【解】 图 2-15 为曝气沉砂池计算图。

（1）曝气沉砂池的有效容积

设最大设计流量时污水的停留时间 $t=6min$，曝气沉砂池的有效容积为：

$$V=60Q_{max}t=60\times1.2\times6=432m^3$$

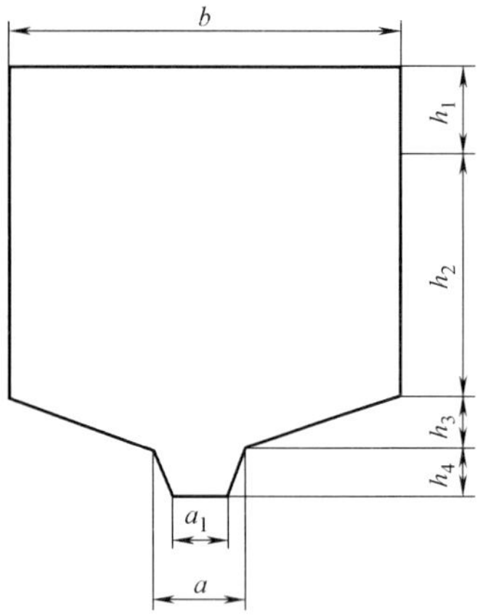

(2) 曝气沉砂池过水断面面积

设最大设计流量时的水平速度 $v=0.10m/s$，则曝气沉砂池过水断面面积为：

$$A=\frac{Q_{max}}{v}=\frac{1.2}{0.1}=12m^2$$

(3) 曝气沉砂池总宽度

设曝气沉砂池有效水深 $h_2=2m$，池总宽度为：$B=\frac{A}{h_2}=\frac{12}{2}=6m$

曝气沉砂池分为三格（$n=3$），每格宽为：$b=\frac{B}{n}=\frac{6}{3}=2m$

图 2-15 曝气沉砂池计算图

宽深比 $b/h_2=2/2=1$，符合设计要求。

(4) 曝气沉砂池长度

$$L=60vt=60\times0.1\times6=36m$$

(5) 曝气沉砂池所需空气量

设污水所需空气量 $q=8L/(m\cdot s)$，则所需空气量为：

$$Q=0.06qL=0.06\times8\times36=17.28m^3/min$$

(6) 沉砂斗所需容积

设两次排砂时间间隔 $T=2d$，污水沉砂量 $X=0.03L/(m^3$ 污水)

$$V=\frac{86400Q_{max}\cdot X\cdot T}{1000}=\frac{86400\times1.2\times0.03\times2}{1000}=6.22m^3$$

曝气沉砂池分为三格，每个沉砂斗所需容积为：$V_1=\frac{V}{3}=\frac{6.22}{3}=2.07m^3$

(7) 沉砂斗各部分尺寸计算

设沉砂槽下底宽 a_1 为 0.5m，沉砂槽斜壁与水平面的夹角为 60°，沉砂槽高度 $h_4=0.4m$，沉砂槽上底宽 a 为：

$$a=2\times0.4\cot60°+0.5\approx0.96m$$

(8) 沉砂斗容积

沉砂斗为沿池长梯形断面渠道，沉砂斗容积为：

$$V_0=\frac{a+a_1}{2}\cdot h_4\cdot L=\frac{0.96+0.5}{2}\times0.4\times36\approx11.2m^3>2.07m^3$$

每格沉砂斗的沉砂量小于沉砂斗容积，满足要求。

(9) 曝气沉砂池总高

设池底坡度为 0.06，坡向沉砂槽，池底斜坡部分的高度为：

$$h_3=0.06(b-a)=0.06\times(2.0-0.96)\approx0.1m$$

设超高 $h_1=0.3m$，曝气沉砂池总高为：

$$H=h_1+h_2+h_3+h_4=0.3+2.0+0.4+0.1=2.8\text{m}$$

（10）排砂方法

采用吸砂机排砂。

2.4.4 沉淀池的设计计算

进行沉淀池设计的基本参数是废水流量、沉降效率、表面水力负荷、水力停留时间、污水在池内的平均流动速度、污泥容重及含水率等。这些参数一般通过实验获得，若无条件，也可根据相似的运行资料，因地制宜地选取经验数据。

1. 设计一般规定

（1）城市污水沉淀池的设计参数宜按表 2-14 采用。生产污水的设计参数，应根据试验或实际生产运行经验确定。

沉淀池主要设计参数　　表 2-14

沉淀池类型		沉淀时间(h)	表面水力负荷[m³/(m²·h)]	每人每日污泥量[g/(人·d)]	污泥含水率(%)	固体负荷[kg/(m²·d)]
初沉池		0.5～2.0	1.5～4.5	16～36	95.0～97.0	—
二沉池	生物膜法后	1.5～4.0	1.0～2.0	10～26	96.0～98.0	≤150
	活性污泥法后	1.5～4.0	0.6～1.5	12～32	99.2～99.6	≤150

（2）沉淀池的超高应大于 0.3m。沉淀池的有效水深宜采用 2～4m，在辐流式沉淀池中指池边水深。

（3）当采用污泥斗排泥时，每个泥斗均应设单独的闸阀和排泥管。泥斗的斜壁与水平面的倾角，方斗宜为 60°，圆斗宜为 55°。坡向泥斗的底坡坡度，在平流式沉淀池中应大于 0.01，在辐流式沉淀池中应大于 0.05。

（4）初沉池的污泥区容积，非机械排泥宜按小于 2d 的污泥量计算。曝气池后的二沉池污泥区容积，宜按小于 2h 的污泥量计算，并应有连续排泥措施。机械排泥的初沉池和生物膜法后的二沉池污泥区容积，宜按 4h 的污泥量计算。

（5）排泥管的直径应大于 0.2m。当采用静水压力排泥时，初沉池的静水压头应大于 1.5m；二沉池的静水压头，生物膜法处理后的应大于 1.2m，活性污泥法处理后的应大于 0.9m。生产污水按污泥性质确定。

（6）沉淀池出水堰最大负荷，在初沉池中应小于 2.9L/(s·m)；在二沉池中应小于 1.7L/(s·m)。可采用多槽出水布置，减轻单位长度出水堰负荷，提高出水水质。

（7）沉淀池应设置撇渣设施。

2. 平流式沉淀池设计要求

平流式沉淀池设计内容包括确定沉淀池的数量，设计入流及出流装置，计算沉淀区和污泥区尺寸，选择排泥和排渣设备。主要设计参数为水平流速、沉淀时间、表面水力负荷、有效池深、池宽、长宽比、长深比等。

（1）沉淀池沉淀区表面积

$$A=\frac{Q_{\max}}{q} \tag{2-35}$$

式中 A——沉淀池沉淀区的表面积，m^2；

$Q_{\max}$——最大设计流量，m^3/h；

q——表面水力负荷，$\text{m}^3/(\text{m}^2\cdot\text{h})$，通过沉淀实验取得，城镇污水可参照表 2-14。

（2）沉淀池有效水深

$$h_2 = q \cdot t \tag{2-36}$$

式中 h_2——沉淀池有效水深，m；

t——沉淀时间，h。

（3）沉淀区有效容积

$$V = A \cdot h_2 \text{ 或 } V = Q_{max} \cdot t \tag{2-37}$$

式中 V——沉淀区有效容积，m^3。

（4）沉淀池长度

$$L = 3.6vt \tag{2-38}$$

式中 L——沉淀池长度，m；

v——最大设计流量时的水平流速，mm/s，初沉池一般水平流速不大于 7mm/s，二沉池一般水平流速不大于 5mm/s。

（5）沉淀池总宽度

$$B = \frac{A}{L} \tag{2-39}$$

式中 B——沉淀池总宽度，m。

（6）沉淀池数量

$$n = \frac{B}{b} \tag{2-40}$$

式中 n——沉淀池数量，沉淀池数量或分格数一般不少于 2 座；

b——每座或每格沉淀池宽度，m；

一般平流式沉淀池长度为 30～50m，不宜大于 60m。为保证污水在沉淀池内流动速度均匀，避免短流。每格长度与宽度比应大于 4∶1，每格宽度或导流墙间距一般采用 3～9m，最大为 12m，池的长度与有效水深比应大于 8∶1。

（7）污泥区所需容积

$$V_w = \frac{S_{湿} \cdot N \cdot T}{1000} \tag{2-41}$$

式中 V_w——污泥区所需容积，m^3；

$S_{湿}$——每人每日湿污泥量，L/(d・人)；

N——设计人口数，人；

T——两次排泥时间间隔，d。

$$S_{湿} = \frac{100S_{干}}{(100 - P_0)\gamma} \tag{2-42}$$

式中 $S_{干}$——每人每日干污泥量，g/(d・人)；

P_0——污泥含水率，%。

γ——污泥密度，kg/m^3，当污泥含水率 $P_0 \geqslant 95\%$时，$\gamma = 1000kg/m^3$。

（8）沉淀池总高度

$$H = h_1 + h_2 + h_3 + h_4 = h_1 + h_2 + h_3 + h_4' + h_4'' \tag{2-43}$$

式中 H——沉淀池总高度，m；

h_1——沉淀池超高，m，一般取 0.3m；

h_2——沉淀区的有效深度，m；

h_3——缓冲层高度，m，一般无机械刮泥设备时为 0.5m；有机械刮泥设备时，缓冲层上缘应高出刮板 0.3m；

h_4——污泥区高度，m；

h_4'——污泥斗高度，m；

h_4''——梯形的高度，m。

【例 2-6】 某城市污水拟采用生物处理法处理，日最大处理流量为 43200m^3/d，设计人口数 $N=250000$，污水沉淀时间 $t=1.5h$。试设计平流式初沉池。

【解】（1）沉淀池沉淀区的表面积

取 $q=2m^3/(m^2 \cdot h)$，则沉淀池沉淀区的表面积为：$A=\frac{Q_{max}}{q}=\frac{43200}{24\times2}=900m^2$

（2）沉淀池有效水深

$$h_2=q \cdot t=2\times1.5=3m$$

（3）沉淀区有效容积

$$V_1=A\times h_2=900\times3=2700m^3$$

（4）沉淀池长度

设水平流速为 3.6mm/s，则沉淀池的总长度为：

$L=3.6vt=3.6\times1.5\times3.6=19.44m$，取 $L=19.5m$。

污水流入口至挡板的距离取 0.5m，流出口至挡板的距离取 0.3m，则沉淀池的总长度

$$L'=L+0.5+0.3=19.5+0.5+0.3=20.3m$$

（5）沉淀池总宽度

$$B=\frac{A}{L'}=\frac{900}{20.3}=44.3m$$

（6）单个沉淀池的宽度

设沉淀池数量为 9 个，则每个沉淀池的宽度为：$b=\frac{B}{n}=\frac{44.3}{9}=4.9m$

验算长宽比 $\frac{L'}{b}=\frac{20.3}{4.9}=4.14>4$，符合规范要求。

（7）污泥区容积

对于初沉池，取污泥量 16～36g/(d·人)，取人均干泥量 $S_{干}=25g/(d \cdot 人)$，污泥含水率 $P_0=95\%$；由此换算成湿污泥量 $S_{湿}$ 为：

$$S_{湿}=\frac{100S_{干}}{(100-P_0)\gamma}=\frac{25\times100}{(100-95)\times1000}=0.5L/(d \cdot 人)$$

污泥贮存时间 T 取 2d，则污泥区的总容积为：

$$V_w=\frac{S_{湿} \cdot N \cdot T}{1000}=\frac{0.5\times250000\times2}{1000}=250m^3$$

每个池污泥部分的容积为：$V'=\frac{V}{n}=\frac{250}{9}=27.8m^3$

（8）沉淀池的总高度

污泥斗底尺寸采用 0.5m×0.5m，上口尺寸采用 4.5m×4.5m，污泥斗斜壁与水平面的夹角为 60°，污泥斗高度为：$h_4'=\frac{(4.5-0.5)}{2}\tan60°=3.46\text{m}$

设池底坡度为 0.01，梯形部分高度 $h_4''=(20+0.3-4.5)\times0.01=0.16\text{m}$

沉淀池的总高度 H 为：

$$H=h_1+h_2+h_3+h_4'+h_4''=0.3+3+0.3+0.16+3.46=7.22\text{m}$$

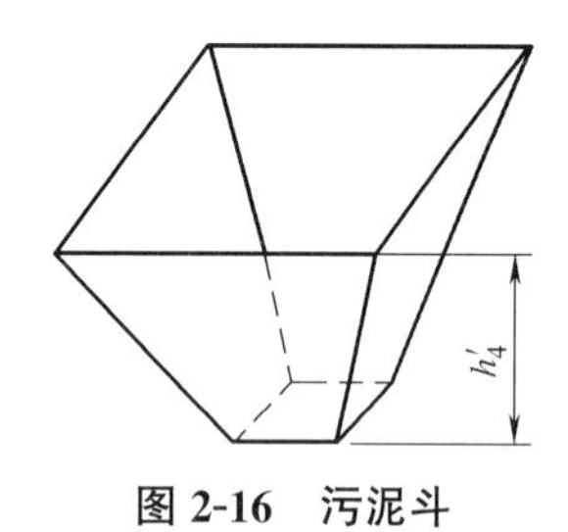

图 2-16　污泥斗

（9）污泥斗的容积

污泥斗为倒正棱台，如图 2-16 所示。

$$V_1=\frac{1}{3}h_4'(S_1+S_2+\sqrt{S_1\times S_2})$$

$$=\frac{1}{3}\times3.46\times(4.5\times4.5+0.5\times0.5+\sqrt{4.5^2\times0.5^2})=26\text{m}^3$$

（10）污泥斗以上梯形部分污泥容积

$$V_2=\left(\frac{L_1+L_2}{2}\right)h_4''\cdot b=\frac{(20+4.5)}{2}\times0.16\times4.9=9.6\text{m}^3$$

（11）污泥斗和梯形部分污泥容积（贮泥区总容积）

$$V=V_1+V_2=26+9.6=35.6\text{m}^3>27.8\text{m}^3$$

（12）计算草图

设计计算草图如图 2-17 所示。

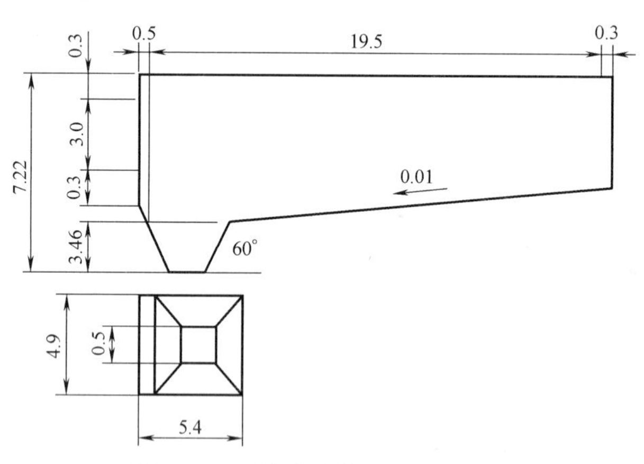

图 2-17　设计计算草图（单位：m）

3. 斜板沉淀池

（1）斜板（管）沉淀池设计要求

1）升流式异向流斜板（管）沉淀池的设计表面水力负荷，一般可按普通沉淀池的设计表面水力负荷的 2 倍计算；但对于二沉池，尚应以固体负荷核算。

2）斜板净距（或斜管孔径）为 0.8～1.0m；斜板（管）斜长为 1.0～1.2m；斜板（管）水平倾角一般宜为 60°；斜板（管）区上部水深，一般为 0.7～1.0m；斜板（管）

区底部缓冲层高度为1.0m。斜板（管）沉淀池应设冲洗设施。斜板沉淀池一般采用重力排泥，每日排泥至少1～2次，或连续排泥。

（2）斜板（管）沉淀池计算

1）斜板沉淀池表面积

$$A=\frac{Q_{max}}{0.91nq} \tag{2-44}$$

式中 A——每座斜板沉淀池表面积，m^2；

Q_{max}——最大设计流量，m^3/h；

q——表面水力负荷，$m^3/(m^2 \cdot h)$；

0.91——斜板区面积利用系数；

n——斜板沉淀池数量，至少2个。

2）斜板沉淀池直径或边长

$$D=\sqrt{\frac{4A}{\pi}} \text{或} B=\frac{A}{L} \tag{2-45}$$

式中 D——斜板沉淀池直径，m；

B——斜板沉淀池宽度，m；

L——斜板沉淀池长度，m。

3）水力停留时间

$$T=\frac{h_2+h_3}{q} \tag{2-46}$$

式中 T——水力停留时间，h；

h_2——斜管区上部清水层高度，m，一般为0.5～1m；

h_3——斜管自身垂直高度，m；

其他设计参数参照前述有关章节。

【例2-7】 某工业园区污水处理厂小时平均处理流量为$500m^3/h$，污水时流量变化系数$K_z=1.2$，初沉池采用升流式斜板沉淀池，进水悬浮物浓度$C_1=300mg/L$，出水悬浮物浓度$C_2=150mg/L$，污泥平均含水率96%，求斜板沉淀池各部分尺寸。

【解】（1）斜板沉淀池表面积

表面水力负荷q取$3.0m^3/(m^2 \cdot h)$，斜板沉淀池数量n取4，则斜板沉淀池表面积为：

$$A=\frac{Q_{max}}{0.91nq}=\frac{1.2\times500}{0.91\times3\times4}=55m^2$$

（2）斜板沉淀池边长

采用正方形池体，则斜板沉淀池边长B为：

$$B=\sqrt{A}=\sqrt{55}=7.4m$$

（3）水力停留时间

斜管区上部清水层高度$h_2=1.0m$，斜管的自身垂直高度$h_3=1.0m$，则：

$$T=\frac{h_2+h_3}{q}=\frac{1+1}{2}=1h$$

（4）污泥部分所需容积

设排泥时间间隔 $T=2\text{d}$，污泥含水率 $P_0=96\%$，斜板沉淀池数量 $n=4$，污泥密度 $\gamma=1000\text{kg/m}^3$，则：

$$V_w=\frac{Q_{max}\cdot 24(C_1-C_2)\cdot 100\cdot T}{1000\gamma(100-P_0)n}=\frac{500\times24\times(300-150)\times100\times2}{1000\times1000\times(100-96)\times4}=22.5\text{m}^3$$

（5）污泥斗容积

设污泥斗下部边长 $a_1=1.0\text{m}$，上部边长 $B=7.4\text{m}$，则污泥斗高度为：$h_5=\frac{(B-a_1)}{2}\tan60°=\frac{(7.4-1.0)}{2}\tan60°=5.54\text{m}$

$$V_1=\frac{h_5}{3}(B^2+a_1^2+a_1\cdot B)=\frac{5.54}{3}\times(7.4^2+1^2+7.4\times1)=116.6\text{m}^3>22.5\text{m}^3$$

满足要求。

（6）斜板沉淀池总高度

设斜板沉淀池超高 $h_1=0.3\text{m}$，缓冲层高度 $h_4=0.8\text{m}$，则斜板沉淀池总高度为：

$$H=h_1+h_2+h_3+h_4+h_5=0.3+1+1+0.8+5.54=8.64\text{m}$$

本章关键词（Keywords）

均化池	Equalization tank
格栅	Screen
沉淀	Precipitation
过滤	Filtration
浮选	Flotation
预处理和初级处理	Pretreatment and primary treatment
混凝剂	Coagulant
助凝剂	Coagulant aids

思考题

1. 某厂每小时排放生产废水 50m^3，废水 COD 变化周期为 8h，一个周期内各小时分别为 20mg/L、80mg/L、90mg/L、140mg/L、60mg/L、40mg/L、7mg/L、140mg/L，欲将污水 COD 平衡到 80mg/L，求需要的均化时间和均化池容积。

2. 某酸性污水的 pH 逐时变化为 5、6.5、4.5、5、7，若水量依次为 $4\text{m}^3/\text{h}$、$4\text{m}^3/\text{h}$、$6\text{m}^3/\text{h}$、$8\text{m}^3/\text{h}$、$10\text{m}^3/\text{h}$，试问完全均化后能否达到排放标准（pH 6～9）。

3. 简述混凝处理法的机理和影响因素。查阅混凝剂的研发进展，归纳对混凝剂改性的途径有哪几类？

4. 重金属的去除涉及哪些物理、化学方法？

5. 曝气沉砂池的工作原理与平流式沉砂池有何区别？设计平流式沉砂池先计算池长而曝气沉砂池先计算池容积，二者的设计思路能否互换？

6. 试说明沉淀有哪几种类型？并讨论各种类型的内在联系与区别，各自体现在哪些场合？

7. 对比平流式沉砂池和平流式沉淀池的停留时间、有效水深等设计参数，说明其异同点。

8. 已知某小型污水处理站设计流量 $Q=400\text{m}^3/\text{h}$，悬浮固体浓度 $SS=250\text{mg/L}$。设沉淀效率为

55%，污泥的含水率为98%。根据性能曲线查得$\mu_0=2.8$m/h，试为该处理站设计竖流式初沉池，并按照计算结果绘制结构草图。

9. 已知某城镇污水处理厂设计平均流量$Q=20000m^3/d$，服务人口100万人，初沉污泥量按25g/(人·d)计算，污泥含水率按97%计算，试为该厂设计曝气沉砂池和平流式沉淀池。

10. 微气泡与悬浮颗粒相黏附的基本条件是什么？有哪些影响因素？

11. 气固比的定义是什么？如何确定适当的气固比？

12. 在废水处理中，气浮法与沉淀法相比较，各有何优缺点？

13. 如何改进及提高沉淀或气浮分离效果？

14. 快滤池、过滤罐、滤元、滤纸分别属于哪种类型的过滤？机理有何不同？

15. 工程设计时过滤罐应放在气浮法系统之前还是之后？为什么？

Chapter 3 Introduction of Biological Oxidation Wastewater Treatment Processes

3.1 Purpose of Biological Oxidation

The overall objectives or purpose of the biological treatment of domestic wastewater are: (1) to transform (i. e., oxidize) dissolved and particulate biodegradable constituents into acceptable end products; (2) to capture and incorporate suspended and non-settleable colloidal solids into a biological floc or biofilm; (3) to transform or remove nutrients, such as nitrogen and phosphorus; and (4) in some cases, to remove specific trace organic constituents and compounds. For industrial wastewater, the objective is to remove or reduce the concentration of organic and inorganic compounds. Because some of the constituents and compounds found in industrial wastewater are toxic to microorganisms, pretreatment may be required before the industrial wastewater is discharged to the municipal collection system. For agricultural irrigation return wastewater, the objective is to remove nutrients, specifically nitrogen and phosphorus, which stimulate the growth of aquatic plants.

In Chapter 2, it was seen that primary sedimentation or settling can remove about 60% of the suspended solids and about 30% of the BOD_5. About 65% of the solids removed (settleable) are organic and the remainder inorganic.

The main purpose of secondary treatment is to reduce the BOD_5 value which does not benefit as much as SS from primary sedimentation or settling. In other words, secondary treatment should be a process that is capable of biodegrading the organic matter into non-polluting end products, e. g. H_2O, CO_2 and biomass (sludge). The end product liquid effluent should be well stabilized or well oxidized so that it does not provide a food source for aerobic bacteria in the receiving water body. Then the discharge to the water body should lead to little or no removal of dissolved oxygen by bacterial action. In order to produce a well-oxidized liquid effluent, a vast array of biological processes exist, some of which are general and some of which are proprietary, that are capable of removing the organic matter from the wastewater.

Secondary treatment systems are broadly categorized as: Suspended growth or activated sludge systems; Attached growth or fixed film systems and dual biological suspended and attached growth systems.

Suspended growth or activated sludge systems are defined as those aerobic processes that achieve a high microorganism concentration through the recycle of biological solids. The bacterial organisms convert biodegradable organic wastewater and certain inorganic fractions into new biomass and other (non-polluting) end products (e. g. water and carbon dioxide). The biomass is removed as sludge and the liquid after settling is removed as clarified effluent. The gases are air stripped. Suspended growth or activated sludge systems and, in particular, the conventional plug flow activated sludge systems are the most common processes for treating both municipal and industrial wastewater.

Attached growth or fixed film systems allow a microbial layer to grow on the surface of the media (stone, plastic) that is exposed to the atmosphere from where it draws its oxygen. The microbial layer is sprayed (from above) with the wastewater. In this way, the microbial layer converts the biodegradable organic wastewater into biomass and byproducts. The trickling or percolating filter is common for treating municipal wastewater that has already passed through primary settling. Rotating biological contactors (RBC) and trickling filters (biofilters) are in operation to handle higher strength wastes than municipal. They are sometimes used as roughing filters with subsequent activated sludge or stone media trickling filters for further treatment.

Dual process systems utilize two stage arrangements of attached growth or fixed film and suspended growth or activated sludge processes with the objective of achieving very high quality effluent standards.

3.2 Mechanisms of Organic Removal by Bio-oxidation

The major organic removal mechanism for most wastewater is bio-oxidation. It should be noted that, in treating industrial wastewater, the active microbial population must be acclimated to the wastewater in question. For the more complex wastewater, acclimation may take up to 6 weeks. When the sludge is being acclimated, the feed concentration of the organic in question must be less than the inhibition concentration or level.

BOD removal from wastewater by a biological sludge may be considered as occurring in two phases. An initial high removal of suspended, colloidal, and soluble BOD are followed by a slow progressive removal of remaining soluble BOD. Initial BOD removal is accomplished by one or more mechanisms, depending on the physical and chemical characteristics of the organic matter. These mechanisms are as following:

(1) Removal of suspended matter or solid by enmeshment in the biological floc. This removal is rapid and depends upon adequate mixing of the wastewater with the sludge.

(2) Removal of colloidal material by physicochemical adsorption on the biological floc.

(3) Biosorption of soluble organic matter by the microorganisms. There is some question as to whether this removal is the result of enzymatic complexing or is a surface phenomenon and whether the organic matter is held to the bacterial surface or will be in

the cell as a storage product or both. The amount of immediate removal of soluble BOD is directly proportional to the concentration of sludge present, the sludge age, and the chemical characteristics of the soluble organic matter.

The type of sludge generated markedly affects its sorptive properties. In general, biomass generated from a batch or plug flow configuration will have better sorptive properties than that generated from a complex mix configuration.

The three mechanisms begin immediately on contact of biomass with wastewater. The colloidal and suspended material must undergo sequential breakdown to smaller molecules in order that it may be made available to the cell for oxidation and synthesis. In an acclimated system, the time required for this breakdown is related primarily to the characteristics of the organic matter and to the concentration of active sludge. In a complex waste mixture at high concentrations of BOD, the rate of synthesis is independent of concentration as long as all components remain and, as a result, there is a constant and maximum rate of cellular growth. With continuing aeration, the more readily removable components are depleted and the rate of growth will decrease with decreasing concentration of BOD remaining in solution.

3.3 Process Selection Based on Reaction Kinetics

In process selection and sizing based on reaction kinetics, particular emphasis is placed on defining the nature of the reactions occurring within the process, the appropriate values of the kinetic coefficients, and the selection of the reactor type.

3.3.1 Nature of the Kinetic Reactions

The nature of the reactions occurring within a process must be known to apply the reaction kinetics approach to design. Selection of reaction rate expressions for the process that is to be designed is typically based on (1) information obtained from the literature, (2) experience with the design and operation of similar systems, or (3) data derived from pilot-plant studies. Fox example, it is of critical importance to know if the reaction is zero-, retarded first-, or second-order, or if the reaction is a saturation type.

The total volume required for various removal efficiencies for first-order kinetics, using complete-mix reactors was compared to the volume required for a plug-flow reactor, greater reactor volume is required for complete-mix reactors to achieve the same removal efficiencies as plug-flow reactors. It should be noted, however, that for zero-order kinetics the volume of the two reactors will be the same.

3.3.2 Selection of Appropriate Kinetic Rate Coefficients

Selection of appropriate kinetic rate coefficient for the process that is to be designed is also based on (1) information obtained from the literature, (2) experience with the design and operation of similar systems, or (3) data derived from pilot-plant studies. In cases where significantly different wastewater characteristics occur or new applications of existing technology or new processes are being considered, pilot-plant testing is recommended. The various rate expressions that have been developed for biological waste treatment, based on the method of analysis presented in this chapter, are considered in Chapter. 5 through 9.

3.3.3 Selection of Reactor Types

Operational factors that must be considered in the type of reactor or reactors to be used in the treatment process include (1) the nature of the wastewater to be treated, (2) the nature of the reaction kinetics governing the treatment process, (3) special process requirements, and (4) local environmental conditions. As noted previously, for biological treatment with the activated sludge process, there is no difference in the size of the reactor required. For example, a complete-mix reactor might be selected over a plug-flow reactor, because of its dilution capacity, if the influent wastewater is known to contain toxic constituents that cannot be removed by pretreatment. Alternatively, a plug-flow or multistage reactor might be selected over a complete-mix reactor to control the growth of filamentous microoganisms. In practice, the construction costs and operation and maintenance costs also affect reactor selection.

3.3.4 Process Selection Based on Mass Transfer

In addition to process selection based on reaction kinetic and loading criteria, a number of treatment processes will be based on mass transfer considerations. The principal operations in wastewater treatment involving mass transfer are aeration, especially the addition of oxygen to water; the drying of biosolids and sludge; the removal of volatile organics from wastewater; the stripping of dissolved constituents such as ammonia from digested supernatant; and the exchange of dissolved constituents as in ion exchange. Fortunately, there is a considerable body of literature on these subjects as well as a vast amount of practical experience. Additional details on these subjects are presented in the subsequent chapters.

3.4 Nutrient Requirements

Several mineral elements are essential for the metabolism of organic matter by microorganisms. All but nitrogen and phosphorus are usually present in sufficient quantity in the carrier water. An exception is process wastewater generated from deionized water or high-strength industrial wastewater. Iron and other trace nutrients may be deficient in this case.

Sewage provides a balanced microbial diet, but many industrial wastes (cannery, pulp and paper, etc.) do not contain sufficient nitrogen and phosphorus and require their addition as a supplement.

The quantity of nitrogen required for effective BOD removal and microbial synthesis has been the subject of much research. These represent average values derived from the treatment of several nitrogen supplemented industrial wastes. When insufficient nitrogen is present, the amount of cellular material synthesized per unit of organic matter removed increases as an accumulation of polysaccharide. At some point, nitrogen-limiting conditions restrict the rate of BOD removal. The rule-of-thumb number is BOD : N : P of 100 : 5 : 1. Nutrient-limiting conditions will also stimulate filamentous growth.

The nitrogen content of sludge as generated in the process has been shown to average 12.3% on the basis of the VSS. The nitrogen content of the sludge will decline during the endogenous phase. The nitrogen content of the non-degradable cellular mass has been shown to average 7%. The phosphorus content of sludge at generation has been found to average 2.6% with the non-degradable cellular mass having a phosphorus content of 1%.

Not all organic nitrogen compounds are available for synthesis. Ammonia is the most readily available form, and other nitrogen compounds must be conversed to ammonia. Nitrite, nitrate, and about 75% of organic nitrogen compounds are also available.

Phosphorus may be fed as phosphoric acid in larger plants and ammonia as an hydrous or aqueous ammonia. In small plants, nutrients may be fed as diammonium phosphate. In many cases, in aerated lagoons treating pulp and paper mill wastewater, nitrogen and phosphorus have not been added, but rather the retention time has been increased.

3.5 Toxicity

Toxicity in biological oxidation systems may have any of several causes:

(1) An organic substance, such as phenol, which is toxic in high concentrations, but biodegradable in low concentrations.

(2) Substances such as heavy metals that have a toxic threshold depending on the operating conditions.

(3) Inorganic salts and ammonia, which exhibit a retardation at high concentration.

The toxic effects of organics can be minimized by employing the complete mixing system, in which the influent is diluted by the aeration tank contents and the microorganisms are in contact only with the effluent concentration. In this way, wastes with concentrations many times the toxic threshold can be successfully treated.

Heavy metals exhibit a toxicity in low concentrations to biological sludge. Acclimation of the sludge to the metal, however, will increase the toxic threshold considerably.

While an acclimated biological process is tolerant of the presence of heavy metals, the metal will concentrate in the sludge by complexing with the cell wall. Metal concentrations in the sludge as high as 4% have been reposed. This in turn creates problems relative to ultimate sludge disposal.

High concentrations of inorganic salts are not toxic in the conventional sense, but rather exhibit progressive inhibition and a decrease in rate kinetics. Biological sludge, however, can be acclimated to high concentration of salt. Processes are successfully operating with as high as 6% salt by weight. Frequently, high salt content will increase the effluent suspended solids. Monovalent ions such as Na^+ and K^+ will disperse the biological flocs while divalent ions such as Ca^{2+} and Mg^{2+} will tend to aid flocculation.

3.6 生物处理方法的类型、过程

3.6.1 生物处理方法的类型

污水生物处理的对象包括污水中呈溶解状态和胶体状态的有机污染物和污水中呈溶解状态的氮和磷。去除BOD或COD是污水生物处理的最初目标，也是主要目标，然而近年来随着对水体富营养化问题的重视，N、P成为污水生物处理的去除对象。

根据参与代谢活动的微生物对溶解氧的需求不同，污水生物处理分为好氧生物处理、缺氧生物处理和厌氧生物处理三种类型。好氧生物处理是在水中有溶解氧（分子氧）存在的条件下进行的生物处理过程；缺氧生物处理是在水中无分子氧存在，但存在化合态氧（如硝酸盐等）的条件下进行的生物处理过程；厌氧生物处理是在水中既无分子氧又无化合态氧存在的条件下进行的生物处理过程。

图3-1归纳了常用生物处理方法的类型，其中人工条件下的污水处理速率比自然条件下的快，是当前主要的污水处理方式。人工生物处理方法中，活性污泥法以其高效、快捷而被广泛应用，大型城市污水处理厂基本上采用活性污泥法。

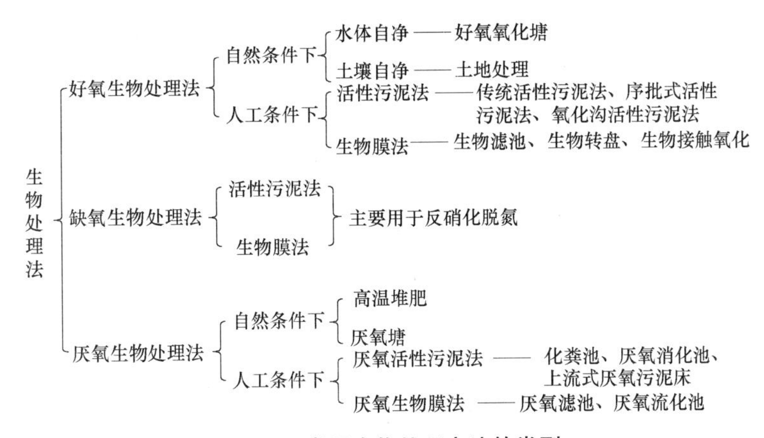

图3-1 常用生物处理方法的类型

3.6.2 生物处理方法的过程

1. 好氧生物处理法

好氧生物处理是在污水中有分子氧存在的条件下，利用好氧微生物（包括兼性微生物，但主要是好氧细菌）降解有机物，使其稳定、无害化的处理方法，本质上是微生物利用污水中存在的有机污染物（以溶解状和胶体状为主）作为底物进行好氧代谢。有机物被微生物吸收后，一部分被氧化分解为简单无机物，同时释放能量，作为微生物生命活动的能源；另一部分作为微生物生长繁殖所需的构造物质，合成新的生物体。于是，这些高能位的有机物经过一系列的生化反应，逐级释放能量，最终以低能位的无机物稳定下来，达到无害化的要求，以便返回自然环境或进一步处置。

污水好氧生物处理的过程可用图3-2表示。

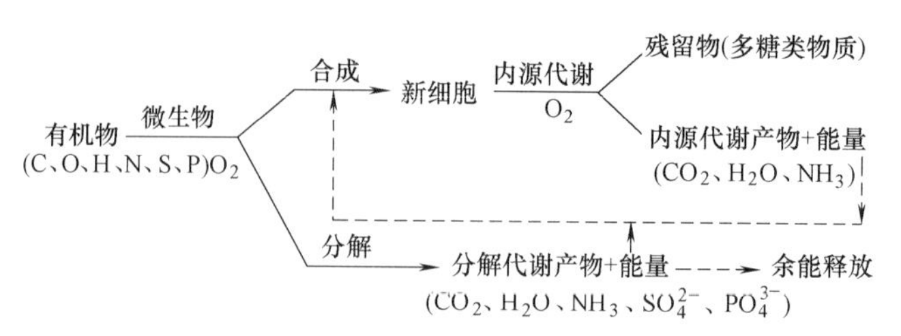

图 3-2　好氧生物处理过程

如图 3-2 所示，有机物被微生物摄取后，通过代谢活动，约有三分之一被分解、稳定，并提供其生理活动所需的能量，约有三分之二被转化，合成新的细胞物质，即进行微生物自身生长繁殖。后者就是污水生物处理中的活性污泥或生物膜的增长部分，通常称其为剩余活性污泥或生物膜，又称生物污泥。在污水的生物处理系统中，生物污泥经固液分离后，需进一步处理和处置，防止二次污染。

好氧生物处理的反应速度较快，所需的反应时间较短，故处理构筑物容积较小，且处理过程中散发的臭气较少。所以，目前对中、低浓度的有机废水，或者 BOD_5 小于 500mg/L 的有机废水，基本上采用好氧生物处理法。

2. 厌氧生物处理法

图 3-3 给出了厌氧生物处理污水的过程图。

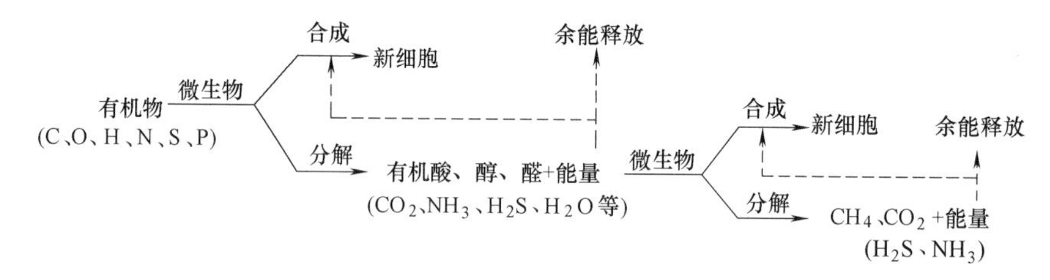

图 3-3　厌氧生物处理过程

如图 3-3 所示，在厌氧生物处理过程中，污水中的有机物可转化为：①可回收利用的甲烷气体；② CO_2、H_2O、NH_3、H_2S 等无机物，并为细胞合成提供能量；③新的细胞物质。其中，仅少量有机物用于新细胞物质的合成。因此，相对于好氧生物处理，厌氧生物处理的剩余污泥量少而且稳定。

由于废水厌氧生物处理过程不需另加氧源，故运行费用低。其主要缺点是反应速度较慢，反应时间较长，处理构筑物容积大等。为维持较快的反应速度，需维持较高的温度，就要消耗能源。

对于有机污泥和高浓度有机废水（一般 $BOD_5 \geqslant 2000$mg/L），可采用厌氧生物处理法。

3. 生物脱氮

水中的含氮化合物在微生物作用下被转化为氮气而从水中脱除的过程称为生物脱氮。氮在污水中主要以有机氮（如氨基酸、蛋白质、尿素等）和氨氮的形式存在，生活污水中有机氮占 40%～50%，氨氮占 50%～60%。在生物处理中，有机氮容易通过微生物转化成氨氮，该过程称为氨化，氨化可在好氧或厌氧条件下进行。

（1）传统的生物脱氮

在氨化的基础上，先通过硝化反应将氨氮转化为亚硝态氮、硝态氮，再通过反硝化反应将亚硝态氮、硝态氮还原成氮气从水中逸出，从而达到脱氮的目的。因此，传统生物脱氮包括硝化和反硝化两个过程。

1）硝化过程

在有溶解氧的条件下，好氧微生物将氨氮氧化成硝态氮的过程称为硝化，首先由亚硝酸盐细菌将氨氮转化为亚硝态氮（也称亚硝化作用），然后由硝酸盐细菌将亚硝态氮进一步转化为硝态氮。亚硝酸盐菌和硝酸盐菌合称为硝化菌，属于专性好氧自养菌。硝酸菌、亚硝酸菌的种类、常见形态见表 3-1。其他硝化菌的形态如图 3-4 所示。

硝酸菌、亚硝酸菌的种类、常见形态　　表 3-1

	亚硝酸菌	硝酸菌
种类	单胞菌属、球菌属、螺菌属、叶菌属	杆菌属、球菌属、囊菌属
形态特点	具有单生鞭毛	没有鞭毛
常见形态		

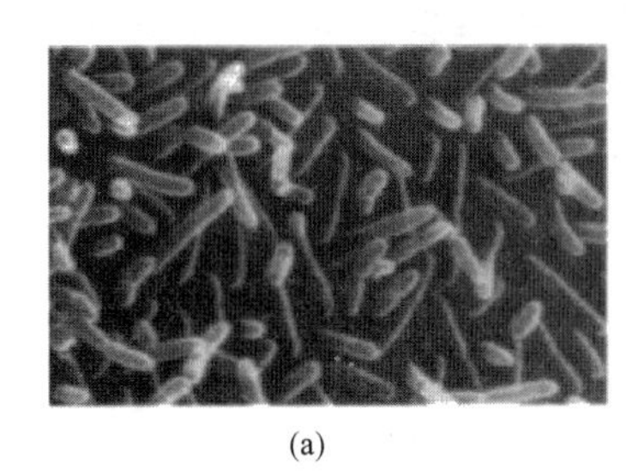
(a)

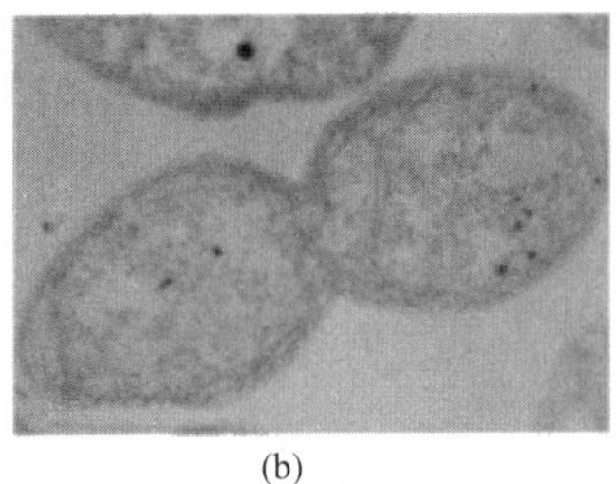
(b)

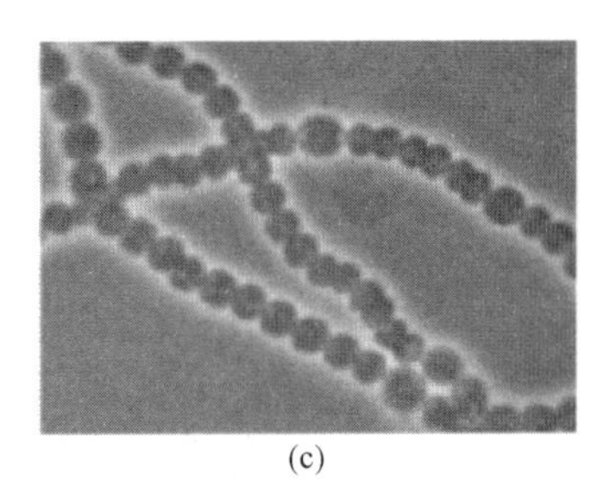
(c)

图 3-4　其他硝化菌的形态

（a）硝化杆菌；（b）硝化球菌；（c）串珠型硝化细菌

硝化过程的总反应式如下：

$$NH_4^+ + 2O_2 \rightarrow NO_3^- + H_2O + 2H^+ \tag{3-1}$$

由上述反应可知，硝化过程需要消耗水中的溶解氧，1g 氨氮完全硝化需氧 4.57g（其中亚硝化反应需氧 3.43g，硝化反应需氧 1.14g），此即硝化需氧量。同时，硝化过程中释放出 H^+，将消耗水中的碱度，1g 氨氮完全硝化消耗 7.11g 碱度（以 $CaCO_3$ 计）。因此，为了保持硝化过程适宜的 pH，污水中应有足够的碱度。

2）反硝化过程

在缺氧的条件下，反硝化菌将 NO_2^- 和 NO_3^- 还原为 N_2 的过程称为反硝化。反硝化菌大多数为兼性菌，在有氧环境下利用分子氧进行有氧呼吸，在无氧环境下，则利用 NO_2^- 或 NO_3^- 中的氧作电子受体、有机物作碳源及电子供体进行无氧呼吸，即反硝化。

反硝化菌的形态如图 3-5 所示。

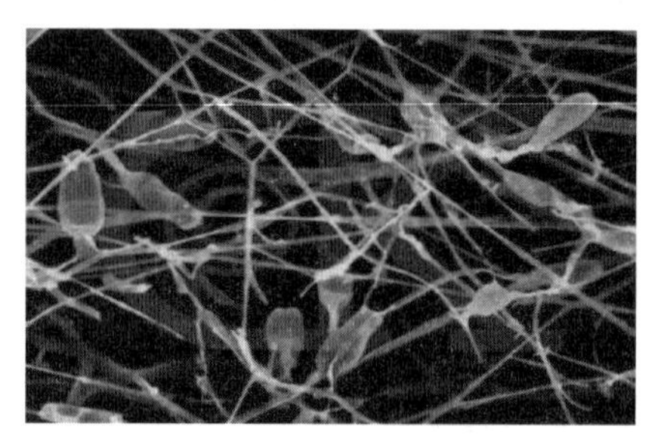

图 3-5 反硝化菌的形态

当以甲醇为碳源时，反硝化过程的反应式如下：

$$6NO_3^- + 5CH_3OH \xrightarrow{\text{反硝化菌}} 3N_2 + 5CO_2 + 7H_2O + 6OH^- \tag{3-2}$$

$$2NO_2^- + CH_3OH \xrightarrow{\text{反硝化菌}} N_2 + CO_2 + H_2O + 2OH^- \tag{3-3}$$

由上述反应可知，在反硝化过程中，不仅能使 NO_2^- 和 NO_3^- 还原，而且还可使有机物分解，并产生 OH^-。每还原 1g 硝态氮需提供有机物（以 BOD_5 计）2.86g，同时产生 3.57g 碱度（以 $CaCO_3$ 计）。

当环境中缺乏有机物时，微生物可通过消耗自身的原生质进行所谓的内源反硝化：

$$C_5H_7NO_2 + 4NO_3^- \rightarrow 5CO_2 + NH_3 + 2H_2\uparrow + 4OH^- \tag{3-4}$$

由上式可见，内源反硝化的结果是细胞物质减少，即反硝化菌减少，并会有氨的生成。因此，污水处理中不希望以此种反应为主，应提供足够的电子供体（碳源）。

此外，在生物处理过程中，污水中的一部分氮（氨氮或有机氮）被同化为微生物细胞的组成成分，并以剩余污泥的形式从污水中去除，此过程称为同化作用，当进水氨氮浓度较低时，同化作用可能成为脱氮的主要途径。

（2）新型生物脱氮

基于硝化-反硝化原理的传统生物脱氮工艺能耗高，还需要有足够的有机碳源作为反硝化电子供体，对于高浓度氨氮污水，上述问题更为突出。因此，国内外学者一直在找高效低耗的生物脱氮工艺，其中短程硝化-反硝化脱氮就是代表性的研究成果。

1）短程硝化-反硝化脱氮

由传统硝化-反硝化脱氮原理可知：硝化过程包括由两类细菌独立催化完成的两个不同反应，这两个反应可以分开；对于反硝化菌，NO_2^- 和 NO_3^- 均可作为最终受氢体。短程硝化-反硝化法就是控制硝化终止于亚硝化阶段，随后以 NO_2^- 作为受氢体进行反硝化，反应见式（3-5）、式（3-6）：

$$NH_4^+ + 1.5O_2 \xrightarrow{\text{亚硝酸盐菌}} NO_2^- + H_2O + 2H^+ \tag{3-5}$$

$$2NO_2^- + 6[H] + 2H^+ \xrightarrow{\text{反硝化菌}} N_2 + 4H_2O \tag{3-6}$$

2）厌氧氨氧化脱氮

厌氧氨氧化（Anaerobic Ammonium Oxidation，ANAMMOX）是 20 世纪末荷兰 Delft 大学开发的一种新型脱氮工艺，其基本原理是在厌氧条件下，以 NO_2^- 和 NO_3^- 为电子受体，将 NH_3-N 氧化成 N_2，或者说以 NH_3-N 作为电子供体，将 NO_2^- 和 NO_3^- 还原成 N_2。参与厌氧氨氧化的细菌是一种自养菌，在厌氧氨氧化过程中无须有机碳源存在。

厌氧氨氧化反应式及反应自由能如下：

$$NH_4^+ + NO_2^- \rightarrow N_2 + 2H_2O \qquad \Delta G = -358kJ/mol(NH_4^+) \tag{3-7}$$

$$5NH_4^+ + 3NO_3^- \rightarrow 4N_2 + 9H_2O + 2H^+ \quad \Delta G = -297kJ/mol(NH_4^+) \tag{3-8}$$

上述反应的 $\Delta G<0$，根据热力学理论可知反应能自发进行，可以提供能量供微生物生长。

3）亚硝酸型完全自养脱氮

亚硝酸盐完全自养脱氮（Completely Autotrophic Nitrogen-removal Over Nitrite，CANON）工艺的基本原理是先将氨氮氧化成亚硝态氮，控制 NH_4^+ 和 NO_2^- 比例为 1∶1，然后通过厌氧氨氧化作为硝化实现脱氮的目的，其反应式如下：

$$NH_4^+ + 1.5O_2 \rightarrow NO_2^- + H_2O + 2H^+ \tag{3-9}$$

$$NH_4^+ + NO_2^- \rightarrow N_2 + 2H_2O \tag{3-10}$$

全过程为自养的好氧亚硝化反应结合自养的厌氧氨氧化反应，无需有机碳源，对氧的消耗比传统硝化-反硝化减少 62.5%，同时减少碱消耗量和污泥生成量。

4）自养反硝化脱氮

自养反硝化脱氮是自养反硝化菌以还原性的无机物（如 S^{2-}、H_2 等）为电子供体，以 NO_2^- 和 NO_3^- 为电子受体进行的反硝化过程，其反应式如下：

$$5S + 6NO_3^- + 8H_2O \rightarrow 3N_2 + 5H_2SO_4 + 6OH^- \tag{3-11}$$

$$5H_2 + 2NO_3^- \rightarrow N_2 + 4H_2O + 2OH^- \tag{3-12}$$

4. 厌氧-好氧生物除磷

污水中磷的存在形态取决于污水的类型，最常见的是磷酸盐、聚磷酸盐和有机磷。常规二级生物处理出水中，90%左右的磷以磷酸盐的形式存在。

污水生物除磷（Biological Phosphorus Removal，BPR）基于这样一种现象：一类特殊的微生物（聚磷菌，Phosphorus Accumulation Organisms，PAOs）在厌氧状态下释放磷，而在好氧状态下可以过量地、超出其生理需要地从环境中摄取磷。其基本原理如图 3-6 所示，在厌氧段，兼性细菌将溶解性有机物通过水解发酵作用转化成乙酸等低分子挥发性脂肪酸（VFA），聚磷菌大量吸收这些脂肪酸，这一过程导致磷酸盐的释放；在好氧段，聚磷菌过量摄取水体中的磷酸盐，并转化为聚磷酸盐储存在菌体内，所需能量来自胞内 PHB 的好氧分解。由于好氧吸磷量远大于厌氧放磷量，故通过剩余污泥的排放可达

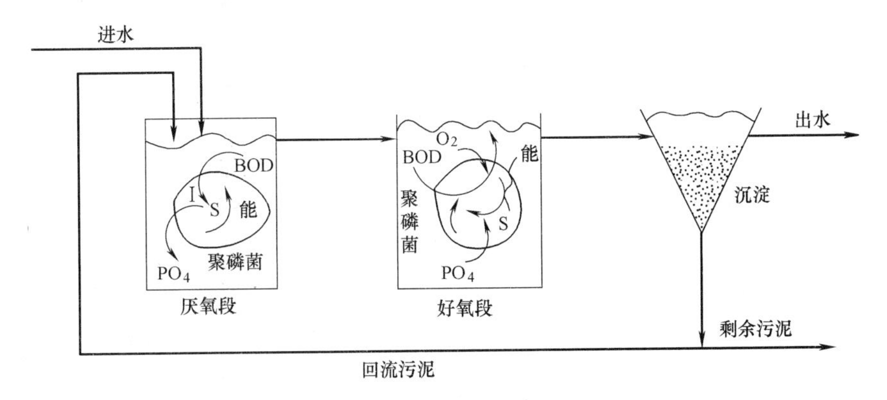

注：I—以 PHB 等形式储存在细胞内的食料；S—储存在菌体内的聚磷酸盐。

图 3-6　污水生物除磷机理

到除磷的目的。

由聚磷菌超量吸磷而形成的富磷活性污泥，含磷量为5%～10%，甚至高达30%，而一般活性污泥含磷量不足3%。实际生物除磷系统中活性污泥的含磷量取决于活性污泥中磷菌的份额，份额越高，除磷能力越强，污泥含磷量越大。此外，在厌氧状态下释放磷越多，合成的PHB越多，则在好氧状态下合成的聚磷量也越多，除磷效果也就越好。聚磷菌形态如图3-7所示。

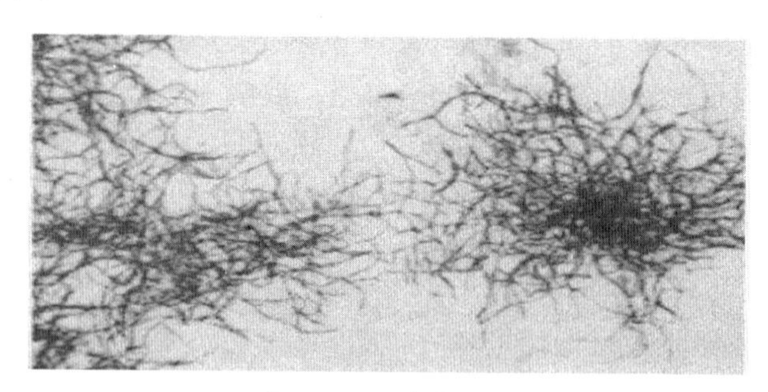

图3-7 聚磷菌形态

好氧条件下，细胞内PHB含量随时间呈指数关系减少；厌氧条件下则呈线性关系增加，且PHB的增加与胞内聚磷的减少呈线性关系。在一定条件下，聚磷菌厌氧有效释磷越彻底，在好氧条件下的吸磷量就越大。

一部分聚磷菌具有脱氮功能，在无游离氧条件下可利用硝酸盐作为电子受体，将硝酸盐还原为N_2或N_xO_y，同时还可大量吸磷。当厌氧段混入硝酸盐时，一部分易降解碳源被反硝化利用，对聚磷菌释磷产生不利影响。

聚磷菌厌氧释磷的程度与基质类型关系很大，一般可直接利用的基质主要为短链挥发性脂肪酸，其他基质则需要转化为短链挥发性脂肪酸后才能被利用。

上述脱氮和除磷是由不同的微生物独立完成的，研究发现，通过反硝化除磷兼性菌的作用，能将反硝化脱氮与生物除磷有机地合二为一，即在无分子氧但有硝态氮的条件下，反硝化除磷菌以硝态氮为电子受体，以简单有机物和自身碳源储存物——聚β羟基丁酸盐为电子供体，通过氧化电子供体产生能量，产生与传统聚磷菌在好氧条件下相同的生物吸磷过程，合成APT、核酸和多聚磷酸盐，在生物吸磷的同时，硝态氮被还原成氮气。显然，被反硝化除磷菌合并后的反硝化除磷过程能够节省碳源和分子氧，同时细胞合成量也较少。

3.7 生物法处理动力学

3.7.1 微生物增长动力学与莫诺特（Monod）方程

当微生物增长所需要的必要条件（如外部电子受体、适宜的物理、化学环境等）都具备时，则对于某一时间增量Δt，微生物浓度的增量ΔX与现存微生物浓度X成正比，即：

$$\Delta X \propto X \cdot \Delta t \tag{3-13}$$

引入比例系数μ，则式（3-13）可改写为等式：

$$\Delta X=\mu \cdot X \cdot \Delta t \tag{3-14}$$

式（3-14）两端同除以 Δt，并取极限 $\Delta t \rightarrow 0$，得到微分：

$$\frac{\mathrm{d}X}{\mathrm{d}t}=\mu \cdot X \tag{3-15}$$

式中 $\frac{\mathrm{d}X}{\mathrm{d}t}$——微生物的增长速度，即单位时间内单位体积反应器中微生物增长量，mg/(L·d)。

式（3-15）两边同除以 X，则有：

$$\mu=\frac{\mathrm{d}X}{\mathrm{d}t} \cdot \frac{1}{X} \tag{3-16}$$

由式（3-16）可知，μ 表示单位微生物的增长速率，称为比增长速度（或称比增长率），单位为 1/d。

微生物增长速度和微生物本身浓度、底物浓度之间关系是污水生物处理中的一个重要课题。1942 年，法国学者 Monod 在研究微生物生长的大量实验数据的基础上，得到微生物比增长速度与底物浓度之间的关系式：

$$\mu=\mu_{\max} \cdot \frac{S}{K_{\mathrm{S}}+S} \tag{3-17}$$

式中 $\mu_{\max}$——微生物最大比增长速度，1/d。

式（3-17）即 Monod 方程，其形式与米氏方程相同，式中的动力学参数 $\mu_{\max}$ 和 K_{S} 可通过试验获得数据后，采用双倒数作图法求取。

3.7.2 底物降解动力学

微生物增长是底物降解的结果。对于某一特定的污水，微生物增长速率与底物降解速率之间存在一定的比例关系：

$$\frac{\mathrm{d}X}{\mathrm{d}t}=Y \frac{\mathrm{d}S}{\mathrm{d}t} \tag{3-18}$$

式中 Y——微生物产率系数，即降解单位重量底物所合成的生物量；

$\frac{\mathrm{d}S}{\mathrm{d}t}$——底物降解（利用）速率。

在式（3-18）的两边同时除以 X，则有：

$$\mu=Y \cdot q \tag{3-19}$$

式中 q——底物比降解（利用）速率，$q=\frac{1}{X} \cdot \frac{\mathrm{d}S}{\mathrm{d}t}$。

将式（3-17）代入式（3-19）中，并定义$q_{\max}=\frac{\mu_{\max}}{Y}$，称为最大底物比降解速率，得

$$q=q_{\max} \frac{S}{K_{\mathrm{S}}+S} \tag{3-20}$$

式（3-20）是 1970 年劳伦斯（Lawrence）和麦卡蒂（Mc Karty）根据莫诺特方程提出的，称为劳-麦方程，反映底物比降解速率与底物降解浓度之间的关系。当底物浓度 S 高，即 $S \gg K_{\mathrm{S}}$ 时，$K_{\mathrm{S}}+S \approx S$，式（3-20）可简化为 $q=q_{\max}$，即底物以最大比降解速率进行降解，不受底物浓度影响，呈零级反应，此时微生物处于对数增殖期，其酶系统全部

被底物所饱和；当底物浓度 S 低，即 $S \ll K_S$ 时，$K_S + S \approx K_S$，式（3-20）可简化为 $q = q_{max}S$，即底物比降解速率与底物浓度成正比，是一级反应，此时微生物处于稳定期或衰亡期，其酶系统已饱和。

3.7.3 微生物净增长与底物降解的关系

以上讨论中的微生物增长量均为合成量，并未考虑因内源代谢而减少的量，而在污水生物处理中，通常控制微生物处于平衡期或内源代谢初期，在新细胞合成的同时，部分微生物也存在内源呼吸而导致微生物产量的减少，实际工程中采用考虑了内源代谢的微生物增长速率，即净增长速率。

微生物的净增长速率为微生物合成（增长）速率与内源代谢速率之差，即：

$$\left(\frac{dX}{dt}\right)_g = \frac{dX}{dt} - \left(\frac{dX}{dt}\right)_t \tag{3-21}$$

式中 $\left(\frac{dX}{dt}\right)_g$——微生物净增长速率；

$\left(\frac{dX}{dt}\right)_t$——微生物内源代谢速率，与现阶段微生物浓度 X 成正比，即：

$$\left(\frac{dX}{dt}\right)_t = K_d X \tag{3-22}$$

式中 K_d——单位重量微生物内源呼吸时的自身氧化速率，1/d，也称衰减系数。

将式（3-18）和式（3-22）代入式（3-21），得：

$$\left(\frac{dX}{dt}\right)_t = Y\frac{dS}{dt} - K_d X \tag{3-23}$$

式（3-23）两边同时除以 X，可得：

$$\mu' = Y \cdot q - K_d \tag{3-24}$$

式中 μ'——微生物比净增长速率，$\mu' = \frac{1}{X}\left(\frac{dX}{dt}\right)_t$。

由式（3-24）可知，当微生物处于高比增长率时，有 $\mu' \gg K_d$，则 $Y \cdot q \gg K_d$，即 $q \gg K_d/Y$，说明底物降解速率远远大于内源呼吸速率，此时微生物处于对数生长期，内源呼吸可以忽略；当微生物比增长率降低，$Y \cdot q$ 相应降低并逐渐接近 K_d，q 也减少并逐渐接近 K_d/Y，表明此时微生物利用的底物大部分用于维持生命活动，而不是用于微生物增长。

式（3-24）表示微生物在低比增长率情况下微生物自身氧化对增长率的影响。在实际工程中，这种影响常用一个实测产率系数 Y_{obs}（或称表观产率系数）表示，即

$$\mu' = Y_{obs} \cdot q \tag{3-25}$$

式（3-24）和式（3-25）均表达生物反应器内微生物的净增长与底物降解之间的基本关系。所不同的是，式（3-24）要求从微生物的理论产量中减去维持生命所自身氧化的量，而式（3-25）描述的是考虑了总的能量需要量之后的实际观测产量。

式（3-17）、式（3-20）及式（3-24）或式（3-25）是污水生物处理工程中常用的基本动力学方程式，式中的 K_S、μ_{max}、Y、K_d、Y_{obs} 等动力学系数可通过实验求得。在实践

中，根据所研究的特定处理系统，通过建立微生物量和底物量的平衡关系，可以建立不同类型生物处理设备的数学模型，用于生物处理工程的设计和运行管理。

3.8 污水的可生化性

污水的可生化性（即可生物处理性）是判断污水能否采用生物处理法处理的主要依据，也是影响污水生物处理效率的关键性能。

污水可生化性是指污水中所含的污染物通过微生物的生命活动所能去除的程度。对污水进行可生化性研究只研究可否采用生物处理，并不研究分解成什么产物，即使有机污染物被生物污泥吸附而去除也是可以的。因为在停留时间较短的处理设备中，某些物质来不及被分解，允许其随污泥进入消化池逐步分解。事实上，生物处理并不要求将有机物全部分解成 CO_2、H_2O 和硝酸盐等，而只要求将水中污染物去除到环境所允许的程度。

研究污水可生化性的目的，在于考察微生物对污染底物的生物降解能力和降解速率，从而确定是否需要在二级处理前采取必要的预处理手段以提高污水的可生化性。根据生物降解难易程度和降解速度，可将有机污染物分为如下四类：

（1）易降解有机物，且无生物毒性或抑制作用。这类物质是可立即被微生物作为碳源和能源利用的有机物，如简单的糖、氨基酸、脂肪酸等。

（2）可降解有机物，但有生物毒性或抑制作用。如对二苯甲酸、聚乙烯醇、烷基苯磺酸钠等，这类有机污染物可逐步被微生物分解利用，但需要对微生物进行驯化。驯化期间基本不发生生物降解，驯化完成后，有机物可作为微生物唯一的碳源和能源而被降解。驯化期长短与有机物的性质、浓度有关，反映污染物生物降解的难易程度。

（3）难降解有机物，但无生物毒性或抑制作用。这类有机物生物降解速度十分缓慢或基本不被降解，如天然高分子有机污染底物木质素、纤维素等，及人工合成的有机污染物，如有机氯化物（如六六六）、多氯联苯、喹啉等。

（4）难降解有机物，且有生物毒性或抑制作用。这类有机物对微生物毒性强，不但不可被微生物降解，而且会抑制微生物的生命活动。因此，一方面可采用物化手段进行预处理，以降低其在污水中的浓度，从而削弱其对微生物的毒性；另一方面可采用厌氧水解酸化预处理，由抗冲击能力较强的水解发酵菌群将污染物质的复杂分子结构打破，转变为简单的小分子结构，提高其可生化性，以利于后续生物处理过程。

表 3-2 总结了可生化性的评价方法。

可生化性评价方法 **表 3-2**

分类	方法	方法要点	方法评价
根据氧化所耗氧量	水质指标法	采用 BOD_5/COD 作为有机物评价指标。方法改进：日本通产省测试法，TOD 代替 COD，采用 BOD 自动测定仪测定有机物 28d 的生化需氧量，并以 BOD_{28}/TOD 评价有机物的生物降解性能	比较简单，但精度不高，可粗略反映有机物的降解性能
	瓦呼仪法	根据有机物的生化呼吸线与内源呼吸线的比较来判断有机物的生物降解性能。测试时，接种物可采用活性污泥，接种量为 1～3gSS/L	较好地反应微生物氧化分解特性，但试验水量少，对结果有影响

续表

分类	方法	方法要点	方法评价
根据有机物的去除效果	静置烧瓶筛选试验	以10mL沉淀后的生活污水上清液作为接种物，90mL含有5mg酵母膏和5mg受试物的BOD标准稀释水作为反应液，两者混合，室温下培养，1周后测受试物浓度，并以该培养液作为下周培养的接种物，如此连续4周，同时进行已知降解化合物的对照试验	操作简单，但在静态条件下混合及充氧不好
	振荡培养试验法	在烧瓶中加入接种物、营养液及受试物等，在一定温度下振荡培养，在不同的反应时间内测定反应液中受试物含量，以评价受试物的生物降解性	生物作用条件好，但吸附对测定有影响
	半连续活性污泥法	测试时，采用试验组及对照组两套反应器间歇运行，测定反应器内COD、TOD或DOC的变化，通过两套反应器结果的比较评价受试物的生物降解性	试验结果可靠，但仍不能模拟处理厂实际运行条件
	活性污泥模型试验	模拟连续流活性污泥法生物处理工艺，采用试验组与对照组，通过两套系统对比和分析来评价	结果最为可靠，但方法较复杂
根据 CO_2 量	斯特姆测试法	采用活性污泥上清液作为接种液，反应时间28d，温度25℃，有机物降解以 CO_2 产量占理论 CO_2 产量的百分率判断	系统复杂，可反映有机物的无机化程度
根据微生物生理生化指标	主要有ATP测试法、脱氢酶测试法、细菌标准平板计数测试法等		试验结果可靠，但测试程序较为复杂

可生化性的评价还需注意生物处理方法、有机污染物浓度、共存污染物的影响，微生物的种属和浓度等。

（1）生物处理方法

上述各方法用于评价好氧生物处理时有机物的可生化性，但好氧微生物难以降解的有机物，对于厌氧微生物而言不一定是难降解的。即使同属好氧活性污泥法的各种方法，由于处理条件不同，也会表现出对同一有机物的不同降解性，如普通活性污泥法不能降解的某些有机物，在延时曝气法系统中却能得到一定程度的降解。

（2）有机污染物浓度

一些有机物在低浓度时毒性较小，可以被微生物降解，但在浓度较高时，则表现出对微生物的强烈毒性，如常见的酚、氰、苯等物质，酚浓度在1%时是一种良好的杀菌剂，但在300mg/L以下，则可被经过驯化的微生物降解。

（3）共存污染物的影响

污水中常含有多种污染物，这些污染物在污水中可能出现复合、聚合等现象，从而增大其抗降解性；有毒物质之间的混合往往会增大毒性作用。因此，对水质成分复杂的污水，不能简单地以某种化合物的存在判断污水生化处理的难易程度。

（4）微生物的种属和浓度

所接种微生物的种属是极为重要的影响因素，不同的微生物具有不同的酶诱导特性，在底物的诱导下，一些微生物能产生相应的诱导酶，而有些微生物则不能，从而对底物的降解能力就不同。目前，国内外的生物处理系统大多采用混合菌种，通过污水的驯化进行自然的诱导和筛选，驯化程度的好坏，对底物降解效率有很大影响，如处理含酚废水，在驯化良好时，酚的可接受浓度可由每升几十毫克提高到500～600mg/L。

此外，进行可生化性试验时，还必须考虑微生物的浓度。如果浓度过低，则培养时间就会很长；浓度过高时，由于微生物的吸附能力强，会因吸附作用使溶液中的有机物浓度降低，难以正确计算有机物的降解率。因此，微生物浓度应尽可能与实际浓度相同。

本章关键词（Keywords）

生物氧化	Biological oxidation
好氧处理	Aerobic treatment
厌氧处理	Anaerobic treatment
生物脱氮	Biological nitrogen removal
生物除磷	Biological phosphorus removal
莫诺特方程	Monod equation
可生化性	Biodegradability

思考题

1. 好氧生物处理常用的方法有哪些？
2. 简述好氧生物处理的过程。
3. 简述厌氧生物处理的过程。
4. 对比分析好氧生物处理与厌氧生物处理的异同点。
5. 简述生物脱氮机理。
6. 简述城镇污水生物脱氮过程的基本步骤。
7. 简述生物除磷的原理。
8. 简述微生物净增长与底物降解的关系。
9. 如何提高污水的可生化性？

Chapter 4 Fixed Film Processes

4.1 Introduction of Fixed Film Processes

Fixed film system is a biological treatment process that supports biomass on its surface and in its porous structure using natural or synthetic materials as media.

In fixed film or attached growth processes, the microorganisms responsible for the conversion of organic material are attached to an inert packing material. The organic material and nutrients are removed from the waste water flowing past the attached growth also known as a biofilm. Packing materials used in fixed film or attached growth processes include rock, gravel, slag, sand, redwood, and a wide range of plastic and other synthetic materials. Fixed film processes can also be operated as aerobic or anaerobic processes. The packing can be submerged completely in liquid or not submerged, with air or gas space above the biofilm liquid layer.

The principal advantages claimed for these aerobic fixed film or attached growth processes over the activated-sludge process are as follows:

(1) Less energy required;

(2) Simpler operation with no issues of mixed liquor inventory control and sludge wasting;

(3) No problems of bulking sludge in secondary clarifiers;

(4) Better sludge thickening properties;

(5) Less equipment maintenance needs;

(6) Better recovery from shock toxic loads.

Compared with the activated sludge process, the disadvantages of the fixed film or biofilm process are as follows:

(1) Active organisms are difficult to control artificially and thus are less flexible in terms of operation.

(2) Since the specific surface area of the carrier material is small, the volumetric load of the device is limited and the space efficiency is relatively low. The investment is usually more than the activated sludge process.

(3) The treated effluent often contains a large peeling of fixed film or biofilm, which reduces the clarity of effluent.

In general, the actual limitations of the fixed film (1) make it difficult to accomplish biological nitrogen and phosphorus removal compared to single-sludge biological nutrient removal suspended growth designs, and (2) result in an effluent with a higher turbidity than activated-sludge treatment. Trickling filters and Rotating biological contactors have also been used in combined processes with activated sludge to utilize the benefits of both processes, in terms of energy savings and effluent quality.

4.2 Classification of Fixed Film Processes

4.2.1 Trickling Filter

The most common aerobic fixed film process used is the trickling filter in which wastewater is distributed over the top area of a vessel containing non-submerged packing material. Historically, rock was used most commonly as the packing material for trickling filters, with typical depths ranging from 1.25 to 2m.

When organic wastewater is sprayed over stones or plastic, a microbial slime layer develops on the surface. This layer has the effect of reducing the BOD_5 of effluents. Traditionally the stones were grouped in a shallow, open-topped cylinder about 1m deep. The stones were of the order of 25 to 100mm in size. In more recent times, plastic media have been used instead of stones to encourage the growth of the microbial layer on media with very high surface area and volume ratios. This type of aerobic process is called the attached growth or fixed film system. The earliest version which is still widely used is the percolating or trickling filter.

Most modern trickling filters vary in height from 5 to 10m and are filled with a plastic packing material for biofilm attachment. The plastic packing material is designed such that about 90% to 95% of the volume in the tower consists of void space. Air circulation in the void space, by either natural draft or blowers, provides oxygen for the microorganisms growing as an attached biofilm. Influent wastewater is distributed over the packing and flows as a non-uniform liquid film over the attached biofilm. Excess biomass sloughs from the attached growth periodically and clarification is required for liquid/solids separation to provide an effluent with an acceptable suspended solids concentration. The solids are collected at the bottom of the clarifier and removed for waste-sludge processing. The packed-bed reactor is filled with some type of packing materials, such as rock, slag, ceramic, or, now more commonly, plastic. With respect to flow, the packed-bed reactor can be operated in either the down-flow or up-flow mode. Dosing can be continuous or intermittent, for example in trickling filter.

4.2.2 Rotation Biological Contactors

The rotating biological contactor (RBC) process consists of a series of closely spaced discs (3~3.5m in diameter) mounted on a horizontal shaft and rotated while about one-half of their surface area is immersed in wastewater. The discs are typically constructed of light-weight plastic. The speed of rotation of the discs is adjustable.

When the process is placed in operation, the microbes in the wastewater begin to adhere to the rotating surfaces and grow there until the entire surface area of the discs is covered with a 1～3mm layer of biological slime. As the discs rotate, they carry a film of wastewater into the air; this wastewater trickles down the surface of the discs absorbing oxygen. As the discs complete their rotation, the film of water is mixed with the reservoir of wastewater adding to the oxygen in the reservoir and mixing the treated and partially treated wastewater. As the attached microbes pass through the reservoir, they absorb other organic compounds for breakdown. The excess growth of microbes is sheared from the discs as they move through the reservoir. These dislodged organisms are kept in suspension by the moving discs. Thus, the discs provide media for the buildup of attached microbial growth, bring the growth into contact with the wastewater and aerate the wastewater and the suspended microbial growth in the reservoir.

As the treated wastewater flows from the reservoir below the discs, it carries the suspended growth out to a downstream settling basin (sedimentation tank) for removal. The process can achieve secondary effluent quality or better. By placing several sets of discs in series, it is possible to achieve even higher degrees of treatment, including biological conversion of ammonia to nitrates.

4.2.3 Submerged Attached Growth Process

Submerged attached growth or biological contact oxidation process, also known as submerged biological aerated filter, is developed on the basis of biological filter. The contact oxidation tank is equipped with biofilm carriers, and the biofilm carriers are submerged in the sewage. In the process of wastewater contact with biofilm, organic matter in water is adsorbed, oxidized and decomposed by microorganisms and transformed into new biofilm. The biofilm falling off from the packing is removed after flowing to the secondary sedimentation tank, and the sewage is purified. The air enters the water flow through the air distribution device at the bottom of the tank, and provides oxygen to the microorganism when the bubble rises.

4.3 生物膜法的净化机理

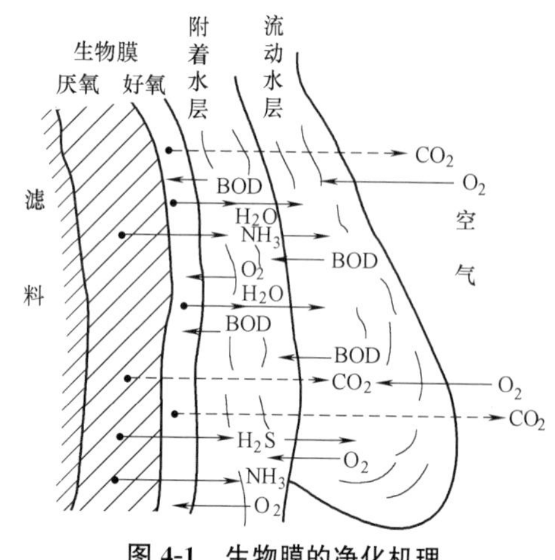

图 4-1 生物膜的净化机理

生物膜法去除水中污染物是一个吸附、稳定的复杂过程，包括污染物在液相中的紊流扩散、在膜中扩散传递、氧向生物膜内部扩散和吸附、有机物的氧化分解和微生物的新陈代谢等过程。生物膜的净化机理如图 4-1 所示。

从图 4-1 可见，生物膜与水层之间进行着多种物质的传递过程，废水进入滤池并在滤料表面流动时，依靠生物膜的吸附作用，废水中的有机物透过生物膜表面的附着水层，从废水主体向生物膜内部迁移；同时，空气中的氧通过膜表面附着水层进入生物膜。生物膜中的微生物在有氧条件下进行新陈代谢，对有机物进行降解，降解产物沿相反方向从生物膜经过附着水层排泄到流动水层或空气中去。

由于新生生物膜的厚度较薄，生物膜内物质传输的阻力较小、速率快，于是大部分污染物在生物膜表面去除，代谢产物从生物膜排出的速率也很快，所以新生的生物膜受产物积累的抑制较弱，生物膜活性高，底物去除率也高。但当生物膜厚度增长到一定程度或有机物浓度较大时，迁移到生物膜的分子氧主要被膜表层的微生物所消耗，导致生物膜内部供氧不足，出现厌氧层，伴随生物膜的增厚，内层微生物不断死亡并解体，大幅降低了生物膜与滤料的黏附力。于是老化的生物膜在自重和过流污水冲刷的共同作用下自行脱落，裸露后的滤料又重新开始新的生物膜生长，这一过程称作生物膜的更新。

在生物膜处理系统中，保持生物膜正常的新陈代谢和生物膜内微生物的活性是保证该系统去除水中污染物的前提条件。

实际运行中，影响污染物去除效果的因素涉及3个方面：一是废水水质特性，如底物浓度、底物可生物降解性等；二是生物膜自身特性，如生物膜厚度、生物膜活性、生物膜内菌群结构等；三是生物膜处理过程控制模式（进水和曝气方式）和特性参数（水力停留时间、污泥负荷、水力负荷等）。

1. 进水底物的组分和浓度

底物浓度的改变会导致生物膜的特性和剩余污泥量的变化。稳定的生物膜系统中，短时底物浓度升高会增加传质推动力，促进生物膜生长；短时底物浓度降低时，底物在生物膜内传质推动力降低，底物多在生物膜表面得到降解，系统的处理效能仍然良好，故生物膜对底物的低浓度耐受力一般高于活性污泥。

由于生物膜较活性污泥吸附能力强，而且成熟生物膜内菌群丰富，好氧菌和厌氧菌在不同区域共存，当难降解污染物被吸附后，可缓慢被厌氧区菌群水解为简单、小分子物质，污水的可生化性得到提高。因此，生物膜较活性污泥更能承受底物的难降解性。

2. 生物膜的厚度与活性

生物膜过厚会阻碍底物向内部传质，使内部菌群活性降低，内层生物膜与载体间黏附减弱而造成生物膜脱落；同时生物膜孔隙内积累的大量杂质或代谢产物，会阻碍底物传质，并对微生物造成毒性或抑制作用，使生物膜活性降低。适宜的生物膜厚度因生物膜过程模式不同而异，如淹没式生物滤池中生物膜厚度一般为300～400μm，而好氧生物转盘生物膜厚度可控制在3mm以内。

生物膜活性与厚度直接相关。研究表明，在考虑了生物膜密度的因素下，厚度小于20μm的生物膜层为高活性区。因此，工程运行中，为保持高活性的生物膜，需采取反冲洗，以维持系统效能稳定。

生物膜系统去除污染物的效能高低还取决膜内的菌群结构，所以工程运行中，需要通过宏观过程控制影响生物膜的微环境条件，进而形成稳定的菌群结构。

3. 控制参数

生物膜反应器过程控制方式主要包括进水方式、供氧方式、反冲洗方式等。一般来讲，进水方式的选择需考虑布水均匀、控制水流剪切力以维持生物膜厚度、冲走脱落的生物膜防止堵塞等，常见的有直流式进水（包括升流式和降流式）和侧流式进水等。供氧方式根据反应器类型不同而异，供氧方式的选择首先满足DO的供给，其次要考虑气、水混合效能，以及气流对生物膜的剪切等因素；由于生物膜内DO传质阻力较活性污泥大，故两者除污效率相同时，生物膜系统要求的DO水平要高于活性污泥系统。目前反冲洗方式主要为气、水联合反冲洗，需考虑反冲洗气、水强度及反冲洗时间等参数，并以更新生物膜但不损伤生物膜为原则。

在实际工程中，需要根据除污效率和反应器类型，确定适合的水力停留时间 *HRT*、有机负荷和水力负荷。负荷是影响生物膜法处理能力的首要因素，是集中反映膜处理系统工作性能的参数。其中污泥负荷直接决定生物膜厚度、活性及生物膜内菌群结构。污泥负荷又称为有机负荷，是指在保证处理水达到水质要求的前提下，单位体积或面积滤料每天所能承受的有机物（通常以 BOD_5 表示）的量，用 N 表示，其中单位体积滤料每天所能承受有机物的量称为 BOD_5 容积负荷，用 N_v 表示，单位是 $kgBOD_5/(m^3 \cdot d)$；单位面积滤料每天所能承受有机物的量称为 BOD_5 面积负荷，用 N_s 表示，单位是 $kgBOD_5/(m^2 \cdot d)$。有机负荷高，则传质速率快，膜内菌群活性高、生长快，生物膜迅速增厚，运行周期短。水力负荷是以污水量来计算的负荷，单位是 $m^3/(m^3 \cdot d)$ 或 $m^3/(m^2 \cdot d)$，后者又可称为滤率 q。水力负荷是运行过程中决定生物膜厚度的主要参数，水力负荷高，则水流剪切力强，老化生物膜可及时脱落，同时生物膜系统不易堵塞，延长运行周期。但水力负荷高也相当于缩短水力停留时间，势必影响系统除污效率，故必要时可采用处理水回流以增加水力负荷，如高负荷生物滤池。

4.4 生物膜法的构造、流程和设计计算

4.4.1 生物滤池

1. 生物滤池的构造

生物滤池由滤床及池体、布水设备和排水系统等部分组成。

（1）滤床及池体

普通生物滤池的池体在平面上多呈方形、矩形或圆形；池壁多用砖石筑造，一般应高出滤料表面0.5～0.9m，具有围护滤料的作用，并防止风力对池表面均匀布水的影响。

滤床由滤料组成。滤料是微生物生长栖息的场所，理想的滤料应具备下述特性：①能为微生物附着提供大量的表面积；②使污水以液膜状态流过生物膜；③有足够的孔隙率，保证通风（即保证氧的供给）和使脱落的生物膜能随水流出滤池；④不被微生物分解，也不抑制微生物生长，有良好的生物化学稳定性；⑤有一定机械强度；⑥价格低廉。早期国内外一般多采用碎石、卵石、炉渣和焦炭等实心拳状无机滤料，其粒径在3～8cm左右，孔隙率在45%～50%左右，比表面积在65～100m^2/m^3之间。但近年来已经广泛使用由聚氯乙烯、聚苯乙烯和聚酰胺等材料制成的呈波形板状、多孔筛状和蜂窝状等人工有机滤料，具有比表面积大（100～200m^2/m^3）和孔隙率高（80%～95%）的优势，可以大幅提高滤池的处理能力。

滤料层一般由底部的承托层和其上的工作层组成，承托层厚为0.2m，无机滤料粒径为60～100mm；工作层厚为1.3～1.8m，无机滤料粒径为30～50mm。对有机物浓度较高的污水，应采用粒径较大的滤料，以防滤料被生物膜堵塞。而塑料滤料因质量轻和孔隙率高，可使滤床高度大幅提高，国外采用的双层滤床高7m左右，国内采用的多层“塔

式”结构，高度常在10m以上。

（2）布水设备

生物滤池布水系统的作用是使污水能均匀地分布在整个滤床表面。布水装置分为两种：一种是固定布水装置，另一种是旋转（回转）布水器。

图4-2为采用固定喷嘴式布水系统的生物滤池，由配水池、虹吸装置、布水管道和喷嘴四部分组成。污水进入配水池，当水位达到一定高度后，虹吸装置开始工作，污水进入布水管路。配水管设有一定坡度以便放空，布水管道敷设在滤池表面下0.5～0.8m，喷嘴安装在布水管上，伸出滤料表面0.15～0.2m，喷嘴的口径为15～20mm。当水从喷嘴喷出，受到喷嘴上部所设倒锥体的阻挡，使水流向四周分散，形成水花，均匀喷洒在滤料上。当配水池水位降到一定程度时，虹吸被破坏，喷水停止。这种布水系统布水不够均匀，而且不能连续冲刷生物膜，所需水头也较大，但它不受生物滤池池形的限制。

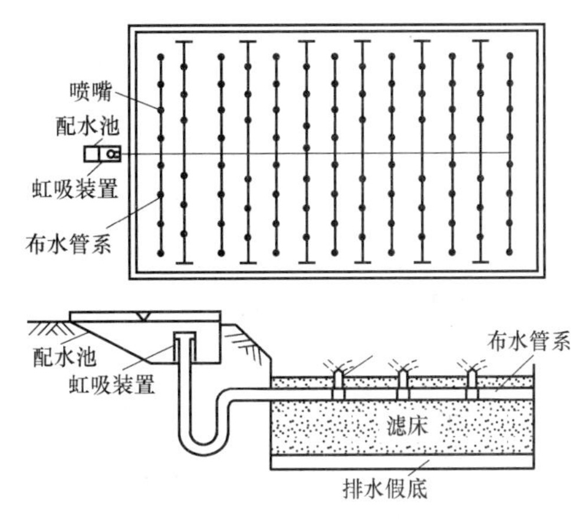

图4-2 采用固定喷嘴式布水系统的生物滤池

旋转布水器的中央是一根空心的立柱，底端与设在池底下面的进水管衔接。布水横管的一侧开有喷水孔口，孔口直径为10～15mm，间距不等，越近池心间距越大，使滤池单位面积接受的污水量基本上相等，如图4-3所示。布水器的横管可为两根或四根，对称布置。污水通过中央立柱流入布水横管，由喷水孔口分配到滤池表面。污水喷出孔口时，作用于横管的反作用力推动布水器绕立柱旋转，转动方向与孔口喷嘴方向相反。所需水头在0.6～1.5m左右。如果水头不足，可用电动机转动布水器。目前旋转布水器因布水均匀，淋水周期短，水力冲刷力强而得到广泛的应用，但由于布水水头和横管上的小孔孔径较小，易产生堵塞问题。

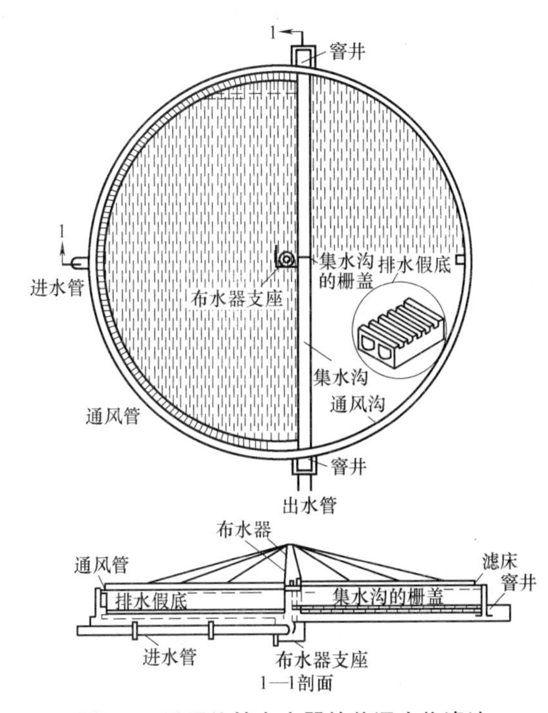

图4-3 采用旋转布水器的普通生物滤池

（3）排水系统

池底排水系统的作用是：①收集滤床流出的污水与生物膜；②保证通风；③支承滤料。池底排水系统由池底、排水假底和集水沟组成，如图4-4所示。排水假底由特制砌块或栅板铺成，滤料堆在假底上面。假底的空隙所占面积不宜小于滤池平面的5%～8%，与池底的距

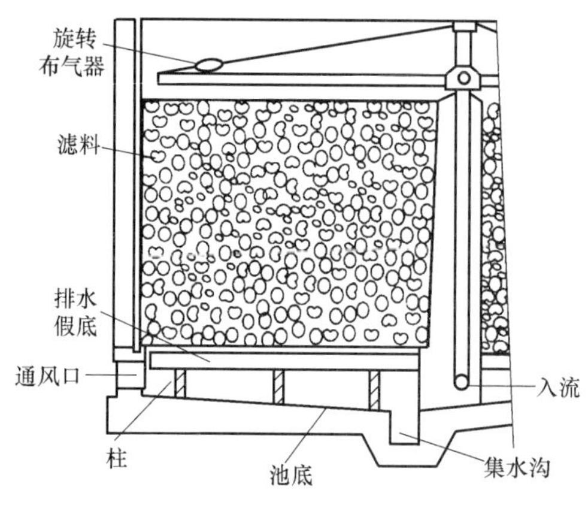

图 4-4 生物滤池池底排水系统示意图

离不应小于 0.6m。

池底除支承滤料外，还要排泄滤床上的来水，池底中心轴线上设有集水沟，两侧底面向集水沟倾斜，池底和集水沟的坡度约 1%～2%。集水沟要有充分的高度，并在任何时候不会满流，确保空气能在水面上畅通无阻，使滤池中空隙充满空气。

2. 生物滤池的工艺流程

生物滤池法的基本流程是由初沉池、生物滤池、二沉池组成。进入生物滤池的污水，必须通过预处理，去除悬浮物、油脂等会堵塞滤料的物质，并使水质均化稳定。一般在生物滤池前设初沉池，但也可以根据污水水质而采取其他方式进行预处理，达到同样的效果。生物滤池后面的二沉池，用于截留滤池中脱落的生物膜，以保证出水水质。

普通生物滤池又称低负荷生物滤池，因其处理废水的负荷较低，一般只有 $1\sim4m^3/(m^2\cdot d)$，有机负荷也仅为 $0.1\sim0.4kgBOD_5/(m^3\cdot d)$。在处理城市污水方面，普通生物滤池曾有长期运行经验。其优点是处理效果好，BOD_5 去除率可在 90%以上，出水 BOD_5 可下降为 25mg/L 以下，硝酸盐含量在 10mg/L 左右，出水水质稳定。缺点是占地面积大，易于堵塞，灰蝇很多，影响环境卫生，故普通生物滤池目前已很少使用。

为缓解普通生物滤池的堵塞问题，可采用交替式二级生物滤池法。图 4-5 是交替式二级生物滤池法工艺流程。运行时，滤池是串联工作的，污水经初沉池后进入一级生物滤池，出水经相应的中间沉淀池去除残膜后用泵送入二级生物滤池，二级生物滤池的出水经过沉淀后排出污水处理厂。工作一段时间后，一级生物滤池因表层生物膜的累积，即将出现堵塞，改作二级生物滤池，而原来的二级生物滤池则改作一级生物滤池。运行中每个生物滤池交替作为一级和二级生物滤池使用。交替式二级生物滤池法流程相比并联流程可提高负荷 2～3 倍。采用交替式二级生物滤池法流程时，两滤池的滤料粒径应相同，构筑物高程上也要考虑水流方向互换的可能性。此外，增设泵站使建设成本增加。

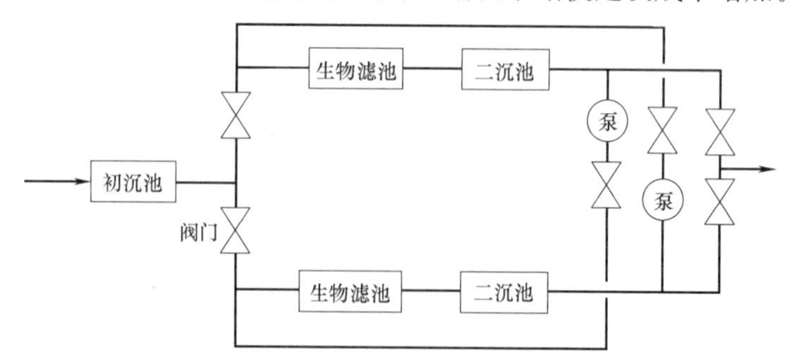

图 4-5 交替式二级生物滤池法工艺流程

目前国内外广泛使用的回流式生物滤池、塔式生物滤池都属于高负荷生物滤池，通过采用新型滤料，革新流程，使负荷比普通生物滤池提高数倍，而池体积大幅缩小。缺点是负荷高时，有机物转化较不彻底，排出的生物膜容易腐化。高负荷生物滤池现场运行工艺比较灵活，可以通过调整负荷和流程，得到不同的处理效率（65%～90%）。如图 4-6 所示，有多种高负荷生物滤池的流程。

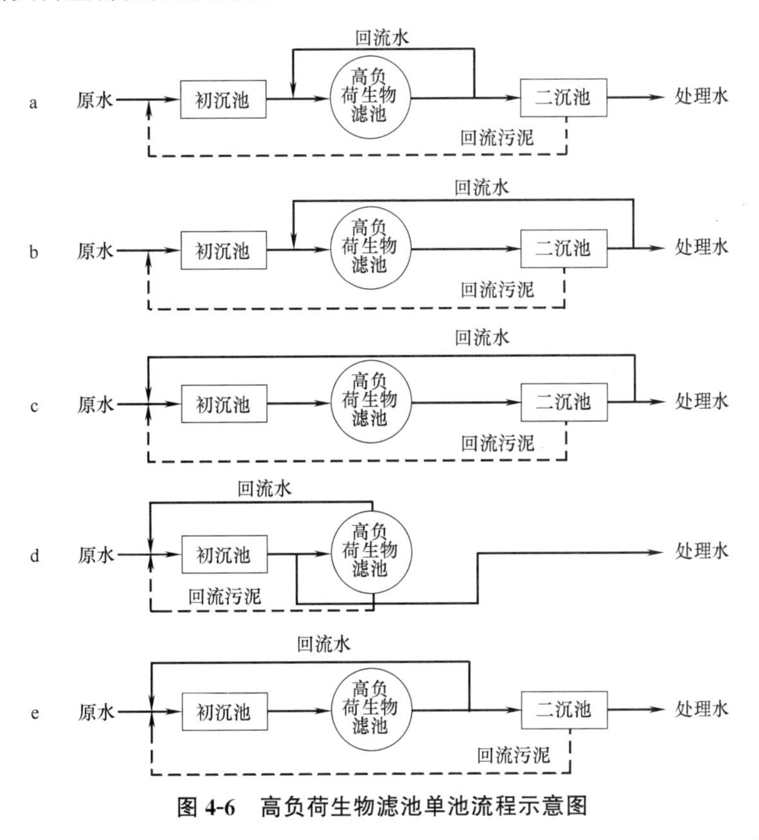

图 4-6　高负荷生物滤池单池流程示意图

如图 4-6 所示，流程 a 中滤池出水直接向滤池回流，并由二沉池向初沉池回流生物污泥，利于生物膜的接种；流程 b 中二沉池出水回流到滤池，加大了滤池的水力负荷可避免加大初沉池的容积；流程 c 中二沉池出水回流到初沉池，生物污泥回流到初沉池，当入流量小于平均流量时，增大回流量，当入流量大于平均流量时，减少或停止回流；流程 d 中滤池出水直接回流到初沉池，初沉池的效果得到提高，并可兼作二沉池；流程 e 中滤池出水回流至初沉池，生物污泥由二沉池回流到初沉池。其中流程 a 和 b 的应用最为广泛。

流程 a、d、e 适合废水浓度低的情况，三种流程中 e 的除污效能最好，但基建费用最高；流程 d 的除污效能最差，但基建费用最低；流程 b、c 适合废水浓度高的情况，其中流程 c 除污效能最好，但基建费用高。

当对处理水质要求较高时，或条件不允许提高滤池高度时，可采用两段滤池系统。工艺流程如图 4-7 所示。两段高负荷生物滤池串联系统不仅可达到有机底物去除率超过 90%的效能，而且滤池中也能发生硝化反应，出水中含有硝酸盐和溶解氧。其存在的主要问题是两段滤池负荷率不均会造成生物膜生长不均衡，一段滤池负荷高，生物膜生长快，脱落后易堵塞滤池；另一段滤池负荷低，生物膜生长不佳，滤池容积利用率不高。

高负荷生物滤池和普通生物滤池存在不同的过程控制。首先，高负荷生物滤池大幅提高了滤池的负荷率，其 BOD_5 容积负荷高于普通生物滤池的 6～8 倍，水力负荷则高达 10 倍；其次，高负荷生物滤池的高滤率是通过限制进水 BOD_5 和运行上采取处理水回流等技术

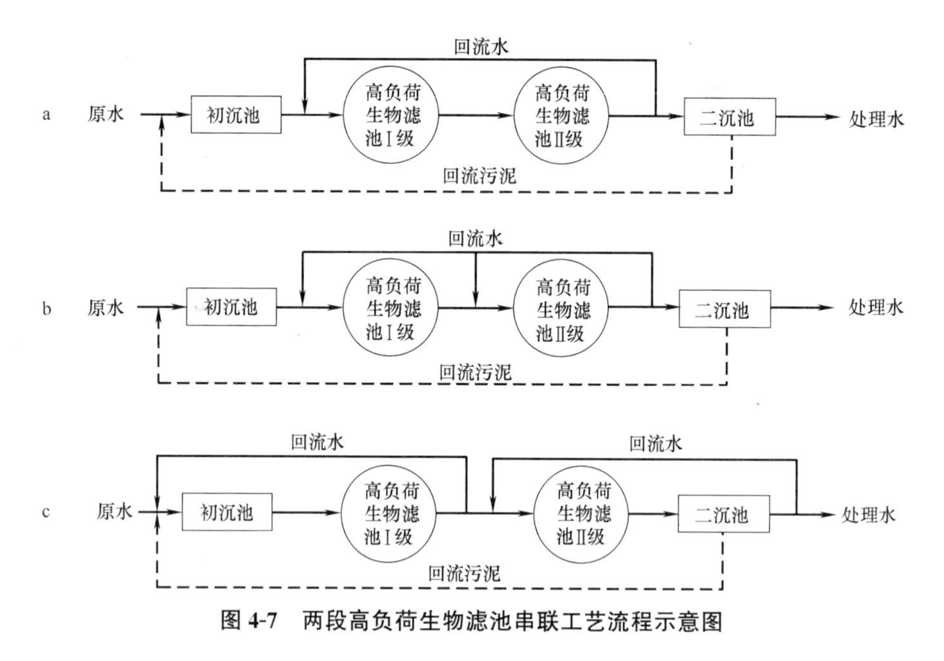

图 4-7　两段高负荷生物滤池串联工艺流程示意图

措施达到；最后，高负荷生物滤池进水 BOD_5 必须低于 200mg/L，否则要以处理水回流加以稀释。回流的处理水量（Q_r）与进入滤池的原污水量（Q）之比称为回流比 R。

塔式生物滤池是由德国化学工程师舒尔茨根据气体洗涤塔原理开发的。因塔内的废水、生物膜、空气三者充分接触，水流紊动剧烈，通风条件改善，氧从空气中经过废水向生物膜内的传质过程得到加强，较高的负荷加快生物的增长和脱落，使塔式生物滤池单位体积填料去除有机物能力显著提高。所以塔式生物滤池的水力负荷比回流式高负荷生物滤池高 2～10 倍，达 30～200$m^3/(m^2 \cdot d)$，有机负荷高达 1～2$kgBOD_5/(m^3 \cdot d)$，进水 BOD_5 可以提高到 500mg/L。

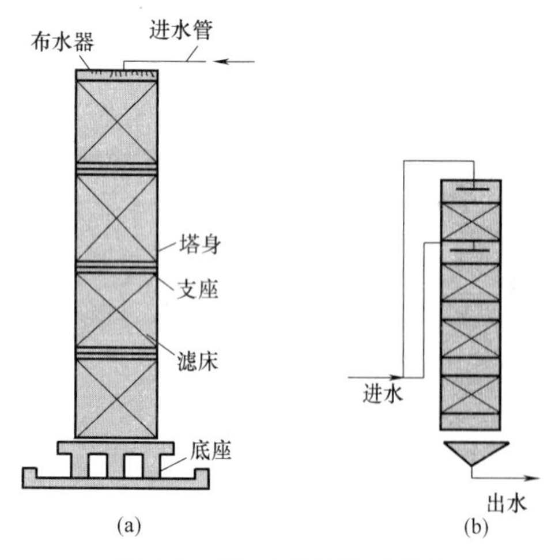

图 4-8　塔式生物滤池示意图

塔式生物滤池在平面上一般呈矩形或圆形（图 4-8），高 8～24m，直径 1～3.5m，直径与高度比为 1∶8～1∶6，这样能使滤池内部形成较强烈的拔风状态，良好的通风满足需氧量大的要求。因塔身高，从承重角度要求滤料的重度小，还要求滤料的孔隙率大以利于通风和排出脱落的生物膜，所以塔式生物滤池大多选用大孔径波纹塑料板滤料或一种玻璃布蜂窝填料。运行中，一旦自然通风供氧不足，出现厌氧状态，必须采用机械通风。

塔式生物滤池占地面积较其他生物滤池大幅减小，对水质、水量适应性强，但废水抽升费用大，且池体过高使运行管理不便，故适宜处理小水量废水。

3. 影响生物滤池性能的主要因素

生物滤池中，有机物的降解过程复杂，影响生物滤池性能的因素很多，主要包括滤池高度、回流、供氧等。

（1）滤池高度

滤床的上层和下层相比，生物膜量、微生物种类和去除有机物的速率均不相同。滤床上层废水中有机物浓度较高，微生物繁殖速率高，种属较低级，以细菌为主，生物膜量较多，有机物去除速率较高。随滤床深度增加，微生物从低级趋向高级，种类逐渐增多，生物膜量从多到少。滤床中的这一递变现象，类似污染河流在自净过程中的生物递变，因为微生物的生长和繁殖同环境因素息息相关，所以当滤床各层的进水水质互不相同时，各层生物膜的微生物就不相同，处理污水（特别是含多种性质相异的有害物质的工业废水）的能力也随之不同。

研究表明，生物滤池的处理效率在一定条件下随滤床高度的增加而增加，但滤床高度超过某一数值后，处理效率的提高很小；滤床不同深度处的微生物不同反映滤床高度对处理效率的影响与废水水质有关。

（2）回流

回流是将生物滤池的一部分出水回流到滤池前与进水混合的工艺操作方式，多用于高负荷生物滤池运行系统。利用处理水回流不但具有加大水力负荷，均化与稳定进水水质、抑制滤池蝇滋长和减轻散发臭味的作用而且能及时冲刷过厚和老化的生物膜，使生物膜迅速更新并保持较高活性。

回流也给生物滤池带来一些不利因素，如增加水力负荷将缩短废水在池中的停留时间；滤池进水被稀释后会降低生物膜吸附有机物的速度；回流水里含有的生物膜微粒导致池中悬浮微生物的增加，可能影响氧向生物膜转移，影响生物滤池效率；冬季采用回流操作会降低滤池温度，导致滤池工作效率降低等。可见，回流对生物滤池的影响是多方面的，在运行管理中要综合分析。

（3）供氧

生物滤池中，微生物所需的氧一般直接来自大气，靠自然通风供给。凡是影响通风条件的因素，如滤料孔隙率、滤池的高度、风力、池内温度与气温之差等，均能影物滤池所需氧量的供给。当入流废水有机物浓度较高时，供氧条件可能成为影响生物工作的主要因素。为保证生物滤池能正常工作，根据试验研究和工程实践，有人建议进水 COD 应小于 400mg/L。当进水浓度高于此值时，可以通过回流的方法降低滤池有机物浓度，以保证生物滤池供氧充足，正常运行。

4. 生物滤池设计规定

（1）进入生物滤池的污（废）水应具有较好的可生化性，BOD_5/COD 宜大于 0.3，pH 宜为 6.5～9.5，水温宜为 12～35℃。污（废）水中 BOD_5：N：P 宜为 100：5：1，且水中不应含对微生物有抑制和毒害作用的污染物。

（2）生物滤池废水处理工艺有除氨氮要求时，进水总碱度（以 $CaCO_3$ 计）与氨氮的比值宜大于 7.14，且好氧池（区）剩余碱度宜大于 70mg/L，不满足上述条件时宜补充碱度。

（3）废水处理有脱总氮要求时，要求反硝化进水的 BOD_5/TKN 应大于 4.0，总碱度（以 $CaCO_3$ 计）与氨氮的比值宜不小于 3.6，不满足上述条件时，应合理补充碳源或碱度。

(4) 废水处理有除磷要求时，进水 BOD_5/TP 宜大于17.0。生物滤池中出水总磷浓度达不到设计要求时，可采用其他方式除磷，如化学除磷等。

(5) 低负荷生物滤池进水的 BOD_5 不应大于 200mg/L，否则将处理出水回流，以稀释进水有机物浓度。

(6) 高负荷生物滤池进水 BOD_5 不应大于 300mg/L，否则处理出水回流。当进水污染物浓度较高或者含有一定对微生物有毒成分时，也应进行回流。

(7) 塔式生物滤池的处理规模不宜超过 10000m³/d，并且根据污（废）水的水质条件，滤池前宜设沉砂池、初沉池或混凝沉淀池、除油池、厌氧水解池等预处理或前处理设施。塔式生物滤池进水的 BOD_5 不应大于 500mg/L，否则处理出水应回流。

(8) 生物滤池废水处理工艺污染物一般去除率及主要设计参数可参考表4-1及表4-2计算。

生物滤池废水处理工艺污染物一般去除率 **表4-1**

污水类别	工艺	污染物去除率(%)					
		SS	BOD_5	COD	氨氮	总氮	总磷
城镇污水	预处理+生物滤池	75～98	80～95	80～90	80～95	50～80(缺氧单元或区域)	40～80(厌氧单元或区域)
工业废水	前处理+生物滤池	75～98	70～90	70～85	—	—	—

生物滤池主要设计参数 **表4-2**

项目	普通生物滤池	高负荷生物滤池	塔式生物滤池
容积负荷[$kgBOD_5/(m^3 \cdot d)$]	0.15～0.3	≤1.8	1.0～3.0
表面水力负荷[$m^3/(m^2 \cdot d)$]	1.0～3.0	10～36	80～200
滤料高度(m)	1.5～2.0	2.0～4.0	8～12
回流	无	无	无
滤料要求	一般采用碎石、卵石、炉渣、焦炭等无机滤料。采用碎石类滤料时，下层滤料粒径宜为60～100mm，上层滤料粒径宜为30～50mm	采用碎石或塑料滤料，采用碎石类滤料时，下层滤料粒径宜为70～100mm；上层滤料粒径宜为40～70mm	宜采用碎石或塑料滤料，采用碎石类滤料时，滤池下层滤料粒径宜为70～100mm；上层滤料粒径宜为40～70mm

5. 生物滤池设计计算

(1) 生物滤池的总体积

$$V=\frac{S_0 Q}{L_v}\times 10^{-3} \tag{4-1}$$

式中 V——生物滤池的总体积，m³；

Q——进入生物滤池流量，m³/d，回流式生物滤池，流量为 $Q(1+R)$；

R——回流比，$R=\frac{Q_R}{Q}$，可根据经验确定；

Q_R——滤池出水回流量，m³/d；

S_0——进入生物滤池的 BOD_5，mg/L；

L_v——生物滤池的容积负荷，$kgBOD_5/(m^3 \cdot d)$。

（2）生物滤池的有效面积

$$A=\frac{V}{h_1} \tag{4-2}$$

式中 A——生物滤池的有效面积，m^2；

h_1——生物滤池滤料层高度，m。

（3）核算生物滤池表面水力负荷

$$q=\frac{Q}{A} \tag{4-3}$$

式中 q——生物滤池的表面水力负荷，$m^3/(m^2 \cdot d)$；

Q——进入生物滤池的总流量，m^3/d。

（4）生物滤池单位容积滤料需氧量

$$R_O=\frac{aQ(S_0-S_e)}{1000V}+bD \tag{4-4}$$

式中 R_O——生物滤池单位容积滤料需氧量，kg/m^3；

a——完全降解 1kgBOD_5 所需的氧量，kgO_2/kgBOD_5，一般城镇污水取 1.46kgO_2/kgBOD_5 左右；

b——单位质量活性生物膜需氧量，一般为 0.18；

D——单位体积滤料上的生物膜量，kg/m^3；

S_0——进入生物滤池的 BOD_5，mg/L；

S_e——生物滤池出水 BOD_5，mg/L；

Q——进入生物滤池的总流量，m^3/d。

6. 生物滤池设计计算例题

【例 4-1】 已知某城镇人口 50000 人，设计当地居民排水量定额为 100L/(人·d)，BOD_5 排放量为 20g/(人·d)。设有一座工厂，废水最大排放量为 3000m^3/d，其排放的 BOD_5 为 1000mg/L。拟将生活污水和工厂废水混合后采用回流式生物滤池进行处理，要求处理后出水 $BOD_5 \leqslant 20$mg/L，回流式生物滤池物料平衡图如图 4-9 所示。当地夏季混合污水平均温度为 25℃，生物滤池池外平均温度为 31℃；冬季混合污水平均温度为 12℃，生物滤池池外平均温度为 4℃。假设脱落的生物膜全部转化为生物污泥，当量人口产泥量为 25g/(人·d)，污泥含水率为 97%。设计回流式生物滤池有关数据。

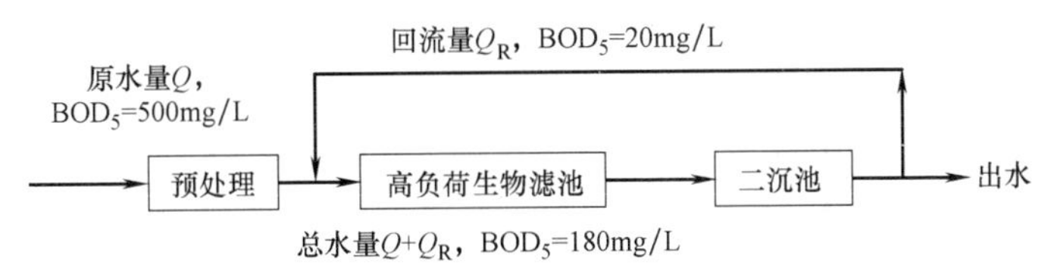

图 4-9 回流式生物滤池物料平衡图

【解】（1）基本设计参数计算

生活污水和工业废水总排放量为：

$$Q=\frac{50000\times100}{1000}+3000=8000m^3/d$$

生活污水和工业废水混合后的 BOD_5 为：

$$S_s=\frac{3000\times1000+50000\times20}{8000}=500mg/L$$

由于生活污水和工业废水混合后的 BOD_5 较高，不满足生物滤池进水要求，应考虑处理后出水回流，设生物滤池处理后出水 BOD_5 为 20mg/L，混合后的污水经出水稀释后进入生物滤池的 BOD_5 为 180mg/L，回流比为：

$$QS_s+Q_RS_e=(Q+Q_R)S_0$$

$$500Q+20Q_R=(Q+Q_R)180$$

$$R=\frac{Q_R}{Q}=\frac{500-200}{180-20}=1.9$$

(2) 生物滤池的个数和滤床尺寸计算

设生物滤池的容积负荷 $L_v=1.8kgBOD_5/(m^3\cdot d)$，生物滤池总体积为：

$$V=\frac{S_0Q(1+R)}{L_v}\times10^{-3}=\frac{8000\times(1.9+1)\times180}{1.8}\times10^{-3}=2320m^3$$

设生物滤池滤料层厚度 h_1 为 2.0m，则滤池的有效面积为：

$$A=\frac{V}{h_1}=\frac{2320}{2}=1160m^2$$

若采用 4 个滤池并联运行，则每个滤池的面积为：

$$A_1=\frac{1160}{4}=290m^2$$

采用圆形滤池，生物滤池的直径为：

$$d=\sqrt{\frac{4A_1}{\pi}}=\sqrt{\frac{4\times290}{\pi}}=19.2m$$

生物滤池超高 h_0 取 0.8m，底部构造层（包括承托层、通风口、排水装置）h_2 取 1.0m，如图 4-10 所示，生物滤池的总高为：

$$H=h_0+h_1+h_2=0.8+2+1=3.8m$$

(3) 校核生物滤池表面水力负荷

$$q=\frac{Q(1+R)}{A}=\frac{8000\times(1.9+1)}{1160}=20m^3/(m^2\cdot d)$$

满足要求。

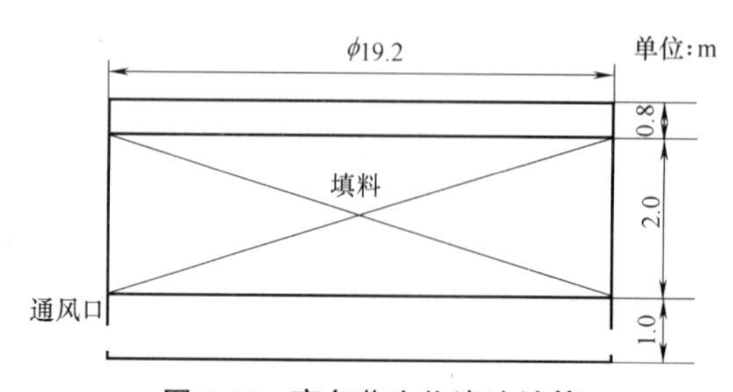

图 4-10 高负荷生物滤池计算

(4) 生物滤池单位容积滤料需氧量

试验测定 $1m^3$ 该滤料平均生物膜量为 2.0kg，则单位容积滤料需氧量为：

$$R_O = \frac{aQ(S_0 - S_e)}{1000V} + bD$$

$$= \frac{1.46 \times 8000 \times (1.9+1) \times (180-20)}{1000 \times 2320} + 0.18 \times 2.0$$

$$= 2.696 \text{kg}/(\text{m}^3 \cdot \text{d})$$

（5）污泥量计算

依据已知条件，当地居民人口数 $N_1 = 50000$ 人。

工业当量人口数 N_2 可依据式（4-5）计算：

$$N_2 = \frac{S_0 Q_{工水}}{S_e} \tag{4-5}$$

式中 $Q_{工水}$——该城镇工业废水总量，m^3/d；

S_0——该城镇工业废水中平均 BOD_5，mg/L；

S_e——处理后废水 BOD_5，mg/L。

代入上述已知数据，工业当量人口数 N_2 为：

$$N_2 = \frac{S_0 Q_{工水}}{S_e} = \frac{3000 \times 1000}{20} = 150000 \text{ 人}$$

因此，当量人口总数 N 为：

$$N = N_1 + N_2 = 50000 + 150000 = 200000 \text{ 人}$$

生物干污泥量 $W_干$ 为：

$$W_干 = \frac{20}{1000} N = \frac{20}{1000} \times 200000 = 4000 \text{kg/d}$$

设生物干污泥占总泥量的 75%，则总干泥量 $W_总$ 为：

$$W_总 = \frac{W_干}{0.75} = \frac{4000}{0.75} \approx 5333.3 \text{kg/d}$$

已知污泥含水率为 97%，则湿污泥量 $W_湿$ 为：

$$W_湿 = \frac{W_总}{(1-0.97) \times 1000} = \frac{5333.3}{(1-0.97) \times 1000} \approx 177.8 \text{m}^3/\text{d}$$

4.4.2 生物转盘

1. 生物转盘的工艺流程

生物转盘工艺流程如图 4-11 所示。根据转盘和盘片的布置形式，生物转盘可分为单轴单级式（图 4-12）、单轴多级式（图 4-13）和多轴多级式。实践表明，处理同一种污水，如果盘片面积不变，相比单级式转盘，将生物转盘分为多级串联运行能显著提高处理水水质和水中溶解氧的含量。通过对转盘上生物相的观察，第一级盘片上的生物膜最厚，随污水中有机物逐渐减少，后面盘片上的生物膜逐级变薄，而后几级生物膜里的微生物种类逐步趋向高级。

二维码 4
纤维转盘滤池
运行视频

与生物滤池相比，生物转盘的优势是不会发生堵塞现象，净化效果好。此外，生物转盘也具有能耗低、抗冲击负荷能力强、无需回流污泥、管理运行方便等生物膜法的优点。缺点是占地面积大、散发臭气、在寒冷地区需做保温处理。

生物转盘过去主要用于水量小的污水处理工程，近年来实践表明，该工艺也可用于一

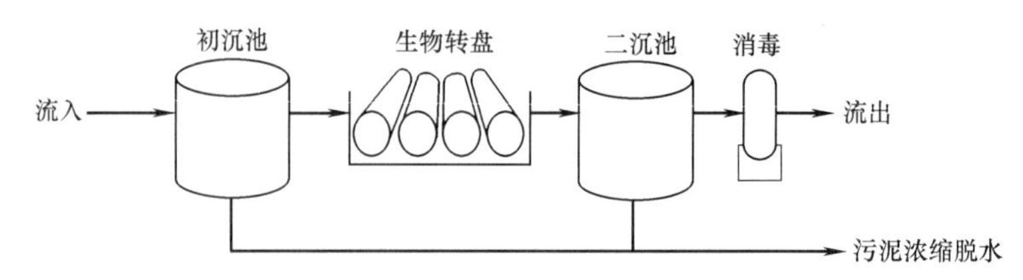

图 4-11　生物转盘工艺流程图

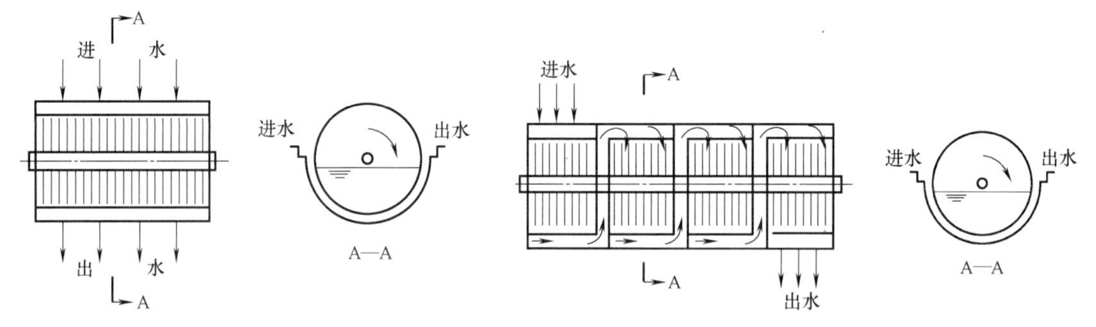

图 4-12　单轴单级式生物转盘　　**图 4-13　单轴多级（四级）式生物转盘**

定规模的污水处理厂。生物转盘可用作完全处理、不完全处理和工业废水的预处理。

2. 生物转盘设计规定

（1）生物转盘池体的接触反应槽应呈半圆形，接触反应槽尺寸根据转盘直径和轴长确定，盘片边缘与接触氧化槽间距应大于 0.1m。接触反应槽槽底应设置放空管，槽的两侧宜采用锯齿堰溢流出水。

（2）盘片转动轴一般采用实心钢轴或无缝钢管，其长度应控制在 0.5～7.0m 之间，直径宜为 0.05～0.08m。

（3）转轴中心应高出槽内水面 0.15m 以上，以保证 30%～40%转盘面积浸没在污水中。转盘盘片直径宜为 1.0～4.0m，转盘净间距宜为 10～30mm，以保证转盘中心部位的通气效果，若为多级转盘，则进水端盘片间距为 20～30mm，出水端一般为 10～20mm，具体可根据工艺需要进行调节。盘片转速一般为 1.0～3.0r/min，盘片外缘线速度宜为 15～18m/min，转盘转速应能根据运行情况调整。

3. 生物转盘设计计算

生物转盘工艺设计的主要内容是计算转盘的总面积，表示转盘处理能力的指标是水力负荷和有机负荷。生物转盘的负荷与废水性质、废水浓度、气候条件及构造、运行等多种因素有关。设计时可以通过试验求得所需的设计参数或根据试验资料和其他方法确定设计负荷。一般盘片表面水力负荷宜为 0.04～0.20m^3 污水/(m^2·d)；无试验资料的情况下，生物转盘盘片 BOD_5 表面有机负荷宜为 0.005～0.20kgBOD_5/(m^2·d)，有脱氮要求时，氨氮表面负荷宜为 0.006～0.007kg/(m^2·d)，同时采用混合液回流，回流比宜为 100%～300%。单位 BOD_5 产泥量按 0.04～0.20m^3/(m^2·d) 计算。

（1）生物转盘总面积

$$A=\frac{(S_0-S_e)Q}{L_A} \tag{4-6}$$

式中　A——生物转盘总面积，m^2；

Q——处理水量，m^3/d；

S_0——进水 BOD_5，mg/L；

S_e——出水 BOD_5，mg/L；

L_A——生物转盘盘片的 BOD_5 面积负荷，$g/(m^2 \cdot d)$。

（2）生物转盘盘片数

$$m=\frac{0.64A}{D^2} \tag{4-7}$$

式中 m——生物转盘盘片数，个；

D——生物转盘直径，m。

（3）废水处理槽有效长度

$$L=m(a+b)K \tag{4-8}$$

式中 L——废水处理槽有效长度，m；

a——盘片净间距，m；

b——盘片厚度，m，根据材料强度确定；

K——系数，一般取 1.2。

（4）废水处理槽有效容积

$$V=(0.294 \sim 0.335)\times(D+2\delta)^2 L \tag{4-9}$$

式中 V——废水处理槽有效容积，m^3；

δ——盘片边缘与处理槽内壁的间距，m。

（5）转轴转速

$$n_0=\frac{6.37}{D}\times\left(0.9-\frac{V_1}{Q_1}\right) \tag{4-10}$$

式中 n_0——转轴转速，r/min；

V_1——每个废水处理槽容积，m^3；

Q_1——每个废水处理槽处理的水量，m^3/d。

4. 生物转盘设计计算例题

【例 4-2】 某居民区最大废水排放量为 $500m^3/d$，废水的 BOD_5 为 150mg/L，废水冬季平均水温为 15℃，要求出水的 BOD_5 不大于 20mg/L，废水处理槽中 VSS/SS 为 0.7，试设计生物转盘的相关技术参数。

【解】（1）生物转盘总面积

$$A=\frac{(S_0-S_e)Q}{L_A}=\frac{(150-20)\times 500}{13}=5000m^2$$

城市污水面积负荷取值范围 $5 \sim 20gBOD_5/(m^2 \cdot d)$，取 $L_A=13gBOD_5/(m^2 \cdot d)$。

（2）生物转盘盘片数

设生物转盘直径 $D=2.5m$，则转盘盘片数为：

$$m=\frac{0.64A}{D^2}=\frac{0.64\times 5000}{2.5^2}=512 \text{ 片}$$

采用二级四轴生物转盘，平均每轴盘片数为 128 片，每级盘片数为 256 片。

（3）废水处理槽有效长度

采用盘片净间距 $a=0.02\text{m}$，盘片厚度 $b=0.005\text{m}$ 的硬聚氯乙烯板制作转盘，$K=1.2$，则

$$L=m(a+b)K=128\times(0.020+0.005)\times1.2=3.84\text{m}$$

（4）废水处理槽有效容积

取生物转盘与氧化槽净间距 $\delta=0.2\text{m}$，废水处理槽有效容积为：

$$\begin{aligned}V&=(0.294\sim0.335)\times(D+2\delta)^2L\\&=0.32\times(2.5+2\times0.2)^2\times3.84\\&=10.3\text{m}^3\end{aligned}$$

废水处理槽净有效容积为

$$\begin{aligned}V_1&=(0.294\sim0.335)\times(D+2\delta)^2\times(L-mb)\\&=0.32\times(2.5+2\times0.2)^2\times(3.84-128\times0.005)\\&=8.6\text{m}^3\end{aligned}$$

（5）废水处理槽有效宽度

$$B=D+2\delta=2.5+2\times0.2=2.9\text{m}$$

采用二级生物转盘，则废水处理槽总宽度为：2.9×2=5.8m。

（6）转轴转速

$$n_0=\frac{6.37}{D}\times\left(0.9-\frac{V_1}{Q_1}\right)=\frac{6.37}{2.5}\times\left(0.9-\frac{8.6}{125}\right)=2.1\text{r/min}$$

（7）污（废）水在废水处理槽内停留时间

$$T_1=\frac{V}{Q}=\frac{8.6\times24}{500/4}=1.7\text{h}$$

（8）污泥量的计算

根据经验，生物转盘的污泥产率系数 $Y=0.3$，则干污泥产生量为：

$$W_{干}=\frac{Y\cdot Q(S_0-S_e)}{VSS/SS}=\frac{0.3\times500\times(150-20)}{0.7\times1000}=27.9\text{kg/d}$$

设污泥的含水率为98%，湿污泥量为：

$$\frac{27.9}{1000}=0.0279\text{t(干泥)/d}$$

$$\frac{0.0279}{100\%-98\%}=1.40\text{m}^3/\text{d}$$

4.4.3 生物接触氧化池

1. 生物接触氧化池的工艺流程

生物接触氧化池由池体、填料和进水布气装置等组成。池体用于设置填料、布水布气装置和支承填料的支架，从填料上脱落的生物膜会有一部分沉积在池底，必要时，池底部可设置排泥和放空设施。

二维码 5
不同填料的安装方式、适用条件及特点

生物接触氧化池填料要求对微生物无毒害、易挂膜、质轻、高强度、抗老化、比表面积大和孔隙率高。目前常采用的填料主要有聚氯乙烯塑料、聚丙烯塑料、环氧玻璃钢等做成的蜂窝状和波纹板状填料、纤维组合填料、立体弹性填料等。近年国内开发的空心塑料体填

料，因其相对密度接近于 1，被称为悬浮填料。在运行时，由于悬浮填料在池内均匀分布，并不断切割气泡，可使氧利用率、动力效率得到提高。

生物接触氧化池中的填料可采用全池布置，底部进水，整个池底安装布气装置，全池曝气，如图 4-14 所示；两侧布置，底部进水，布气管布置在池子中心，中心曝气，如图 4-15 所示。或单侧布置，上部进水，侧面曝气，如图 4-16 所示。填料全池布置、全池曝气的形式，由于曝气均匀，填料不易堵塞，氧化池容积利用率高等优势，是目前生物接触氧化法采用的主要形式。但无论哪种形式，曝气池的填料都应分层安装。辫带式纤维滤料及应用效果如图 4-17 所示，斜管填料及应用效果如图 4-18 所示。

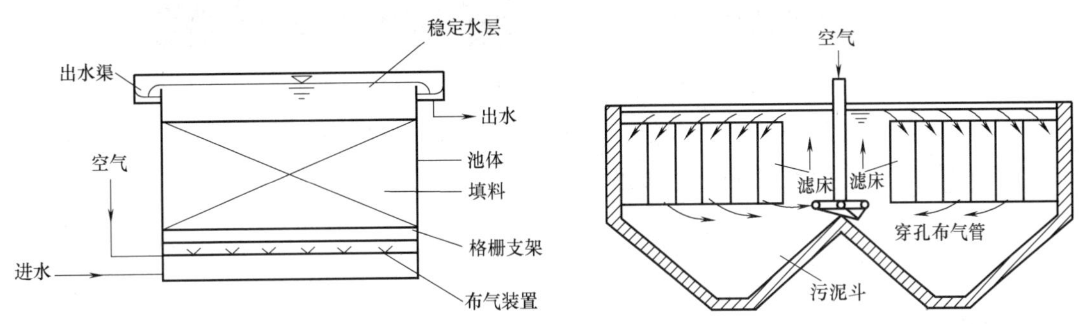

图 4-14　接触氧化池构造示意图

图 4-15　中心曝气的生物接触氧化池

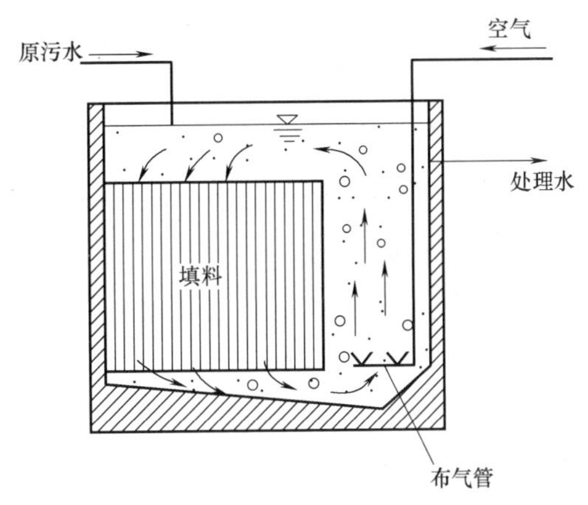

图 4-16　侧面曝气的生物接触氧化池

图 4-17　辫带式纤维滤料及应用效果

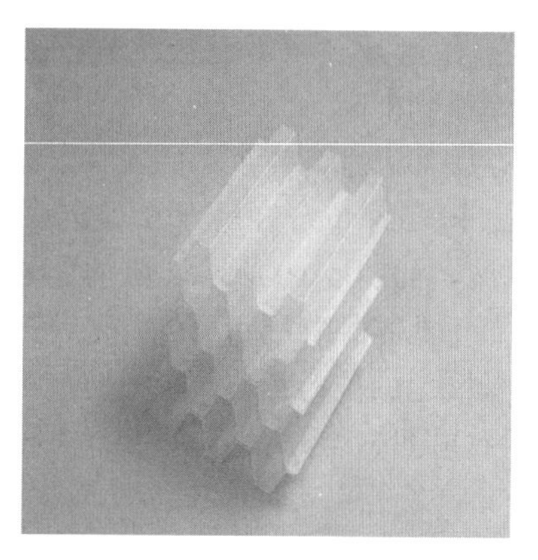

图 4-18　斜管填料及应用效果

生物接触氧化池应根据进水水质和处理程度确定采用单级式、二级式或多级式，图 4-19、图 4-20 是生物接触氧化法的两种基本流程。

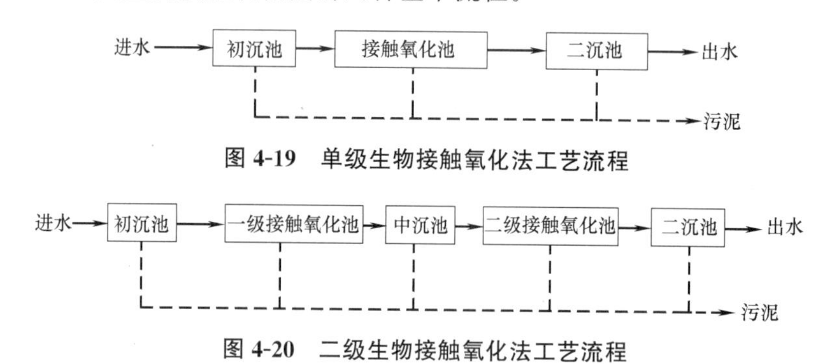

图 4-19　单级生物接触氧化法工艺流程

图 4-20　二级生物接触氧化法工艺流程

在一级处理流程中，原废水经预处理（主要为初沉池）后进入接触氧化池，出水经过二沉池分离脱落的生物膜，实现泥水分离。在二级处理流程中，两段接触氧化池串联运行，可根据实际需要进行调整，如将氧化池分格，不设中沉池等；多级处理流程中连续串联三座或以上的接触氧化池。第一级接触氧化池内的微生物处于对数增长期和减速增长期的前段，生物膜增长较快，有机负荷较高，有机物降解速率也较大；后续的接触氧化池内微生物处在生长曲线的减速增长期后段或生物膜稳定期，生物膜增长缓慢，处理水水质逐步提高。

生物膜挂膜方法主要有以下 3 种：

（1）自然挂膜法

将带有自然菌种的工业废水用泵慢速打入空的生物滤池中，废水中的自然菌种和空气微生物附着在滤料上，逐渐形成一层黏性的微生物薄膜。

（2）活性污泥挂膜法

取处理后的活性污泥回收作菌种。

（3）优势菌种挂膜法

从自然环境或废水处理中筛选分离得到对某种工业废水有强降解能力的菌株，也可以通过遗传育种获得优良菌种或构建基因工程菌。

一级和二级工艺流程相比较，一级法生物膜生长快、活性大，降解有机物速度快，操作方便，投资少，但氧化池有时会引起短路；二级法更适应原水水质变化，使出水水质趋于稳定和改善，氧化池的流态属于完全混合型，能提高生化效率，缩短生物氧化时间，但由于二级法需增加工艺流程及设施设备，使得投资费用比一级法高。一般来说，当有机负荷较低而水力负荷较大时，采用一级法；当有机负荷较高时，需采用二级法。

生物接触氧化法是介于活性污泥法和生物滤池二者之间的废水生物处理技术，兼有活性污泥法和生物膜法的优点，但仍以生物膜除污染过程为主，具体特点如下：

（1）由于填料的比表面积大，池内的充氧条件良好。生物接触氧化池内单位容积的生物固体量高于活性污泥法曝气池及生物滤池。因此，生物接触氧化池具有较高的容积负荷。

（2）生物接触氧化法不需要污泥回流，不存在污泥膨胀问题，运行管理简便。

（3）由于生物固体量多，水流又属完全混合型，因此，生物接触氧化池对水质水量的骤变有较强的适应能力。

（4）生物接触氧化池有机容积负荷较高时，其F/M保持在较低水平，污泥产率较低。

目前，生物接触氧化法在国内的废水处理领域，特别在有机工业废水生物处理、小型生活污水处理中得到广泛应用，成为污（废）水处理的主流工艺之一。其中所用填料种类繁多、发展迅速。

2. 生物接触氧化池的设计规定

（1）城镇污水集中处理工程和工业园区集中式废水处理工程应设置沉砂池。进水悬浮物浓度高于 BOD_5 设计值 1.5 倍时，城镇污水处理工程应设置初沉池。一般情况下工业废水处理工程可不设初沉池。

（2）接触氧化池的进水应符合下列条件：

水温宜为12～37℃、pH宜为6.0～9.0、营养组合比（BOD_5：氨氮：磷）宜为100：5：1，当氮磷比例不满足要求时，应适当进行调整；去除氨氮时，进水总碱度（以 $CaCO_3$ 计）与氨氮（NH_3-N）的比值应大于7.14，不满足时应补充碱度；脱总氮时，进水易降解碳源 BOD_5/TN应大于4.0，不满足时应补充碳源。

（3）多级接触氧化工艺的第一级生物接触氧化池的水力停留时间应占总水力停留时间的55％～60％。

（4）接触氧化池池体设计应符合下列条件：

接触氧化池的长宽比宜取1：1～2：1，有效水深宜取3～6m，超高应大于0.5m。接触氧化池采用悬挂式填料时，应由下至上布置曝气区、填料层、稳水层和超高。其中，曝气区高宜采用1.0～1.5m，填料层高宜取2.5～3.5m，稳水层高宜取0.4～0.5m。接触氧化池进水端宜设导流槽，其宽度不宜小于0.8m。导流墙下缘至填料底面的距离宜为0.3～0.5m，至池底的距离不宜小于0.4m。竖流式接触氧化池宜采用堰式出水，过堰负荷宜为2.0～3.0L/(s·m)。

（5）工艺参数应符合下列条件：

去除碳源有机污染物时，对于城镇污水和性质相似的工业废水，接触氧化池可参照表4-3所列参数取值，当水质指标相差较大时，应通过试验或参照类似工程取得设计参数。

同时去除碳源有机污染物和脱氮，应分别设置缺氧池和接触氧化池并参照表4-4所列参数取值。

去除碳源有机污染物的接触氧化池设计参数（设计水温20℃） **表4-3**

项目	单位	参数值
接触氧化池填料容积负荷	$kgBOD_5/(m^3$ 填料·d)	0.5～3.0
悬挂式填料填充率	%	50～80
悬浮式填料填充率	%	20～50
污泥产率	$kgVSS/kgBOD_5$	0.2～0.7
水力停留时间	h	2～6

同时去除碳源有机污染物和脱氮的接触氧化池设计参数（设计水温10℃） **表4-4**

项目	单位	参数值
接触氧化池填料容积负荷	$kgBOD_5/(m^3$ 填料·d)	0.4～2.0
硝化池填料容积负荷	$kgTKN/(m^3$ 填料·d)	0.5～1.0
好氧池悬挂式填料填充率	%	50～80
好氧池悬浮式填料填充率	%	20～50
缺氧池悬挂式填料填充率	%	50～80
缺氧池悬浮式填料填充率	%	20～50
污泥产率	$kgVSS/kgBOD_5$	0.2～0.6
污泥回流比	%	100～300
水力停留时间	h	好氧池:4～16;缺氧池:0.5～3.0

注：此参数仅适用于城镇污水和生活污水。

3. 生物接触氧化池的设计计算

（1）去除有机污染物的生物接触氧化池有效容积

$$V=\frac{(S_0-S_e)Q}{1000L_v\eta} \tag{4-11}$$

式中 V——生物接触氧化池有效容积，m^3；

Q——生物接触氧化池进水水量，m^3/d；

S_0——生物接触氧化池进水BOD_5，mg/L；

S_e——生物接触氧化池出水BOD_5，mg/L；

L_v——生物接触氧化池去除有机污染物BOD_5的容积负荷，$kgBOD_5/(m^3$ 填料·d)；

η——生物接触氧化池填料的填充比，%。

（2）脱氮反应的生物接触氧化池有效容积

1）硝化好氧池有效容积

$$V=\frac{(N_0-N_e)Q}{1000L_N\eta} \tag{4-12}$$

式中 V——硝化好氧池有效容积，m^3；

Q——生物接触氧化池进水水量，m^3/d；

N_0——生物接触氧化池进水凯氏氮，mg/L；

N_e——生物接触氧化池出水凯氏氮，mg/L；

L_N——生物接触氧化池的硝化容积负荷，$kgTKN/(m^3$ 填料·d)；

η——生物接触氧化池填料的填充比，%。

2）反硝化缺氧池有效容积

$$V=\frac{(N_0-N_e)Q}{1000L_{DN}\eta} \tag{4-13}$$

式中 V——反硝化缺氧池有效容积，m^3；

Q——生物接触氧化池进水水量，m^3/d；

N_0——反硝化池进水的硝态氮，mg/L；

N_e——反硝化池出水的硝态氮，mg/L；

L_{DN}——缺氧池的反硝化容积负荷，$kgNO_3$-N/(m^3 填料·d)；

η——生物接触氧化池填料的填充比，%。

（3）同时去除碳源有机污染物和氨氮的生物接触氧化池有效容积

同时去除碳源有机污染物和氨氮时，应分别计算去除碳源有机污染物的生物接触氧化池的有效容积和去除氨氮的生物接触氧化池的有效容积，取两者较大值或者取两者计算之和作为生物接触氧化池的有效容积。

（4）生物接触氧化池容积校核

采用水力停留时间对生物接触氧化池容积进行校核：

$$V=\frac{Qt}{24} \tag{4-14}$$

式中 V——生物接触氧化池容积，m^3；

Q——生物接触氧化池进水水量，m^3/d；

t——污水在生物接触氧化池的水力停留时间，h。

（5）生物接触氧化池所需空气量的计算

$$D=D_0Q \tag{4-15}$$

式中 D_0——$1m^3$ 污（废）水所需空气量，根据污（废）水水质、试验资料或参考类似工程实际运行经验参数确定。生物接触氧化池的需氧量，需同时满足微生物降解污染物的需氧量和氧化池搅拌强度的要求，D_0 宜大于 10，一般取 15～20。

（6）生物接触氧化池污泥量计算

$$W_{干}=YQ(S_0-S_e)+(X_0-X_h-X_e)Q \tag{4-16}$$

式中 $W_{干}$——污泥干重，kg/d；

Y——污泥产率系数，$kgVSS/kgBOD_5$；

Q——生物接触氧化池进水水量，m^3/d；

S_0——生物接触氧化池进水 BOD_5，mg/L；

S_e——生物接触氧化池出水 BOD_5，mg/L；

X_0——生物接触氧化池进水悬浮物浓度，mg/L；

X_e——生物接触氧化池出水悬浮物浓度，mg/L；

X_h——生物接触氧化池进水悬浮物中可生物降解物质所占百分比，%。

4. 生物接触氧化池的设计计算例题

【例 4-3】 某工业废水日最大排放量为 $800m^3/d$，废水的 BOD_5 为 180mg/L，进水悬

浮物（SS）浓度为120mg/L，冬季废水平均水温为16℃。拟采用生物接触氧化法处理，要求出水的BOD_5不大于20mg/L，出水悬浮物浓度不大于15mg/L。试设计生物接触氧化池的相关技术参数。

【解】（1）生物接触氧化池的有效容积（即填料体积）

取填料容积负荷$L_v=2.0kgBOD_5/(m^3 \cdot d)$，采用悬浮式填料，充填率取40%，接触氧化池有效容积为：

$$V=\frac{(S_0-S_e)Q}{1000L_v\eta}=\frac{(180-20)\times 800}{2\times 0.4\times 1000}=160m^3$$

（2）生物接触氧化池的总面积和座数

设生物接触氧化池填料高度h_0采用3.0m，则生物接触氧化池总面积为：

$$A=\frac{V}{h_0}=\frac{160}{3}=53.3\text{，取 }54m^2$$

采用3座生物接触氧化池，单池面积为：

$$A_1=\frac{A}{n}=\frac{54}{3}=18m^2\leqslant 25m^2$$

每座接触氧化池面积为$18m^2$，满足单池面积小于$25m^2$要求。

（3）生物接触氧化池深度

生物接触氧化池超高$h_1=0.5m$；填料层上水深$h_2=0.5m$；填料层至池底的高度$h_3=0.5m$，则接触氧化池总深度为：

$$H=h_0+h_1+h_2+h_3=3+0.5+0.5+0.5=4.5m$$

（4）污水在生物接触氧化池内的有效停留时间

$$t=\frac{V}{Q}=\frac{160}{800}\times 24=4.8h$$

（5）生物接触氧化池所需空气量的计算：

采用多孔管鼓风曝气供氧，取水气比$D_0=15\sim 20m^3/m^3$，所需总空气量为：

$$D=D_0Q=15\times 800=12000m^3/d$$

（6）污泥量计算

根据《生物接触氧化法污水处理工程技术规范》HJ 2009—2011，该工艺去除有机物产生的污泥量宜按去除每千克BOD_5产生0.2～0.4kgVSS计，污泥含水率为96%～98%。

因此，污泥产率系数Y可以取$0.3kgVSS/kgBOD_5$，污泥含水率为97%，设该废水进水悬浮物中可生物降解物质为60%，污泥干重$W_干$为：

$$\begin{aligned}W_干&=YQ(S_0-S_e)+(X_0-X_h-X_e)Q\\&=0.3\times 800\times(0.18-0.02)+(0.12-0.12\times 0.6-0.015)\times 800\\&=64.8kg/d\end{aligned}$$

设污泥含水率为97%，湿污泥量为：

$$\frac{64.8}{1000}=0.0648t/d$$

湿污泥体积为：

$$\frac{0.0648}{(100\%-97\%)}=2.16m^3/d$$

本章关键词（Keywords）

生物膜法	Fixed film or attached growth processes，Biofilm process
生物滤池	Trickling filter
生物转盘	Rotation biological Contactors
生物接触氧化法	Submerged attached growth，Biological contact oxidation process
填料	Packing material
转盘	Discs
处理水槽	Reservoir

思考题

1. 什么叫生物膜?

2. 影响生物滤池处理效率的因素有哪些? 简述不同生物滤池的工艺特点和适用条件。

3. “处理同一种污水，如果盘片面积不变，相比单级式转盘，将生物转盘分为多级串联运行能显著提高处理水水质和水中溶解氧的含量”。试分析原因。

4. 当接触氧化池处理率下降时，是否需要添加营养物质?

5. 接触氧化装置生物膜培养过程中发现生物膜形成又脱落，如何解决和避免?

6. 采用浸没式生物膜法处理废水，在 C/N 较低的情况下能否提高脱氮效果?

7. 接触氧化池如何处理高 BOD 废水?

8. 生化装置采用接触氧化工艺时，沉淀池应该设回流系统吗?

9. 在接触氧化池挂膜过程中，有哪些方法有利于污泥黏附至填料上以及让填料上的污泥快速成长?

10. 接触氧化池在进水时曝气是否不利于生物膜的形成? 营养物质在进水前或后各个生化池均加? 进水时投加是否会造成营养缺失?

11. 某小区人口为 1000 人，全年平均气温大于 10℃，污水排放量按 150L/(人·d) 计算，BOD_5 为 300mg/L，要求出水 BOD_5 不大于 50mg/L，拟采用普通生物滤池，试计算滤池的结构尺寸。

12. 一座 $5000m^3/d$ 的污水处理站，工艺为生物接触氧化法，进水水质不稳定，COD 在 300～1500mg/L 之间，出水始终不能达到排放标准，该如何处理?

Chapter 5 Activated Sludge Processes

The activated sludge process (or suspended growth process) was developed around 1913 at the Lawrence Experiment Station in Massachusetts by Clark and Gage, and by Ardern and Lockett at the Manchester Sewage Works in Manchester, England. The activated sludge process was so named because it involved the production of an activated mass of microorganisms capable of stabilizing waste under aerobic conditions. In the aeration tank, contact time is provided for mixing and aerating influent wastewater with the microbial suspension, which is generally referred to as the mixed liquor suspended solids (MLSS) or mixed liquor volatile suspended solids (MLVSS). Mechanical equipment is used to provide the mixing and transfers oxygen into the process. The mixed liquor then flows to a sedimentation tank (clarifier) where the microbial suspension is settled and thickened. The settled biomass, described as activated sludge because of the presence of active microorganisms, is returned to the aeration tank to continue biodegradation of the influent organic material. A portion of the thickened solids are removed daily or periodically because the process produces excess biomass to accumulate with the non-biodegradable solids contained in the influent wastewater. If the accumulated solids are not removed, they will eventually find their way to the system effluent.

Many activated sludge processes or suspended growth processes used in municipal and industrial wastewater treatment are operated with a positive dissolved oxygen concentration (aerobic), but applications exist where suspended growth anaerobic (no oxygen present) reactors are used, such as for high organic concentration industrial wastewater and organic sludge.

5.1 Basic Flowsheet of Activated Sludge Processes

Activated sludge process is a sewage treatment method that uses suspended activated sludge in aeration tank to adsorb and degrade organic pollutants. Its basic flowsheet or shematic diagram is shown in Figure 5-1.

It can be seen from Figure 5-1 that the properly pretreated sewage enters the aeration tank together with the return sludge to form a mixed liquid. In the aeration tank, the microorganisms in the return sludge, the organic matter in the sewage and the oxygen injected into the aeration tank through the aeration equipment are fully mixed and contacted. The microorganisms metabolize on the biodegradable organic matter in the sewage water, and the dissolved oxygen is consumed at the same time, BOD_5 of sewage is reduced, then

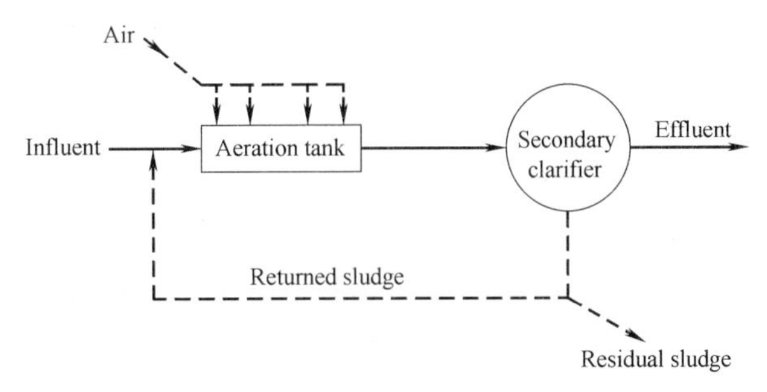

Figure 5-1 Basic flowsheet of activated sludge process

the mixed liquid flows into the secondary sedimentation tank (clarifier) for solid-liquid separation, and the clean water flows out through the overflow weir at the upper part. Most of the sludge at the bottom of the secondary sedimentation tank after sedimentation and concentration is returned to the aeration tank through the return sludge system, and the rest is discharged in the form of surplus sludge and sent to another sludge treatment system for further treatment to eliminate secondary pollution. As a biochemical reactor, the aeration tank keeps a certain amount of microorganisms by returning activated sludge and discharging excess sludge to bear the amount of organic pollutants entering the reactor; As an important part of the activated sludge process system, the secondary sedimentation tank separates activated sludge and water, and is closely connected with the aeration tank through reflux to provide activated sludge microorganisms required by the aeration tank and operate together with an organic whole formed by the aeration tank.

An important feature of the activated sludge process is the formation of floc particles, ranging in size from 50 to 200μm, which can be removed by gravity settling, leaving a relatively clear liquid as treated effluent. Typically, more than 99% of the suspended solids can be removed in the clarification step.

The operating characteristics of the sedimentation tank or clarifier may also affect sludge settling characteristics. Poor settling is often a problem in center-feed circular tanks where sludge is removed from the tank directly under the point where the mixed liquor enters. Sludge may actually be retained in the tank for many hours rather than the desired 30min and cause localized septic conditions. If this is the case, then the design is at fault, and changes must be made in the inlet feed well and sludge withdrawal equipment.

5.2 Development of Activated Sludge Processes

5.2.1 Complete Mix Activated Sludge

In the complete-mix reactor, it is assumed that complete mixing occurs instantaneously and uniformly throughout the reactor as fluid particles enter the reactor. Fluid particles leave the reactor in proportion to their statistical population. Complete mixing can be accomplished in round or square reactors if the contents of the reactor are uniformly and continuously redistributed. The actual time required to achieve completely mixed conditions will depend on the reactor geometry and the power input.

To obtain complete mixing in the aeration tank, proper choices of tank geometry, feeding arrangement, and aeration equipment are required. Through the use of complete mixing, with either diffused or mechanical aeration, it is possible to establish a constant oxygen demand as well a uniform mixed liquor suspended solids (MLSS) concentration throughout the tank or basin volume. Hydraulic and organic load transients are dampened in these systems, giving a process that is very resistant to upset from shock loadings. Influent wastewater and recycle sludge are introduced to the aeration tank or basin at different points. An activated sludge system is shown in Figure 5-2.

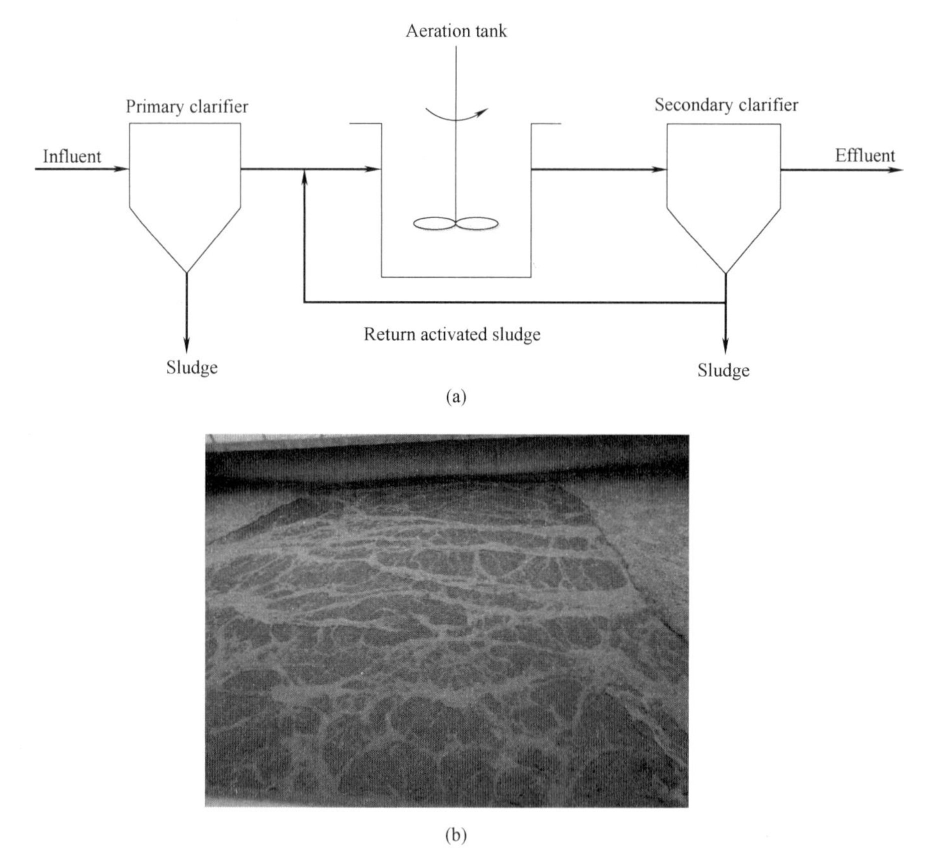

Figure 5-2 Complete Mix Activated Sludge System

(a) Process; (b) Operation

5.2.2 Plug-Flow Activated Sludge

Fluid particles pass through the reactor with little or no longitudinal mixing and exit from the reactor in the same sequence in which they entered. The particles retain their identity and remain in the reactor for a time equal to the theoretical detention time. This type of flow is approximated in long open tanks with a high length-to-width ratio in which longitudinal dispersion is minimal or absent or closed tubular reactors. A Plug-Flow activated sludge system is shown in Figure 5-3.

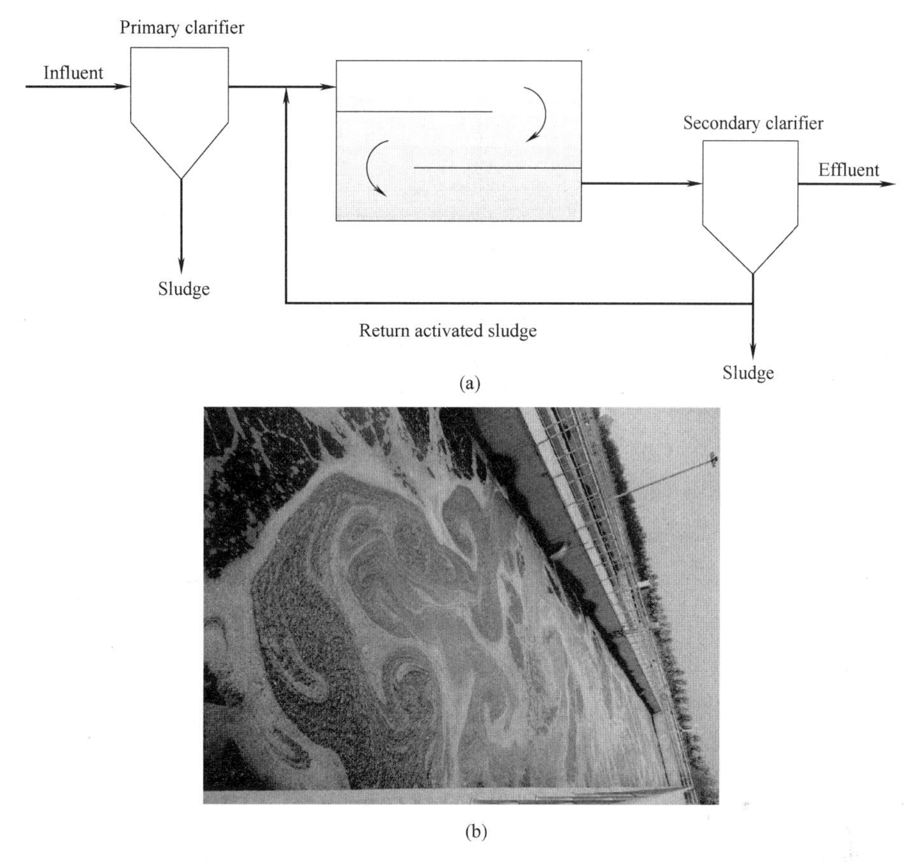

Figure 5-3 Plug-Flow activated sludge system

(a) Process; (b) Operation

5.2.3 Extended Aeration

In this process, sludge wasting is minimized. This results in low growth rates, low sludge yields, and relatively high oxygen requirements by comparison with the conventional activated sludge processes. Extended aeration is a reaction defined mode rather than a hydraulically defined mode, and can be nominally plug flow or complete mix. Design parameters typically include a food/microorganisms (F/M) ratio of 0.05 to 0.15, a sludge age of 15 to 35d, and mixed liquor suspended solids concentrations of 3000 to 5000 mg/L. The extended aeration process can be sensitive to sudden increases in flow due to resultant high MLSS loadings to the final sedimentation tank or clarifier, but is relatively insensitive to shock loads in concentration due to the buffering effect of the large biomass volume. While the extended aeration process can be used in a number of configurations, a significant number is installed as loop-reactor systems where aerators of a specific type provide oxygen and establish a unidirectional mixing to the basin contents. The application and modifications of loop reactor systems in wastewater treatment have been significant over recent years. An extended aeration system is shown in Figure 5-4.

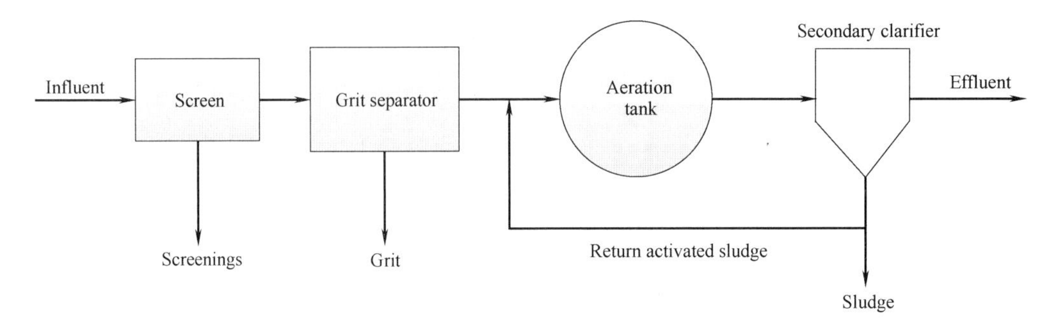

Figure 5-4 Extended aeration system

5.2.4 Oxidation Ditch Systems

A number of loop-reactor or ditch system variants are now available. In any ditch system, it is necessary to adequately match tank or basin geometry and aerator performance in order to yield an adequate channel velocity for mixed liquor solids transport. The key design factors in these systems relate to the type of aeration that is to be provided. It is normal to design for a 0.3m/s mid channel velocity in order to prevent solids deposition. The ditch system is particular amenable to those cases where both BOD and nitrogen removal are desired. Both reactions can be achieved in the same tank or basin by alternating aerobic and anoxic zones. An Oxidation Ditch system is shown in Figure 5-5.

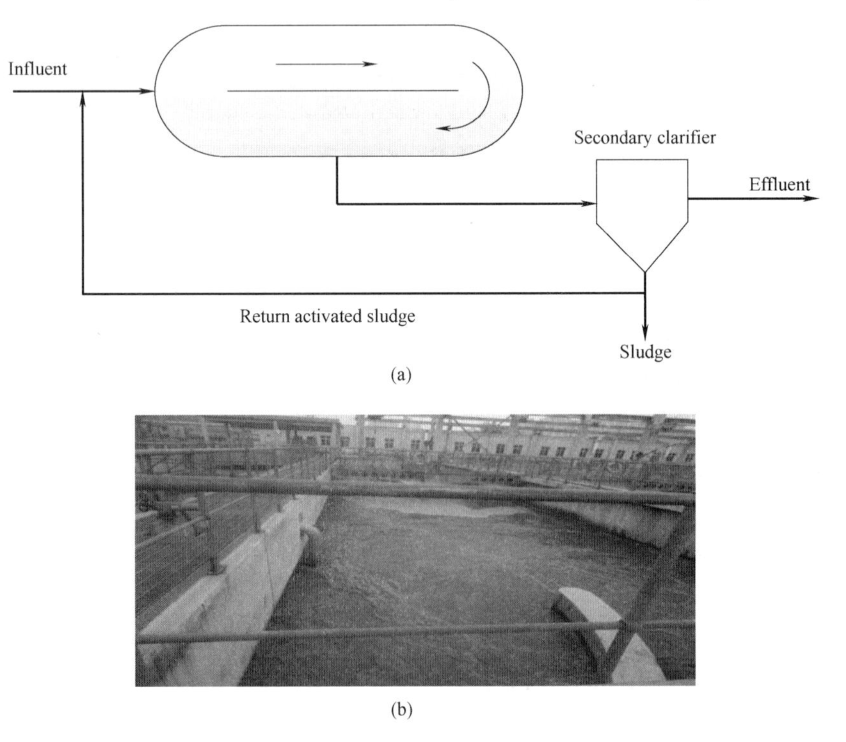

Figure 5-5 Oxidation ditch system

(a) Process; (b) Operation

5.2.5 Oxygen Activated Sludge

The high-purity oxygen system is a series of well-mixed reactors employing concurrent gas-liquid contact in a covered aeration tank. The process has been used for the treatment of municipal, pulp and paper mill, and organic chemical wastewater. Feed wastewater, recycle sludge, and oxygen gas are introduced into the first Stage. Gas-liquid contacting can be done by submerged turbines or surface aeration. An oxygen activated sludge system is illustrated on Figure 5-6.

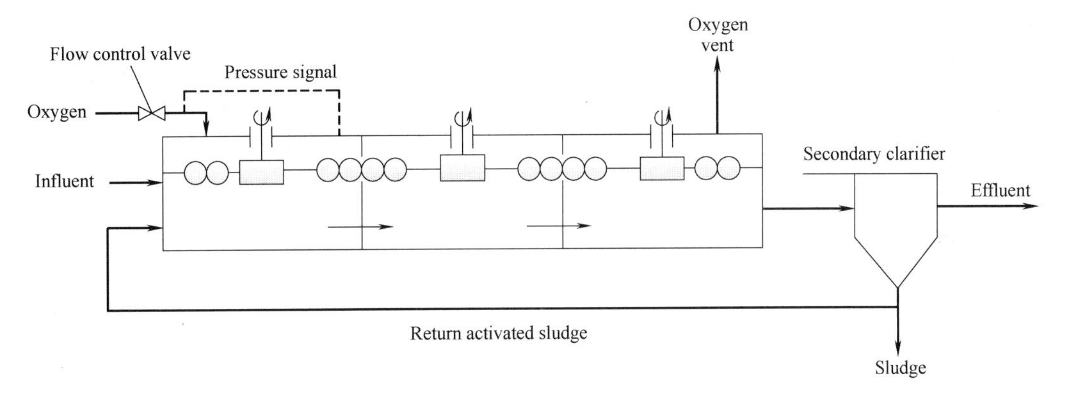

Figure 5-6 Oxygen activated sludge system

Oxygen gas is automatically fed to either system on a pressure demand basis with the entire unit operating, in effect, as a respirator a restricted exhaust line from the final stage vents the essentially odorless gas to the atmosphere. Normally the system will operate most economically with a vent gas composition of about 50% oxygen. For economic considerations, about 90% of oxygen utilization with on-site oxygen generation is desired. Oxygen may be generated by a traditional cryogenic air-separation process for large installations or a pressure-swing adsorption process for smaller installations. At peak load conditions, the oxygen system is usually designed to maintain 6.0mg/L dissolved oxygen in the mixed liquor.

Since high dissolved oxygen concentrations are maintained in the mixed liquor, the system can usually operate without filamentous bulking problems. The maintenance of all aerobic floc with high zone settling velocities also permits high MLSS concentrations in the aeration tank. Solids levels will usually range from 4000 to 9000mg/L, depending on the BOD of the wastewater and design volume of the system.

Pure oxygen also is employed in open aeration tanks or basins in which oxygen under high pressure is mixed with the influent wastewater. When introduced into the aeration tanks, the supersaturated gas comes out of solution in the form of microscopic bubbles.

5.2.6 Combined Aerobic Treatment Process

Several treatment process combinations have been developed that couple trickling filter with the activated sludge process. The combined biological processes are known as dual processes or coupled trickling filter/activated sludge systems. Combined processes have resulted as part of plant upgrading where either a trickling filter or activated sludge process is added; they have also been incorporated into new treatment plant designs. Combined processes have the advantages of the two individual processes, which can include: (1) the stability and resistance to shock loads of the fixed film (attached growth process), (2) the volumetric efficiency and low energy requirement of fixed film for partial BOD removal, (3) the role of fixed film pretreatment as a biological selector to improve activated sludge settling characteristics, and (4) the high-quality effluent possible with suspended growth (activated sludge) process.

The first group of the combined treatment processes, is shown in Figure 5-7, is commonly referred to as the trickling filter/solids contact (TF/SC) or trickling filter/activated sludge (TF/AS) process. The principal difference between these processes is the shorter aeration period in the TF/SC process of minutes versus hours for the TF/AS process. Both processes use a trickling filter, an activated sludge aeration tank, and a final sedimentation tank or settling tank. In both processes, the trickling filter effluent is fed directly to the activated sludge process without clarification and the return activated sludge from the secondary sedimentation tank is fed to the activated- sludge aeration tank or basin.

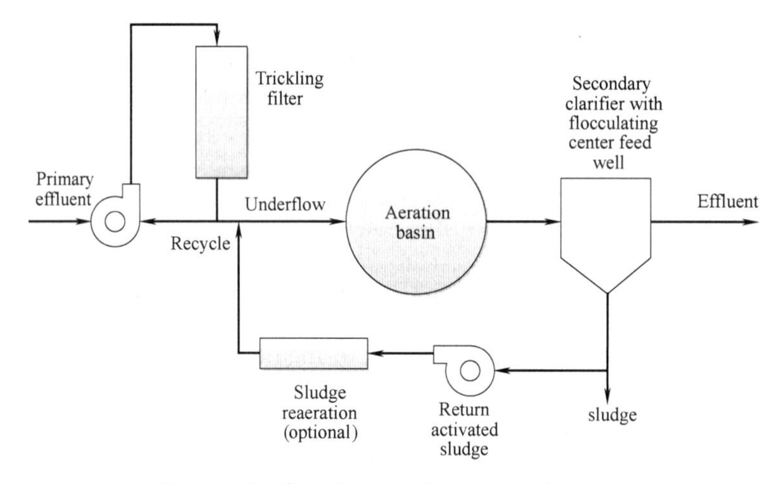

Figure 5-7 Combined aerobic treatment process

The most common application for the TF/AS process is where the trickling filter is designed as a roughing filter for 40% to 70% BOD removal and may be referred to as a roughing filter/activated sludge (RF/AS) process. The trickling filter loading is about 4 times that used for the TF/SC process. The aeration tank hydraulic retention time may be 50% to 70% of that used in the conventional activated sludge or suspended growth process. The RF/AS process is attractive for treating higher-strength industrial wastewater because of the relatively low energy use per quantity of BOD removed on the trickling filter. The use of this filter also results in good SVI values for the activated-sludge mixed liquor, as it acts as a biological selector in removing soluble BOD.

The main differences between the TF/SC and RF/AS processes are in their trickling filter loadings and activated sludge *SRT* values. A relatively low organic load for the trickling filter is used for the TF/SC process, and the purpose of the aeration tank is to remove remaining soluble BOD and to develop a flocculent activated sludge mass that incorporates dispersed solids from trickling filter sloughing. The process is able to produce an advanced treatment effluent quality, which is low in TSS and BOD concentrations, near 10mg/L and typically less.

5.3 活性污泥的性质及净化反应过程

5.3.1 活性污泥的性质

活性污泥是由细菌、真菌、原生动物、后生动物等微生物群体与废水中的悬浮物质、胶体物质混杂在一起所形成的具有很强的吸附分解有机物能力和良好沉降性能的絮绒状泥颗粒，因具有生物化学活性，所以被称为活性污泥。

活性污泥含水率很高，一般在99%以上。从外观上看，活性污泥是絮绒颗粒，又称生物絮凝体，其相对密度约在1.002～1.006，静置时可立即凝聚成较大的绒粒而下沉。活性污泥的比表面积可达20～100cm^2/mL。污泥的颜色因废水水质而异，一般为黄色或茶褐色，供氧不足或出现厌氧状态时呈黑色，供氧过多或营养不足时呈灰白色，略显酸性，稍具土壤的气味并夹带一些霉臭味。

活性污泥的组成包括下列4部分：具有代谢功能活性的微生物群体M_a；微生物（主要是细菌）自身氧化残留物M_e；来自原污水的难降解有机物M_i；来自原废水的无机物M_{ii}。

5.3.2 活性污泥净化反应过程

活性污泥净化反应过程比较复杂，既有活性污泥本身对有机污染物的吸附、絮凝等物理、化学或物理化学过程，也有活性污泥内微生物对有机污染物的生物转化、吸收等生物或生物化学过程，大致可以分为初期吸附去除阶段与代谢稳定阶段。

1. 初期吸附去除阶段

在废水与活性污泥接触、混合后的较短时间（5～10min）内，废水中的有机污染物，尤其是呈悬浮态和胶体态的有机物，表现出高的去除率，此时的去除并非降解，而是被污泥吸附，黏着在生物絮体的表面，这种由物理吸附和生物吸附交织在一起的初期高速去除现象叫初期吸附。在此过程中，混合液中有机底物迅速减少，BOD迅速降低，如图5-8中吸附区曲线。这是由于活性污泥的表面积大，并且在表面上富集大量的微生物，外部覆盖多糖类的黏质层，当废水中悬浮态、胶体态的有机底物与活性污泥絮体接触时，便被迅速凝聚和吸附去除。

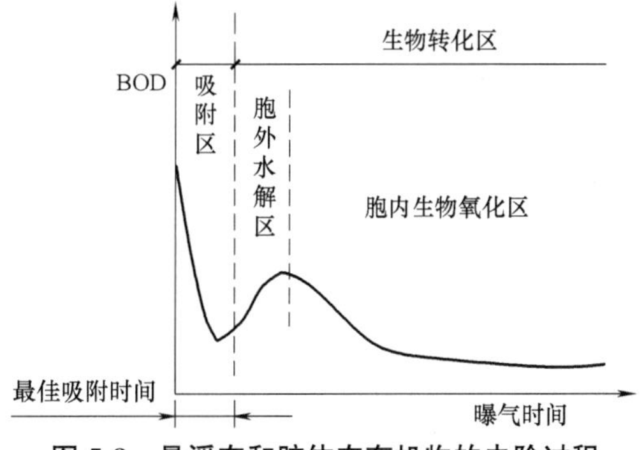

图5-8 悬浮态和胶体态有机物的去除过程

初期吸附过程进行得很快，一般在30min内便能完成，废水BOD的吸附去除率可达 70%，对于含悬浮态和胶体态有机物较多的废水，BOD 可下降 80%～90%。初期吸附速度主要取决于微生物的活性程度和反应器内水力扩散程度与水力动力学规律，前者决定活性污泥微生物的吸附、凝聚效能，后者则决定活性污泥絮体与有机底物的接触程度。活性污泥微生物的高吸附活性取决于较大的比表面积和适宜的微生物增殖期，一般而言，处于“饥饿”态的内源呼吸期微生物，其吸附活性最强。

2. 代谢稳定阶段

被吸附在活性污泥微生物细胞表面的有机污染物，在透膜酶的作用下，溶解态和小分子有机物直接透过细胞壁进入细胞体内，而胶体态和悬浮态的大分子有机物（如淀粉、蛋白质等）则先在细胞外酶——水解酶的作用下，被水解为溶解态小分子后再进入细胞体内，此水解产生的部分溶解性简单有机物会扩散到混合液中，造成混合液 BOD 升高，如图 5-8 中胞外水解区曲线所示。

进入细胞体内的有机污染物，在各种胞内酶（如脱氢酶、氧化酶等）的催化作用下，被氧化分解为中间产物，有些中间产物合成为新的细胞物质，另一些则氧化为稳定的无机物，如 CO_2 和 H_2O 等，并释放能量供合成细胞所需，这个过程即物质的氧化分解过程，也称稳定过程。在此过程中，不稳定的高分子有机物质通过生化反应被转化为简单稳定的低分子无机物质，混合液 BOD 逐渐降低。稳定过程所需时间取决于有机物的转化程度，要比吸附过程长得多。

5.4 活性污泥法的重要指标

5.4.1 活性污泥的评价指标

活性污泥微生物群体是一个以好氧细菌为主的混合群体，其他微生物包括酵母菌、放线菌、霉菌以及原生动物、后生动物等。在污水处理系统或曝气池内微生物群体的增殖规律与纯菌种的增殖规律相同，即停滞期（适应期）、对数期、静止期（也减速增殖期）和衰亡期（内源呼吸期）。活性污泥的核心菌胶团是由成千上万细菌相互黏附形成的生物絮体。细菌在对数增长期时，个体处于旺盛生长，其运动活性大于范德华力，菌体不能结合；但到了衰亡期，动能低微，范德华力大，菌体相互黏附，形成生物絮体，因此，静止期与衰亡期个体是活性污泥的重要微生物。

不同增殖期对应于不同微生物组合及不同生物处理工艺。所以利用光学显微镜或电子显微镜，观察活性污泥中的细菌、真菌、原生动物、后生动物等微生物的种类、数量、优势度等状况，在一定程度上可反映整个系统的运行状况。

1. 污泥浓度

混合液悬浮固体浓度（MLSS）也称为污泥浓度，指曝气池中单位体积混合液中活性污泥悬浮固体的质量（mg/L），包括如前所述的 M_a，M_e，M_i 及 M_{ii} 四者在内的总量。混合液挥发性悬浮固体浓度（MLVSS）指混合液悬浮固体中有机物的质量，包括 M_a，M_e 及 M_i 三者，不包括污泥中无机物质。

采用具有活性的微生物的浓度作为活性污泥浓度从理论上更加准确，但测定活性微生物的浓度非常困难，无法满足工程应用要求。而 MLSS 测定简便，工程上往往以它作为评价活性污泥量的指标，同时用 MLVSS 代表混合液悬浮固体中有机物的含量，比 MLSS

更接近活性微生物的浓度，测定也较为方便，且对某一特定的污水处理系统，MLVSS/MLSS 相对稳定，因此，可用 MLVSS 表示污泥浓度，一般生活污水处理厂曝气池混合液 MLVSS/MLSS 为 0.6～0.7。

2. 污泥沉降比

为了在后续沉淀池中进行泥水分离和提供更高的回流污泥浓度，在设计二沉池时必须考虑混合液污泥的沉降或浓缩特性，通常使用污泥沉降比（SV%）和污泥体积指数 SVI 表示活性污泥的沉降性能。

污泥沉降比是曝气池混合液静止 30min 后沉淀污泥的体积分数，标准采用 1L 的量筒测定污泥沉降比。由于正常的活性污泥在静沉 30min 后可接近它的最大密度，故可反映污泥的沉降性能。污泥沉降比与所处理的污水性质、污泥浓度、污泥絮体颗粒大小及污泥絮体性状等因素有关，混合液污泥浓度在 3000mg/L 左右时，正常曝气池污泥沉降比在 30%左右。在培菌阶段也用 SV 评价污泥的培养程度。

3. 污泥体积指数

污泥体积指数 SVI 指曝气池混合液沉淀 30min 后，每单位质量干泥形成的湿污泥的体积，常用单位为 mL/g。SVI 通常按下述方法测定：①在曝气池出口处取混合液样品；②测定 MLSS；③测定样品的 SV%，读取 1L 混合液沉淀污泥的体积（mL）；④按下式计算 SVI：

$$\mathrm{SVI}=\text{沉淀污泥的体积}(\mathrm{mL/L})/\mathrm{MLSS}(\mathrm{g/L}) \tag{5-1}$$

SVI 是判断污泥沉降浓缩性能的一个重要参数，比 SV%能更准确反映污泥的沉降性能。通常认为 SVI 为 100～150 时，污泥沉降性能良好；SVI>200 时，污泥沉降性能差，易发生污泥膨胀；若 SVI 过低，如小于 50 时，则说明污泥絮体细小紧密，含无机物较多，污泥的活性差。

5.4.2 活性污泥工艺运行技术指标

1. BOD 负荷

BOD 负荷是活性污泥处理系统设计、运行最基本的参数之一，可用污泥负荷和容积负荷两种方法表示，而污泥负荷是影响有机污染物降解、活性污泥增长的重要因素，与污水处理效率、活性污泥特性、污泥生成量、氧的消耗量密切相关。

在活性污泥法中，一般将有机物量 F 与活性污泥量（M）的比值（F/M）称为污泥负荷，即单位质量活性污泥在单位时间内所承受的有机物量，以 N_s 或 L_s 表示，单位是 $kgBOD_5$/(kgMLSS・d) 或 $kgBOD_5$/(kgMLVSS・d)，即：

$$N_s=F/M=QS_0/XV \tag{5-2}$$

式中 Q——污水平均流量，m^3/L；

S_0——曝气池入流污水的 BOD_5，mg/L；

X——曝气池混合液浓度，MLSS 或 MLVSSmg/L；

V——曝气池有效容积，m^3。

而单位曝气池有效容积在单位时间内所承受的有机物量称为容积负荷，以 N_v 或 L_v 表示，单位是 $kgBOD_5/(m^3 \cdot d)$，即：

$$N_v=QS_0/V=N_sX \tag{5-3}$$

此外，工程上采用去除负荷 N_r 或 L_r 表示有机物的去除情况，单位为 $kgBOD_5$/

(kgMLSS·d)：

$$N_r=Q(S_0-S_e)/XV \text{ 或 } N_r=Q(S_0-S_e)/VX_v \quad (5\text{-}4)$$

式中　X_v——曝气池混合液 MLVSS，mg/L。

式（5-4）更精准地体现了去除负荷的本质，但由于 MLSS 容易监测，在进行设计计算时使用式（5-3）更方便。

（1）BOD 负荷与处理效率的关系

在一定的污泥负荷范围内，随着污泥负荷 N_s 的升高，处理效率 η 将下降，出水的 BOD_5 将升高。但对不同的底物，η 随 N_s 的变化关系有很大差别。粪便污水、食品工业废水等所含底物是糖类、有机酸、蛋白质等易降解的一般性有机物，即使污泥负荷升高，BOD 去除率下降的趋势也较缓慢；相反，醛类、酚类的分解需要特种微生物，当污泥负荷超过某一值后，BOD 去除率显著下降。对同一种废水，在不同的污泥负荷范围内，其 BOD 去除率变化速度也不同。

（2）BOD 负荷对污泥容积指数的影响

如果在运行时 BOD 负荷波动进入高 SVI 负荷区，污泥沉淀性差，将会出现污泥膨胀。所以一般在高负荷时应选择在 1.5～2.0$kgBOD_5$/(kgMLSS·d) 范围内，中负荷时为 0.2～0.4$kgBOD_5$/(kgMLSS·d)，低负荷时为 0.03～0.05$kgBOD_5$/(kgMLSS·d)。

（3）BOD 负荷对污泥生成量的影响

在曝气池中，活性污泥微生物的增殖是微生物合成反应与内源代谢两项生理活动的综合结果，即是微生物合成与内源呼吸的差值，工程上采用式（5-5）计算：

$$\Delta X=aQ(S_0-S_e)-bVX \quad (5\text{-}5)$$

式中　ΔX——活性污泥微生物净增殖量，kg/d；

S_0——进入曝气池污水含有的有机污染物量，kgBOD/d；

S_e——经活性污泥处理后出水的有机污染物量，kgBOD/d；

X——混合液活性污泥浓度，kg/m^3；

a——污泥产率（降解单位 BOD 增殖的微生物的量，kg/kg），生活污水取 0.07～0.49；

b——微生物内源代谢的自身氧化率（也叫分解系数），生活污水取 0.07～0.075。

等式两边除以 VX 得：

$$\Delta X/VX=aQ(S_0-S_e)/VX-b$$

由于 $N_r=Q(S_0-S_e)/VX$，故：

$$\Delta X/VX=aN_r-b \quad (5\text{-}6)$$

式（5-6）表明活性污泥比增长率 $\Delta X/VX$ 与去除负荷 N_r 相关，如果 N_r 降低，则污泥生成量减少，且当 N_r 低到 b/a 时，污泥生产率为零，即系统基本不排剩余污泥。

（4）BOD 负荷对需氧量的影响

曝气池总需氧量包括有机物去除的需氧量以及微生物有机体自身内源呼吸需氧量之和，在工程上表示为：

$$m_{O_2}=a'VX_vN_r+b'VX_v \quad (5\text{-}7)$$

式中　m_{O_2}——曝气池混合液每日的需氧量，kg/d；

a'——氧化分解有机物过程的需氧率，即微生物每代谢 1kgBOD 所需氧的质量数，$kgO_2/kgBOD_5$；

b'——内源分解自身氧化过程的需氧率，即每 kg 活性污泥自身氧化所需氧的质量数，$kgO_2/(kgMLVSS \cdot d)$。

$$m_{O_2}/VX_v = a'N_r + b' \tag{5-8}$$

从式（5-8）可见，当活性污泥系统在较高的污泥去除负荷下运行时，去除单位质量 BOD 的需氧量较小，因为底物在高负荷系统的停留时间短，一些只被吸附而未经氧化的有机物随剩余污泥排出，同时微生物自身的氧化分解程度较弱；反之，在低负荷情况下，有机物能被彻底氧化，微生物自身氧化作用也强，因此，需氧量消耗大。从需氧量方面看，高负荷系统比低负荷系统经济。

（5）BOD 负荷对曝气池容积的影响

采用高值的污泥负荷，将加快有机污染物的降解速度和活性污泥的增长速度，底物在曝气池中的停留时间短，曝气池的容积减小，在经济上比较适宜，但处理水水质未必能够达到预定的要求。而采用低值的污泥负荷，有机污染物的降解速度和活性污泥的增长速度都将降低，底物在曝气池中的停留时间长，曝气池的容积加大，建设费用有所增加，但处理水的水质可能提高，并达到要求。

（6）BOD 负荷对营养比要求的影响

采用不同污泥负荷时，微生物处于不同生长阶段。在低负荷时，污泥自身氧化程度较高，在有机体氧化过程中释出氮、磷成分，所以氮、磷的需要量减小，如在延时曝气法中，BOD_5 ∶ N ∶ P＝100 ∶ 1 ∶ 0.2 时即可使微生物正常生长，而在一般污泥负荷下，则要求 BOD_5 ∶ N ∶ P＝100 ∶ 5 ∶ 1。

2. 污泥龄 θ_c

曝气池内活性污泥总量与每日排放的污泥量之比称为污泥龄（简称泥龄），即活性污泥在曝气池内的平均停留时间，故又称“生物固体平均停留时间”，用 θ_c 表示，即：

$$\theta_c = VX/\Delta X = VX/Q_w X_r \tag{5-9}$$

式中 ΔX——曝气池内每日增长的活性污泥量，即要排放的活性污泥量，kg/m^3；

Q_w——排放的剩余污泥流量，m^3/d；

X_r——剩余污泥浓度。

污泥龄是活性污泥处理系统设计、运行的重要参数，与污泥去除负荷、混合液污泥浓度 X、出水底物浓度 S_e 等参数密切相关。污泥龄还影响活性污泥的絮凝沉淀性能、污泥产量、活性污泥微生物状况等。因此，适宜的污泥龄对活性污泥处理系统至关重要。一般而言，为使溶解性有机物去除率高，可选较小的 θ_c；为使活性污泥有较好的絮凝沉淀性，宜选用中等大小的 θ_c；为使微生物净增殖量最小，就选较大的 θ_c。

在活性污泥系统设计中，既可采用污泥负荷，也可采用污泥龄作设计参数，但在实际运行时，控制污泥负荷比较困难，需要测定有机物量和污泥量，而用污泥龄作为运转控制参数只要求调节每日的排泥量，过程控制简单。

5.5 氧转移原理和曝气设备

为满足峰值条件下的持续有机负荷，曝气设备的峰值因子应至少为平均 BOD 负荷的 1.5～2.0 倍。曝气设备的尺寸也应基于平均负荷时曝气池中的残余溶解氧（DO）量，当平均负荷DO值为2mg/L时，峰值负荷为1.0mg/L。曝气设备的设计必须具有足够的灵活性，以（1）满足最低需氧量，（2）防止过度曝气和节约能源，（3）满足最高需氧量。

5.5.1 氧转移原理及其影响因素

氧由气相转入液相的原理常用双膜理论解释。双膜理论的基本点为：在气-液界面两侧存在着气膜和液膜；气膜和液膜对气体分子的转移产生阻力；氧在膜内以分子扩散方式转移，其速度慢于在混合液内发生的对流扩散的转移方式；氧是难溶气体，其阻力主要来自液膜。

根据气液传质扩散的双膜理论，推导出氧转移速率的公式：

$$dC/dt = K_{La}(C_s - C) \tag{5-10}$$

式中 dC/dt——液相溶解氧浓度变化速率，即单位体积液相的氧转移速率，$kgO_2/(m^3 \cdot h)$；

K_{La}——氧总传质系数，1/h；

C_s——氧在液相中的饱和浓度，kg/m^3；

C——氧在液相中的实际浓度，kg/m^3。

由式（5-10）可知，要提高氧转移速率dC/dt，可从以下两个方面考虑：加速气液界面更新，增大气液接触面积，降低液膜厚度，以减小传质阻力，提高K_{La}；或提高气相中的氧分压，如采用纯氧曝气、深井曝气等，以提高C_s。

影响氧传递的因素主要有以下几个方面：

（1）废水水质：水中各种杂质会对氧的转移造成影响，其中主要是溶解性有机物，特别是某些表面活性物质，会在气液界面处集中，形成一层分子膜，增加氧传递的阻力，使K_{La}降低，阻碍氧分子的扩散转移。

（2）水温：水温对氧的转移影响较大，水温上升，水的黏度降低，扩散系数提高，液膜厚度减小，K_{La}增高；反之，水温下降则扩散系数降低。

（3）氧分压：气相中氧分压增大，使C_s增大，则传递速率加快；反之，则速率降低。

此外，氧转移速率还受气泡大小、水流的紊流程度、气泡与液体的接触时间、液相中氧的浓度梯度等因素的影响，可以通过曝气设备的选择、系统运行方式的改变等人为措施使氧转移速率得以强化。

5.5.2 曝气设备

曝气设备按供气方式主要分为鼓风曝气和机械曝气两大类。

1. 鼓风曝气

鼓风曝气系统是由进风空气过滤器、鼓风机、空气输配管系统和浸没于混合液下的扩散器组成。鼓风机须提供满足生化反应所需的氧量，这样的风量才能基本保持混合液悬浮固体呈悬浮状态。风压则要满足克服管道系统和扩散器的摩阻损失以及扩散器上部的静水压。鼓风机进口空气过滤器是为了防止灰尘进入扩散器内部造成阻塞。

扩散器是整个鼓风曝气系统的关键部件，它的作用是将空气分散成不同尺寸的气泡。气泡在扩散装置的出口处形成，气泡尺寸越小，与周围混合液的接触面积越大。气泡在上升过程中随水流循环运动，气泡中的氧随之转移溶解于混合液中。由于气泡内压力越来越小，其在上升过程至液面间破裂，变成更小的气泡。

根据分散气泡的大小，扩散器又可分成微气泡扩散器、小气泡扩散器、中气泡扩散器等几种类型。

微气泡扩散器形成的气泡直径在 100μm 左右，气液接触面大，氧利用率高，但缺点是压力损失较大，易堵塞，对送入的空气必须进行过滤处理。微气泡扩散器制造材料一般分为两大类：一种为多孔性刚性材料，如刚玉、陶粒、锌粉、粗瓷等掺以适当的如酚醛树脂一类的黏合剂，在高温下烧结定形而成，产生的微孔易被沉积物堵塞；另一种材料由柔性橡胶和多孔塑料膜制成，可形成管式（图 5-9）、圆盘式（图 5-10）等形状，膜上用激光均匀制成微孔，鼓风时，空气进入膜片与支撑管或支撑底座之间，使膜片微微鼓起，孔眼张开，空气从孔眼逸出，达到空气扩散的目的。供气停止，压力消失，在膜片的弹性作用下，孔眼自动闭合，并且由于水压的作用，膜片压实在底座之上，曝气池混合液不会倒流，孔眼不会堵塞。

为了便于维护管理，可以将微孔曝气管制成成组的可提升设备，需要维护时，随时可以将扩散器提出水面进行清理。这类扩散设备的氧转移效率可达 30%，具体安装要求及性能参数可参照生产厂家提供的数据。

小气泡扩散器是采用多孔材料（陶瓷、砂粒、锌粉、塑料等）制成的扩散板或扩散管，分散气泡直径可小于 1.5mm（图 5-11）。

中气泡扩散器常用穿孔管和莎纶管。穿孔管由管径介于 25～50mm 之间的钢管或塑料管制成，在管壁两侧向下呈 45°角方向开有直径为 2～3mm 的孔眼，孔眼间距为 50～100mm，两边错开排列，孔口的气体流速不小于 10m/s，以防堵塞（图 5-12）。

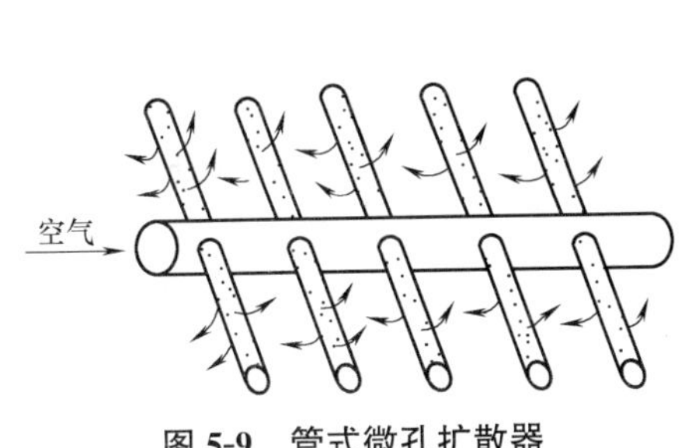

图 5-9　管式微孔扩散器

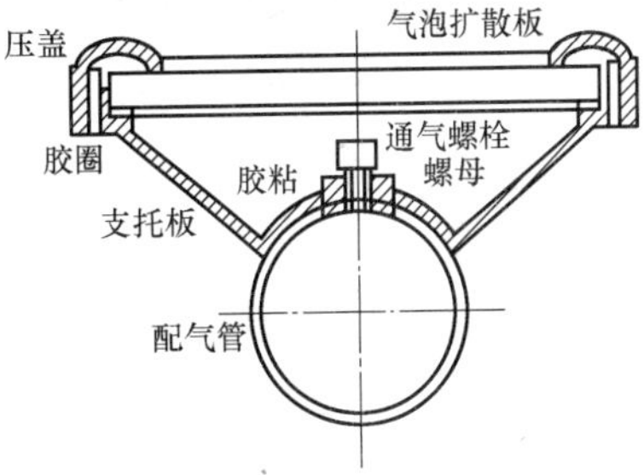

压盖
气泡扩散板
胶圈
通气螺栓
螺母
胶粘
支托板
配气管

图 5-10　圆盘式微孔扩散器

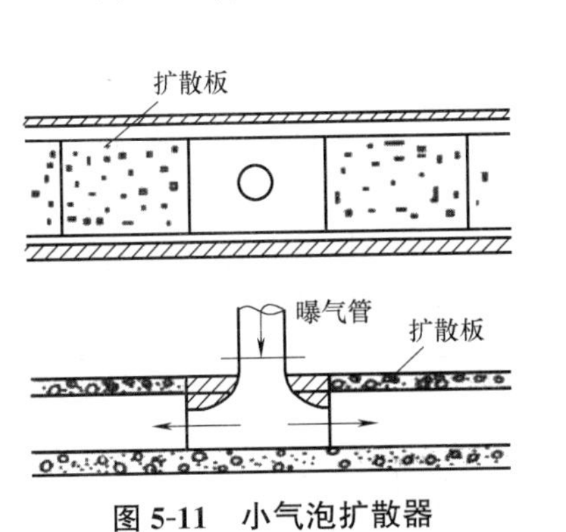

图 5-11　小气泡扩散器

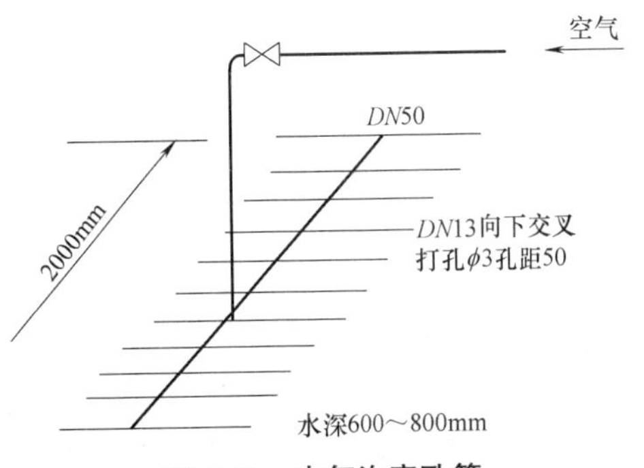

图 5-12　中气泡穿孔管

大、中气泡空气扩散器堵塞的可能性小，空气净化要求低，养护管理比较方便，但其氧的传递速率较低；微小气泡扩散器氧的传递速率高，反应时间短，曝气池的容积可以缩小，但对空气净化要求高、管理维护难度较大。因此，选择何种扩散器要因地制宜。

扩散器可以布满整个曝气池底，或沿曝气池横断面的一侧布置，使混合液中的悬浮固体呈悬浮状态，沿一侧布置时可以在曝气池断面上形成旋流，增加气泡和混合液的接触时间，有利于氧的传递。

鼓风曝气常用的风机有罗茨鼓风机和离心式鼓风机。罗茨鼓风机造价便宜，国产单机风量在 $80m^3/min$ 以下，一般适用于中、小型污水处理厂，但运行时噪声大，必须采取消声、隔声措施。离心式鼓风机又可分为单级高速离心风机和多级离心风机。单机风量大，风量调节方便，运行噪声小，工作效率高，但进口离心风机价格较贵，一般适用于大、中型污水处理厂。

2. 机械曝气

鼓风曝气是采用液面下曝气，机械曝气则是通过安装于池面的表面曝气器实现的。机械曝气器按传动轴的安装方向可分竖轴式和卧轴式两类。

竖轴式曝气器的传动轴与液面垂直，装有叶轮，其基本充氧途径是：当叶轮快速转动时，把大量的混合液以液幕、液滴的形式抛向空中，在空中与大气接触进行氧的转移，然后挟带空气形成气液混合物回到曝气池中，由于气液接触界面大，从而使空气中的氧很快溶入水中；随着曝气器的转动，在曝气叶轮的后侧形成负压区，卷吸部分空气；曝气叶轮的转动具有提升、输送液体的作用，使混合液连续上下循环流动，气液接触界面不断更新，不断使空气中的氧向液体中转移，同时池底含氧量小的混合液向上环流和表面充氧区发生交换，从而提高了整个曝气池混合液的溶解氧含量。因为混合液的流动状态同池型有密切的关系，故曝气的效率不仅决定于曝气器的性能，还与曝气池的池形有密切关系。

曝气叶轮的淹没深度一般在 10～100mm，可以调节。淹没深度大时提升水量大，但所需功率亦会增大，叶轮转速一般为 20～100r/min，也可以进行二挡或三挡调速，以适应进水水量和水质的变化。常用的这类曝气器叶轮有泵型、倒伞型和平板型，如图 5-13 所示。

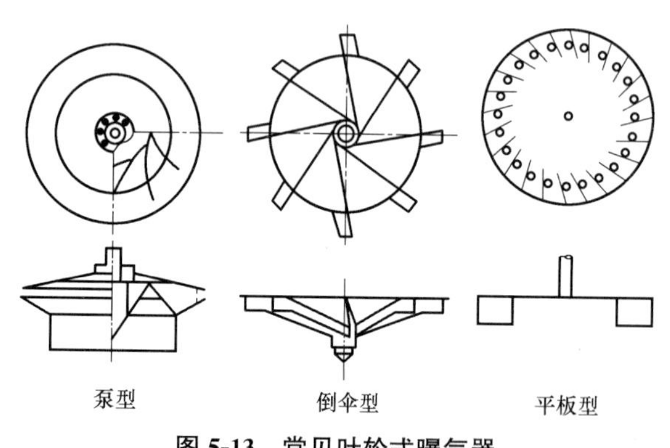

图 5-13　常见叶轮式曝气器

卧轴式曝气器的转动轴与水面平行，主要用于氧化沟系统。在转动轴上安装开有鳞片孔的转碟，或在垂直于转轴的方向装有不锈钢丝（转刷）或塑料板条，由电机驱动，转速在 50～70r/min，淹没深度为转刷直径的 1/4～1/3。转动时，转碟或转刷把大量液滴抛向空中，并使液面剧烈波动，促进氧的溶解；同时推动混合液在池内流动，促进曝气器附近的混合液紊流，便于溶解氧的扩散（图 5-14）。

用于衡量曝气设备性能的主要指标有：①动力效率（*EP*），即每消耗 1kWh 的动力能传递到水中的氧量，单位为 kgO_2/kWh；②氧的利用效率（*EA*），通过鼓风曝气系统转移到混合液中的氧量占总供氧的百分比，单位为%；③氧转移效率（*EL*），也称为充氧能力，单位为 kgO_2/h。对鼓风曝气系统，其性能评价按第一、二项指标评定，其中主要为第一项；对于机械曝气装置，则按第一、三项指标评定，其中主要为第一项。

上面所提及的各类曝气设备除了要满足充氧要求外，还应满足如下的最低混合强度要求：采用鼓风曝气器时，按规范规定，处理 $1m^3$ 废水的曝气量不应小于 $3m^3$，如果曝气池水位较深，则可以按最低曝气强度（单位池底面积、单位时间内的曝气量）$1.2m^3/(m^2 \cdot h)$（中气泡曝气）～$2.2m^3/(m^2 \cdot h)$（小气泡曝气）控制。采用机械曝气器时，混合全池废水所需功率不宜小于 $25W/m^3$；氧化沟不宜小于 $15W/m^3$。

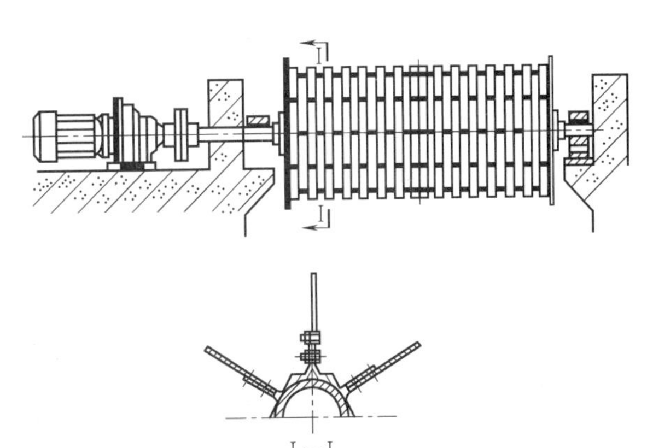

图 5-14　卧轴式曝气器

板式和管式曝气器与曝气效果如图 5-15 所示。

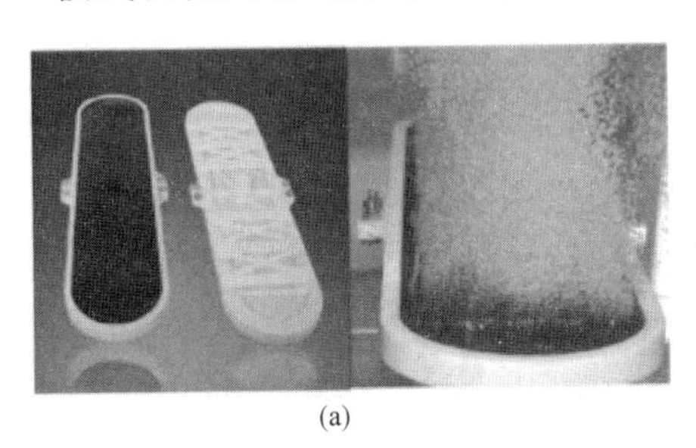

(a)

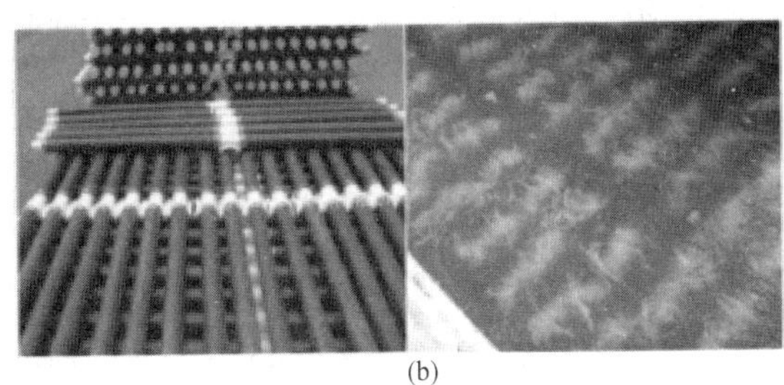

(b)

图 5-15　板式和管式曝气器与曝气效果

（a）板式曝气器；（b）管式曝气器

二维码 6
其他曝气设备
曝气效果、气泡
特点及工作场合

5.6　活性污泥法的设计计算

5.6.1　设计一般规定

（1）根据去除碳源污染物、脱氮、除磷、好氧污泥稳定化等不同要求和外部环境条件，选择适宜的活性污泥处理工艺。

（2）生物反应池的超高，当采用鼓风曝气时为 0.5～1.0m；当采用机械曝气时，其设备操作平台宜高出设计水面 0.8～1.2m。

（3）污（废）水中含有大量产生泡沫的表面活性剂时，应有除泡沫措施。

（4）廊道式生物反应池的池宽与有效水深之比宜采用 1∶1～2∶1。有效水深应结合流程设计、地质条件、供氧设施类型和选用风机压力等因素确定，可采用 4.0～6.0m。在条件许可时，水深尚可加大。每组生物反应池在有效水深一半处宜设置放水管。

（5）生物反应池中的好氧区（池），采用鼓风曝气器时，处理每立方米污水的供气量应大于 $3m^3$。好氧区采用机械曝气器时，混合全池污（废）水所需功率应大于 $25W/m^3$；氧化沟工艺应大于 $15W/m^3$。缺氧区（池）、厌氧区（池）应采用机械搅拌，混合功率宜采用 $2～8W/m^3$。机械搅拌器布置的间距、位置，应根据试验资料确定。

（6）生物反应池的设计，应充分考虑冬季低水温对去除碳源污染物、脱氮和除磷的影响，必要时可采取降低负荷、增长污泥龄、调整厌氧区（池）及缺氧区（池）水力停留时间和保温或增温等措施。

（7）生物反应池的始端可设缺氧或厌氧选择区（池），水力停留时间宜采用0.5～1.0h。

（8）阶段曝气生物反应池宜采取在生物反应池始端1/2～3/4的总长度内设置多个进水口。

（9）吸附再生生物反应池的吸附区和再生区可在一个反应池内，也可分别由两个反应池组成，并应符合：吸附区的容积不应小于生物反应池总容积的1/4，吸附区的停留时间应大于0.5h。当吸附区和再生区在一个反应池内时，沿生物反应池长度方向应设置多个进水口；进水口的位置应适应吸附区和再生区不同容积比例的需要；进水口的尺寸应按通过全部流量计算。

（10）完全混合生物反应池分为合建式和分建式。合建式生物反应池宜采用圆形，曝气区的有效容积应包括导流区。沉淀区的表面水力负荷宜为0.5～1.0$m^3/(m^2 \cdot h)$。

（11）污（废）水、回流污泥进入生物反应池的厌氧池（区）、缺氧池（区），宜采用淹没入流方式。

5.6.2 传统活性污泥法去除污（废）水中碳源污染物的主要设计参数

传统活性污泥法去除污（废）水中碳源污染物的主要设计参数参考表5-1。

传统活性污泥法去除污（废）水中碳源污染物的主要设计参数　　表5-1

运行方式		污泥龄(d)	污泥负荷(N_s)[$kgBOD_5$/(kgMLSS·d)]	容积负荷(L_v)[$kgBOD_5$/(m^3·d)]	污泥浓度(mg/L)	停留时间(h)	回流比(%)
推流式		3～5	0.2～0.4	0.4～0.9	1500～2500	4～8	25～75
阶段曝气		3～5	0.2～0.4	0.4～1.2	1500～3000	3～5	25～75
高负荷		0.2～0.5	1.5～5.0	1.2～1.4	200～500	1.5～3	5～15
完全混合		3～5	0.25～0.5	0.5～1.8	2000～4000	3～5	100～400（合建）
延时曝气		20～30	0.05～0.15	0.1～0.4	3000～6000	18～36	75～150
AB法	A级	0.5～1	2～6	0.1～0.4	3000～3000	0.5	50～80
	B级	15～20	0.1～0.3	0.1～0.4	3000～5000	2～4	50～80

5.6.3 计算公式

1. 曝气池容积设计计算

（1）活性污泥负荷法

$$V=\frac{Q(S_0-S_e)}{XN_s} \tag{5-11}$$

式中　V——曝气池容积，m^3；

Q——曝气池平均进水流量，m^3/d；

S_0——曝气池进水BOD_5，mg/L或kg/m^3；

S_e——曝气池出水 BOD_5，mg/L 或 kg/m³（当去除效率大于 90%时可不考虑）；

X——曝气池混合液悬浮固体浓度，mgMLSS/L；

N_s——曝气池 BOD_5 污泥负荷，kgBOD_5/(kgMLSS·d)。

（2）容积负荷法

$$V=\frac{Q(S_0-S_e)}{L_v} \tag{5-12}$$

式中 L_v——曝气池 BOD_5 容积负荷，kgBOD_5/(m^3·d)。

（3）污泥龄法

$$V=\frac{QY\theta_c(S_0-S_e)}{X(1+K_d\theta_c)} \tag{5-13}$$

式中 Y——污泥产率系数，kgVSS/kgBOD_5，根据试验资料确定，一般取 0.4～0.8；

θ_c——污泥龄，d；

K_d——衰减系数，d^{-1}，应以当地夏季和冬季的污水温度按式（5-14）进行修正。

$$K_T=K_{20}\theta_T^{T-20} \tag{5-14}$$

式中 K_T——T℃时的衰减系数，d^{-1}；

K_{20}——20℃时的衰减系数，d^{-1}，一般为 0.04～0.075；

T——设计温度，℃；

θ_T——温度系数，一般采用 1.02～1.06。

2. 剩余污泥量计算

（1）按污泥龄计算

$$\Delta X=\frac{VX}{\theta_c} \tag{5-15}$$

式中 ΔX——曝气池每天排出的总固体量，kgSS/d；

（2）按污泥产率系数或污泥表观产率系数计算

$$\Delta X_v=YQ(S_0-S_e)-K_dVX_v \tag{5-16}$$

式中 ΔX_v——每天排出的以 VSS 计的总固体量，kgVSS/d；

式（5-16）中产率系数 Y 没有扣除内源呼吸而消耗的微生物量，因此，也称总产率系数，也可以用表观产率系数 Y_{obs} 计算剩余污泥量。

$$\Delta X_v=Y_{obs}Q(S_0-S_e) \tag{5-17}$$

$$Y_{obs}=\frac{Y}{1+K_d\theta_c} \tag{5-18}$$

式中 Y_{obs}——污泥表观产率系数，kgVSS/kgBOD_5，即扣除内源呼吸而消耗的微生物量，又称观测产率系数或净产率系数。

式中其他各项参数意义同前。

3. 需氧量计算

$$m_{O_2}=\frac{Q(S_0-S_e)}{0.68}-1.42\Delta X_v \tag{5-19}$$

式中 m_{O_2}——曝气池混合液每天需氧量，kg/d。

1.42——污泥的氧当量系数，即完全氧化 1 单位的细胞（以 $C_5H_7NO_2$ 表示细胞），需要 1.42 单位的氧。

式中其余各项参数意义同前。

5.6.4 计算例题

【例 5-1】 某城镇污水处理厂日处理流量为 $10000m^3/d$，最大小时流量为 $Q_{max}=500m^3/h$，污水流量总变化系数 K_z 为 1.5，污水经一级预处理后 BOD_5 为 170mg/L，要求二级生物处理出水水质 $BOD_5<20mg/L$。已知该地区的大气压为 1.013×10^5Pa，水温为 14～25℃，曝气池中 MLVSS/MLSS 为 0.7，污泥龄 $\theta_c=5d$，曝气池中悬浮固体（MLSS）浓度为 3000mg/L，二沉池回流污泥浓度为 4070mg/L。求合建式完全混合曝气池有关数据、剩余污泥量和所需空气量。

【解】（1）合建式完全混合曝气池结构示意图如图 5-16 所示。

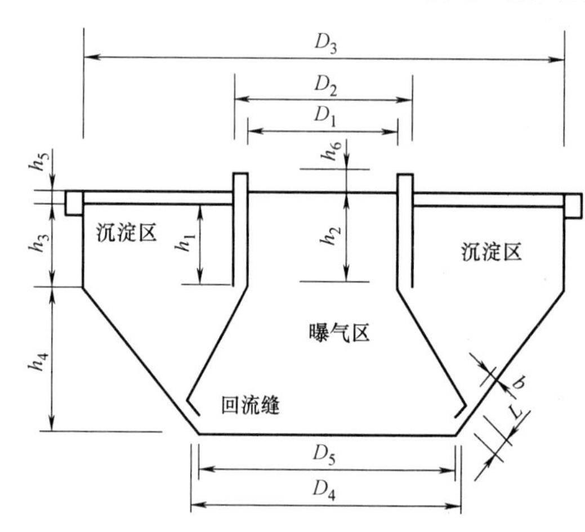

图 5-16 合建式完全混合曝气池结构示意图

（2）BOD_5 处理效率

已知进入生化池的 BOD_5 为 170mg/L，出水 BOD_5 小于 20mg/L，则 BOD_5 处理效率为：

$$\eta=\frac{S_0-S_e}{S_0}\times100\%=\frac{170-20}{170}\times100\%=88\%$$

（3）计算曝气池容积

1）按污泥负荷计算

取曝气池污泥负荷 N_s 为 $0.3kgBOD_5/(kgMLSS\cdot d)$，曝气池中悬浮固体（MLSS）浓度为 3000mg/L，则曝气池容积为：

$$V=\frac{Q(S_0-S_e)}{XN_s}=\frac{10000\times(170-20)}{0.3\times3000}=1666.7m^3$$

设计 6 座完全混合曝气池，则每座曝气池曝气区的容积为 $277.8m^3$。

2）按污泥龄计算

取活性污泥产率系数 $Y=0.6kg\ MLVSS/kgBOD_5$，污泥龄 $\theta_c=5d$，衰减系数 $K_d=0.06d^{-1}$，MLVSS/MLSS=0.7，则曝气池容积为：

$$V=\frac{QY\theta_c(S_0-S_e)}{X(1+K_d\theta_c)}=\frac{10000\times0.6\times5\times(170-20)}{3000\times0.7\times(1+0.06\times5)}=1648.4m^3$$

经比较，选取曝气池曝气区容积为 $1666.7m^3$。设计 6 座完全混合曝气池，每座容积为 $277.8m^3$。

3）复核曝气池容积负荷

$$L_v=\frac{Q(S_0-S_e)}{V}=\frac{10000\times(170-20)}{1666.7}=0.9kgBOD_5/(m^3\cdot d)$$

计算结果表明曝气池容积负荷在规范许可范围内。

（4）计算水力停留时间

$$t=\frac{V}{Q}=\frac{1666.7\times24}{10000}=4.0h$$

(5) 计算每天排除的剩余污泥量

1) 按表观产率系数计算

$$Y_{obs}=\frac{Y}{1+K_d\theta_c}=\frac{0.6}{1+0.06\times5}=0.46$$

计算系统每天排除的以 VSS 计的干污泥量

$$\begin{aligned}\Delta X_v &=Y_{obs}Q(S_0-S_e)\\ &=0.46\times10000\times(170-20)\times10^{-3}=690\text{kg/d}\end{aligned}$$

计算以 SS 计的总排泥量：$\frac{690}{0.7}=985.7\text{kg/d}$

2) 按污泥龄计算剩余污泥排放量

$$\Delta X=\frac{VX}{\theta_c}=\frac{1666.7\times3}{5}=1000\text{kg/d}$$

3) 湿污泥量计算

剩余污泥按含水率 99.2%计算，计算每天排放的湿污泥量：

$$\frac{985.7}{1000}\approx1\text{t(干泥)/d}$$

$$\frac{1}{100\%-99.2\%}=125\text{m}^3/\text{d}$$

(6) 计算污泥回流比

曝气池悬浮固体（MLSS）浓度为 3000mg/L，回流污泥浓度为 4070mg/L，曝气池混合液中微生物主要来自回流污泥，因此，污泥回流比为：

$$4070Q_R=3000(Q+Q_R)$$

$$R=\frac{Q_R}{Q}=\frac{3000}{4070-3000}=280\%$$

(7) 曝气池设计需氧量计算

$$\begin{aligned}m_{O_2} &=\frac{Q(S_0-S_e)}{0.68}-1.42\Delta X_v\\ &=\left[\frac{10000(170-20)}{0.68}\times10^{-3}-1.42\times690\right]\\ &=1226\text{kg/d}\end{aligned}$$

(8) 标准需氧量计算

采用竖轴式泵型叶轮机械曝气，曝气叶轮浸没深度为 10～100mm，可以调节，相关参数选择如下：

机械曝气设备氧转移系数 α 在 0.7～1 之间，取 0.85，氧在污水中的饱和溶解度修正系数 β 取 0.95，所在地区为一个标准大气压，压力修正系数 ρ 为 1，20℃清水中溶解氧饱和度 $C_{s(20)}$ 为 9.17mg/L，25℃清水中溶解氧饱和度 $C_{s(25)}$ 为 8.4mg/L，曝气池内平均溶解氧浓度 C 取 2mg/L。

曝气池混合液需氧量按最不利温度 25℃考虑，最不利温度 25℃换算成标准条件下（20℃，脱氧清水）充氧量为：

$$O_s = \frac{m_{O_2} \cdot C_{s(20)}}{\alpha[\beta\rho C_{s(T)} - C] \cdot 1.024^{(T-20)}}$$

$$= \frac{1266 \times 9.17}{0.85 \times (0.95 \times 1 \times 8.4 - 2.0) \times 1.024^{(25-20)}} = 2028.5\text{kg/d} = 84.5\text{kg/h}$$

(9) 机械曝气设备选型

查阅有关厂家泵型叶轮表面曝气机相关参数，选用 6 台泵型叶轮表面曝气机，每座曝气池布置一台，每台曝气机标准充氧量为 8.6kg/h，叶轮直径为 0.76m，电机额定功率为 7.5kW，采用变频驱动，叶轮转速在 88～126r/min 之间，叶轮周边线速度在 3～5m/s 之间，可以调节。曝气区直径以 3～5.3 倍叶轮直径为宜，水深不宜超过 3.5 倍叶轮直径。若叶轮下加置导流筒（上口位置应在叶轮进水口高度之间），池深可在 5m 以上，最多深至 7m。

(10) 曝气区直径

曝气区直径取 5 倍叶轮直径 D，则曝气区直径为：

$$D_1 = 5D = 5 \times 0.76 = 3.8\text{m}$$

(11) 导流室设计

导流室污水下降速度 v_2 取 15mm/s，污泥回流比 R 取 280%，导流室面积为：

$$F_1 = \frac{(K_z + R)Q}{86.4nv_2} = \frac{(1.5 + 2.8) \times 10000}{86.4 \times 6 \times 15} = 5.53\text{m}^2$$

曝气区外壁厚度 δ_1 取 0.2m，则导流室外径为：

$$D_2 = \sqrt{(D_1 + \delta_1)^2 + 4F_1/\pi}$$

$$= \sqrt{(3.8 + 0.2)^2 + (4 \times 5.53)/\pi} = 4.8\text{m}$$

导流室宽度 b 为：

$$b = \frac{D_2 - D_1 - \delta_1}{2} = \frac{4.8 - 3.8 - 0.2}{2} = 0.4\text{m}$$

(12) 完全混合曝气池沉淀区设计

完全混合曝气池沉淀区表面水力负荷 q 取 0.8m^3/(m^2 · h)，沉淀区面积为：

$$F_2 = \frac{K_z Q}{24nq} = \frac{1.5 \times 10000}{24 \times 6 \times 0.8} = 130.2\text{m}^2$$

导流室外壁采用 UPVC 板，板厚度 δ_2 取 0.01m，则沉淀区直径为：

$$D_3 = \sqrt{(D_2 + \delta_2)^2 + 4F_2/\pi}$$

$$= \sqrt{(4.8 + 0.01)^2 + (4 \times 130.2)/\pi} = 13.7\text{m}$$

沉淀池周边设出水堰，出水堰水力负荷为：

$$q_1 = \frac{K_z Q}{86.4n\pi D_3} = \frac{1.5 \times 10000}{86.4 \times 6\pi \times 13.7} = 0.67\text{L/(s} \cdot \text{m)}$$

(13) 其他部位尺寸计算

曝气区超高 h_6 取 0.5m，沉淀区超高 h_5 取 0.3m，沉淀时间 t 取 1.5h，则沉淀区高度为：

$$h_1 = qt = 0.8 \times 1.5 = 1.2\text{m}$$

曝气区直壁高度为：

$$h_2=h_1+0.414b=1.2+0.414\times0.4=1.4\text{m}$$

曝气区深度 H 取 4.5m，沉淀区直壁高度 $h_3=1.5$m，斜壁高度为：

$$h_4=H-h_3=4.5-1.5=3\text{m}$$

池底直径为：

$$D_5=D_3-2h_4=13.7-2\times3=7.7\text{m}$$

回流窗流速 v_3 取 80mm/s，设计 6 座完全混合曝气池，回流窗面积为：

$$f_1=\frac{(K_z+R)Q}{86.4nv_3}=\frac{(1.5+2.8)\times10000}{86.4\times6\times80}=1.04\text{m}^2$$

回流窗沿曝气池外壁均匀布置，尺寸为 0.2m×0.2m，共 26 个。

污泥回流缝宽度 b_1 取 0.1m，顺流圈长度 L 取 0.4m，顺流圈内径 D_4 取 7.5m，回流缝面积为：

$$f_2=b_1\pi\left(D_4+\frac{L+b_1}{\sqrt{2}}\right)=0.1\pi\times\left(7.5+\frac{0.4+0.1}{\sqrt{2}}\right)=2.47\text{m}^2$$

回流缝流速为：

$$v_4=\frac{RQ}{86.4nf_2}=\frac{2.8\times10000}{86.4\times6\times2.47}=21.9\text{mm/s}$$

回流缝流速 v_4 大于 20mm/s，小于 40mm/s，符合要求。

（14）复核曝气区容积

曝气沉淀池总体积为：

$$V=\frac{\pi D_3^2h_3}{4}+\frac{\pi h_4(D_3^2+D_3D_5+D_5^2)}{12}$$

$$=\frac{\pi\times13.7^2\times1.5}{4}+\frac{3\pi\times(13.7^2+13.7\times7.7+7.7^2)}{12}=497.7\text{m}^3$$

沉淀区容积为：

$$V_2=\frac{\pi(D_3^2-D_2^2)h_1}{4}=\frac{\pi\times(13.7^2-4.8^2)\times1.2}{4}=155.1\text{m}^3$$

曝气区及导流区实际有效容积为：

$$V_3=0.95(V-V_2)=0.95\times(497.7-155.1)=325.5\text{m}^3$$

曝气区及导流区实际有效容积（325.5m^3）大于曝气池曝气区计算容积（277.8m^3），满足要求。

【例 5-2】 某城市污水处理厂日处理流量为 50000m^3/d，小时最大处理能力 Q_{max}＝2500m^3/h，污水经一级预处理后 BOD_5 为 220mg/L，要求生物处理出水 BOD_5＜20mg/L。已知该地区的大气压为 1.013×10^5Pa，水温为 14～25℃，曝气池中 MLVSS/MLSS＝0.7，污泥龄 θ_c＝5d，曝气池中混合液悬浮固体（MLSS）浓度为 2000mg/L，二沉池回流污泥浓度为 8000mg/L。求普通推流式曝气池有关数据、剩余污泥量和所需空气量。

【解】 （1）BOD_5 处理效率

已知进入生化池的 BOD_5 为 220mg/L，出水 BOD_5 小于 20mg/L，则 BOD_5 处理效率为：

$$\eta=\frac{S_0-S_e}{S_0}\times100\%=\frac{220-20}{220}\times100\%=90.9\%$$

(2) 计算曝气池容积

1) 按污泥负荷计算

取污泥负荷 $N_s=0.3$kgBOD$_5$/(kgMLSS · d)，曝气池 MLSS 为 2000mg/L，则曝气池容积为：

$$V=\frac{Q(S_0-S_e)}{XN_s}=\frac{50000\times(220-20)}{0.3\times2000}=16666.7\text{m}^3$$

2) 按污泥龄计算

取活性污泥产率系数 $Y=0.6$kgMLVSS/kgBOD$_5$，污泥龄 $\theta_c=5$d，衰减系数 $K_d=0.06\text{d}^{-1}$，MLVSS/MLSS=0.7，则曝气池容积为：

$$V=\frac{QY\theta_c(S_0-S_e)}{X(1+K_d\theta_c)}=\frac{50000\times0.6\times5\times(220-20)}{2000\times0.7\times(1+0.06\times5)}=16483.5\text{m}^3$$

经比较，选取曝气池容积为 16666.7m^3，取整为 16700m^3。

(3) 确定曝气池各部分尺寸

设计 3 个推流式曝气池，每个曝气池容积为：$V_{单}=\frac{16700}{3}=5566.7\text{m}^3$

取曝气池有效水深 $h_1=4.0$m，每个曝气池面积为：$F=\frac{5566.7}{4}=1391.7\text{m}^2$

取曝气池宽 $b=6$m，$b/h_1=6/4=1.5$，介于 1～2 之间，符合规定。

设每个曝气池采用 4 个廊道，则曝气池长为：$L=\frac{F}{b}=\frac{1391.7}{4\times6}=58\text{m}$

曝气池长度介于 50～70m 之间，符合规定。

取曝气池超高 $h_2=0.5$m，则池总高度 H 为：$H=h_1+h_2=4+0.5=4.5$m。

曝气池平面布置如图 5-17 所示。

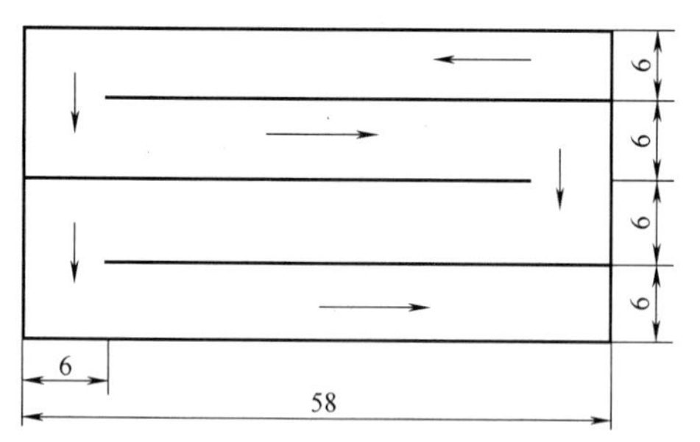

图 5-17 曝气池平面布置示意图（单位：m）

(4) 计算曝气池水力停留时间

$$t=\frac{V}{Q}=\frac{16700\times24}{50000}=8.0\text{h}$$

(5) 计算污泥回流比

曝气池悬浮固体（MLSS）为 2000mg/L，回流污泥浓度为 8000mg/L，曝气池混合液中微生物主要来自回流污泥，因此，污泥回流比为：

$$8000Q_R=2000(Q+Q_R)$$

$$R=\frac{Q_R}{Q}=\frac{2000}{8000-2000}=33.3\%$$

(6) 计算每天排除的剩余污泥量

1) 按表观产率系数计算

$$Y_{obs}=\frac{Y}{1+K_d\theta_c}=\frac{0.6}{1+0.06\times5}=0.46$$

系统排除的以 VSS 计的干污泥量为：

$$\Delta X_{v}=Y_{obs}Q(S_{0}-S_{e})$$
$$=0.46\times50000\times(220-20)\times10^{-3}=4600\text{kg/d}$$

以 SS 计的总干污泥量为：$\frac{4600}{0.7}=6571.4\text{kg/d}$

2）按污泥龄计算

$$\Delta X=\frac{VX}{\theta_{c}}=\frac{16700\times2}{5}=6680\text{kg/d}$$

3）湿污泥量计算

剩余污泥按含水率 99.2%计算，每天排放的湿污泥量为：

$$\frac{6680}{1000}=6.68\text{t(干泥)/d}$$

$$\frac{6.68}{100\%-99.2\%}=835\text{m}^{3}/\text{d}$$

（7）曝气池需氧量计算

$$m_{O_2}=\frac{Q(S_{0}-S_{e})}{0.68}-1.42\Delta X_{v}$$
$$=\left[\frac{50000(220-20)}{0.68}\times10^{-3}-1.42\times4600\right]$$
$$=8174\text{kg/d}$$

（8）所需空气量计算

采用鼓风曝气，设曝气池有效水深 $h_{1}=4.0\text{m}$，曝气扩散器安装在距池底 0.2m 处，则扩散器上有 $H=3.8\text{m}$ 的静水压，其他相关参数选择：

氧总转移系数 $\alpha=0.7$，氧在污水中的饱和溶解度修正系数 $\beta=0.95$，因所在地区为一个标准大气压 $p=1.013\times10^{5}\text{Pa}$，压力修正系数 $\rho=1$，扩散设备堵塞系数 $F=0.9$，采用管式微孔曝气设备，氧转移效率 $E_{A}=20\%$，扩散器压力损失为 4kPa，查有关手册可知，20℃清水中溶解氧饱和度 $C_{s(20)}$ 为 9.17mg/L，25℃清水中溶解氧饱和度为 $C_{s(25)}$ 为 8.4mg/L，曝气池内平均溶解氧浓度 $C=2\text{mg/L}$。

扩散器出口绝对压力 p_{d} 为

$$p_{d}=p+9.8\times10^{3}H=(1.013\times10^{5}+9.8\times10^{3}\times3.8)=1.39\times10^{5}\text{Pa}$$

空气离开曝气池面，气泡含氧体积分数 φ_{0}，按下式计算：

$$\varphi_{0}=\frac{21(1-E_{A})}{79+21(1-E_{A})}\times100=\frac{21(1-20\%)}{79+21(1-20\%)}\times100=17.5$$

曝气池混合液平均溶解氧饱和度按最不利温度 $T=25℃$考虑，因此，25℃时曝气池混合液中平均溶解氧饱和度为：

$$C_{sb(25)}=C_{s(25)}\left(\frac{p_{d}}{2.026\times10^{5}}+\frac{\varphi_{0}}{42}\right)=8.4\times\left(\frac{1.39\times10^{5}}{2.026\times10^{5}}+\frac{17.5}{42}\right)=9.3\text{mg/L}$$

最不利温度 $T=25℃$换算成标准条件下（20℃，脱氧清水）充氧量为：

$$O_s = \frac{m_{O_2} \cdot C_{s(20)}}{\alpha[\beta\rho C_{sb(25)} - C] \cdot 1.024^{(T-20)} \cdot F}$$

$$= \frac{8174 \times 9.17}{0.7 \times (0.95 \times 1 \times 9.3 - 2.0) \times 1.024^{(25-20)} \times 0.9} = 15460.6\text{kg/d} = 644.2\text{kg/h}$$

最不利温度 $T=25$℃时，曝气池所需空气量为：

$$G_S = \frac{O_s}{0.28E_A} = \frac{644.2}{0.28 \times 20\%}$$

$$= 11503.6\text{m}^3/\text{h} = 192\text{m}^3/\text{min}$$

如果选用 5 台风机，3 用 2 备，则单台风机风量为 3834.5m^3/h（64m^3/min）。

冬季需氧量小于夏季，故空气量计算从略。

5.7 二 沉 池

5.7.1 废水悬浮固体的控制

二沉池出水中存在悬浮固体可能有以下几种原因：

（1）高功率曝气将絮凝体破坏；

（2）二沉池水力性能差；

（3）废水中 TDS 过高；

（4）混合液温度过低或过高；

（5）混合液温度的快速变化；

（6）混合液表面张力过低。

涡轮式或机械式表面曝气器产生的混合液湍流水平过高可导致絮凝体被破坏，从而导致高出水悬浮固体。可以通过降低曝气池的功率水平或在曝气池和二沉池之间加设絮凝区解决该问题。

出水悬浮固体浓度高通常是由于二沉池水力性能不佳，导致急流或返混。这些条件导致絮凝体固体在二沉池周边的出水堰上涌。可以通过安装一个斯坦福德（Stamford）挡板，将絮凝体向上流动的方向导离出水堰。

但是，对于工业废水，废水中悬浮固体分散性高可能是由以下几种原因造成的：

（1）总溶解盐（TDS）含量高可能导致不可沉降悬浮固体增加。虽然絮凝体分散的具体原因尚未确定，但增加 TDS 通常会导致不可沉降分散固体增加。高 TDS 还会增加液体的相对密度，从而降低生物污泥的沉降速率。在驯化条件下，盐含量似乎对过程动力学几乎没有影响。

（2）悬浮固体的分散性随着曝气池温度的降低而增加。

（3）悬浮固体的分散性随着表面张力的降低而增加。

（4）有机物的性质可能增加出水悬浮固体。

二沉池之前，可通过添加混凝剂减少出水悬浮固体。一个重要的条件是需要保证有足够的时间进行絮凝。这可以通过曝气池和二沉池之间的絮凝室或二沉池内的絮凝实现。

阳离子聚电解质、明矾或铁盐可用作混凝剂。混凝剂的选择取决于处理过程以选择最经济的混凝剂。使用阳离子聚合物时，应避免过量，否则将导致电荷反转和固体的再分散。

5.7.2 二沉池设计考虑因素

尽管二沉池是生物滤池和活性污泥法的组成部分，但环境工程师特别关注在活性污泥法后使用的二沉池。

图 5-18 运行中的辐流式二沉池

由于活性污泥生物絮体的高固体含量和蓬松性质，二沉池显得非常重要。此外，污泥在此也可以很好地循环，充分浓缩。

二沉池可分为平流式、竖流式和辐流式沉淀池（多采用），如图 5-18 所示。

二沉池在设计构造上要注意以下特点：

（1）二沉池的进水部分，应使布水均匀并形成有利于絮凝的条件，使絮体结大。

（2）二沉池中污泥絮体较轻，容易被出流水挟走，要限制出流堰处的流速，使单位堰长的出水量不超过 $10m^3/(m \cdot h)$。

（3）污泥斗的容积，要考虑污泥浓缩的要求。在二沉池内，活性污泥中的溶解氧只有消耗，没有补充，容易耗尽。缺氧时间过长可能影响活性污泥中微生物的活力，并可能因反硝化而使污泥上浮，故浓缩时间一般不超过 2h。

5.7.3 辐流式二沉池设计

1. 辐流式沉淀池设计要求

（1）辐流式沉淀池直径（或正方形的一边）与有效水深的比值宜为 6～12；

（2）一般采用机械排泥，当辐流式沉淀池直径（或正方形的一边）小于 20m 时也可采用空气提升或多斗排泥，采用机械排泥时刮泥机旋转速度宜为 1～3r/h，刮泥板的外缘线速度不超过 3m/min，一般采用 1.5m/min。

（3）非机械排泥时缓冲层高度宜为 0.5m；机械排泥时，应根据刮泥板高度确定且缓冲层上缘宜高出刮泥板 0.3m；

（4）坡向泥斗的池底坡度应大于 0.05。

（5）周边进水周边出水辐流式沉淀池应保证进水渠的均匀配水。

（6）辐流式沉淀池单池直径不宜小于 16m 且不宜大于 50m；池径小于 20m 时，一般采用中心传动的刮泥机，其驱动装置设在池中心；池径大于 20m 时，一般采用周边传动的刮泥机，其驱动装置设在桁架的外缘。

（7）在进水口的周围应设置整流板，整流板的开口面积为过水断面积的 6%～20%。

（8）出水堰前应设置浮渣挡板收集浮渣，刮渣板安装在刮泥机桁架的一侧。

（9）辐流式二沉池设计参数可按表 5-2 选用。

二沉池主要设计参数　　表 5-2

沉淀池位置	沉淀时间 (h)	表面水力负荷 [$m^3/(m^2 \cdot h)$]	污泥含水率 (%)	每人每日污泥量 [g/(人·d)]	固体表面负荷 [$kg/(m^2 \cdot d)$]	堰口负荷 [L/(s·m)]
活性污泥法后	1.5～4.0	0.6～1.5	99.2～99.6	12～32	≤150	≤1.7
生物膜法后	1.5～4.0	1.0～2.0	96～98	10～26	≤150	≤1.7

注：当二沉池采用周边进水周边出水辐流式沉淀池时，固体负荷不宜超过 200kg/(m^2·d)。

2. 中心进水周边出水辐流式沉淀池设计

（1）每座沉淀池表面积

$$A=\frac{Q_{max}}{nq} \tag{5-20}$$

式中　A——每座沉淀池表面积，m^2；

Q_{max}——最大设计流量，m^3/h；

q——表面水力负荷，$m^3/(m^2 \cdot h)$；

n——沉淀池的数量，个。

（2）沉淀池直径

$$D=\sqrt{\frac{4A}{\pi}} \tag{5-21}$$

式中　D——沉淀池直径，m。

（3）沉淀池有效水深

$$h_2=qt \tag{5-22}$$

式中　h_2——沉淀池的有效水深，m；

t——沉淀时间，h。

（4）校核堰口负荷

$$q'=\frac{Q_0}{3.6\pi D} \tag{5-23}$$

式中　q'——沉淀池堰口负荷，L/(s·m)；

Q_0——单池设计流量，m^3/h。

（5）校核固体负荷

$$G=\frac{24(1+R)Q_0X}{A} \tag{5-24}$$

式中　G——沉淀池固体负荷，$kg/(m^2 \cdot d)$；

A——每座沉淀池表面积，m^2。

（6）污泥部分所需容积

$$V=\frac{SNT}{1000n} \tag{5-25}$$

式中　V——污泥部分所需容积，m^3；

S——每人每日湿污泥量，L/(人·d)；

N——设计人口数，人；

T——排泥时间间隔，d，活性污泥法后二沉池按 2h 计算，生物膜法后二沉池按 4h 计算。

（7）污泥斗容积

$$V_1=\frac{\pi}{3}h_5(r_1^2+r_2^2+r_1\cdot r_2) \tag{5-26}$$

式中 V_1——污泥斗容积，m^3；

r_1——污泥斗上口半径，m；

r_2——污泥斗下口半径，m；

h_5——污泥斗高度，m。

（8）污泥斗以上圆锥体部分污泥区容积

$$V_2=\frac{\pi}{3}h_4(R^2+r_1^2+r_1\cdot R) \tag{5-27}$$

式中 V_2——污泥斗以上圆锥体部分污泥区容积，m^3；

R——辐流式沉淀池的半径，m；

h_4——沉淀池池底带有坡度部分的高度，m。

（9）沉淀池总高度

$$H=h_1+h_2+h_3+h_4+h_5 \tag{5-28}$$

式中 H——沉淀池总高度，m；

h_1——沉淀池超高，一般取 0.3m；

h_2——有效水深，m；

h_3——缓冲层高度，m；

h_4——底坡高度，m；

h_5——污泥斗高度，m。

3. 例题

【例 5-3】 某城镇污水处理厂最大设计处理废水量 $Q_{max}=2500m^3/h$，设计人口 $N=300000$，污水经生物处理后进入二沉池，悬浮固体浓度为 3000mg/L，二沉池池底污泥回流至生物池，污泥回流比 R 为 50%，二沉池采用机械刮泥，设计中心进水周边出水辐流

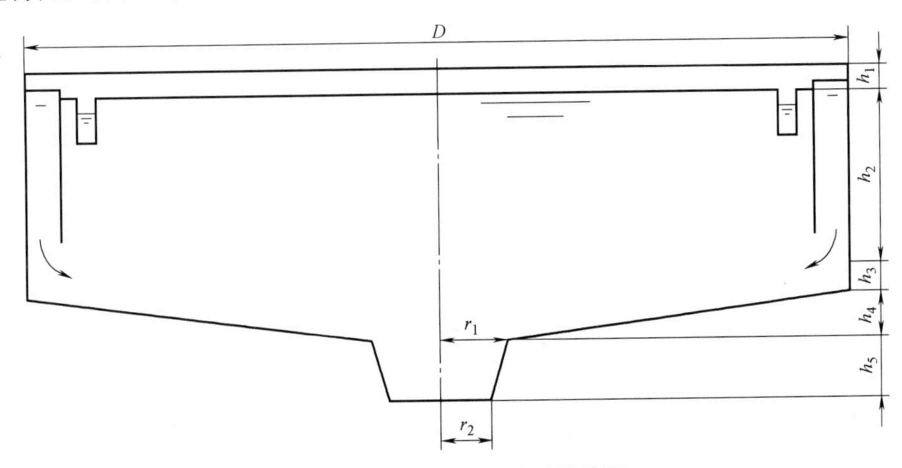

图 5-19 辐流式二沉池计算草图

式二沉池各部分尺寸。

【解】 辐流式二沉池计算草图如图 5-19 所示。

（1）每座二沉池的表面积

$$A=\frac{Q_{\max}}{nq}=\frac{2500}{5\times1.0}=500\text{m}^2$$

二沉池表面水力负荷 $q=1.0\text{m}^3/(\text{m}^2\cdot\text{h})$，二沉池座数 $n=5$。

（2）二沉池直径

$$D=\sqrt{\frac{4A}{\pi}}=\sqrt{\frac{4\times500}{3.14}}=25.2\text{，取 }26\text{m。}$$

（3）二沉池有效水深

取水力停留时间 $t=3\text{h}$，则二沉池有效水深 h_2 为：

$$h_2=qt=1.0\times3=3\text{m}$$

径深比＝26/3＝8.7，符合规范要求。

（4）校核堰口负荷

设计 5 座辐流式二沉池，则单池设计流量 Q_0 为 $500\text{m}^3/\text{h}$，因此，二沉池堰口负荷为：

$$q'=\frac{Q_0}{3.6\pi D}=\frac{500}{3.6\pi\times26}=1.7\text{L}/(\text{s}\cdot\text{m})$$

二沉池出水堰口负荷规范要求。

（5）校核固体负荷

设计 5 座辐流式二沉池，污泥回流比 R 为 50%，悬浮固体浓度为 3kg/m^3，因此，二沉池出口固体负荷为：

$$G=\frac{24(1+R)Q_0X}{A}=\frac{24\times(1+0.5)\times500\times3}{500}=108\text{kg}/(\text{m}^2\cdot\text{d})$$

二沉池出水堰口固体负荷规范要求。

（6）污泥部分所需容积

取每人每日干污泥量 $S_{干}=25\text{g}/(人\cdot\text{d})$，污泥含水率 P 为 99.5%，则每人每日的湿污泥量为：

$$S=\frac{100S_{干}}{(100-P)\gamma}=\frac{25\times100}{(100-99.5)\times1000}=0.5\text{L}/(人\cdot\text{d})$$

活性污泥法后二沉池，排泥周期取 $T=2\text{h}$，污泥部分所需容积为：

$$V=\frac{SNT}{1000n}=\frac{0.5\times300000\times2}{1000\times24\times5}=2.5\text{m}^3$$

（7）污泥斗容积

取污泥斗上部半径 $r_1=2\text{m}$，污泥斗下部半径 $r_2=1\text{m}$；污泥斗倾角 $\alpha=60°$，污泥斗高度为：

$$h_5=(r_1-r_2)\tan\alpha=(2-1)\tan60°=1.73\text{m}$$

因此，污泥斗容积为：

$$V_1=\frac{\pi}{3}h_5(r_1^2+r_2^2+r_1\cdot r_2)=\frac{3.14}{3}\times1.73\times(2^2+1^2+2\times1)=12.7\text{m}^3$$

（8）污泥斗以上圆锥体部分污泥区容积

设池底径向坡度为 0.05，二沉池半径为 13m，则圆锥体的高度为：

$$h_4=(R-r_1)\times0.05=(13-2)\times0.05=0.55\text{m}$$

因此，污泥斗以上圆锥体部分污泥区容积为

$$V_2=\frac{\pi}{3}h_4(R^2+r_1^2+r_1\cdot R)=\frac{3.14}{3}\times0.55\times(13^2+2^2+2\times13)=114.6\text{m}^3$$

(9) 污泥斗总容积

污泥斗总容积为污泥斗容积（V_1）与污泥斗以上圆锥体部分污泥区容积（V_2）之和，因此，污泥斗总容积为：

$$V_1+V_2=114.6+12.7=127.3\text{m}^3>2.5\text{m}^3$$

(10) 二沉池的总高度

取二沉池的超高 $h_1=0.3$m；缓冲层高度 $h_3=0.5$m，则二沉池总高度为：

$$H=h_1+h_2+h_3+h_4+h_5=0.3+3+0.5+0.55+1.73=6.1\text{m}$$

其他设计从略。

5.8 活性污泥法处理系统的运行管理

活性污泥法系统的运行管理需要认识和理解一系列对系统产生重要影响的问题，管理过程主要体现为对下列指标的监控：水力负荷、有机负荷、污泥浓度、污泥龄、污泥回流比、回流污泥浓度、曝气量、曝气时间、氧传递速率、曝气设备、溶解氧浓度、pH 和碱度；另外还包括保障曝气设备正常运行、预防控制污泥膨胀等。

5.8.1 水质和水力负荷

为了保证活性污泥里微生物的生长代谢，污水需含有足量的营养物质以及产生生命活动需要的能源物质，包括有机物、N、P、Na、K、Ca、Mg、Fe、Co、Ni 等，普通活性污泥处理系统要求 BOD∶N∶P=100∶5∶1；通常城镇生活污水的 BOD_5、N、P 的组成能满足微生物生长要求，但对工业废水，上述营养比例一般不满足，甚至缺乏某些微量元素，此时需补充相应组分，缺氮可投加氨水、铵盐或含氮高的工业废水；缺磷可投加磷酸盐。城镇污水中氮、磷元素往往超过微生物合成代谢所需的营养比，需采取辅助措施脱氮、除磷，当碳源不足时，可利用初沉或剩余污泥发酵产酸作为补充碳源，或投加外购商品碳源。

大部分污水的流量变化是不易控制的因素，当地的生活方式和集水范围影响污水处理厂的流量变化。在一般的设计中，高峰值约为平均流量的 200%，最低值约为平均流量的 50%。水力负荷的变化对活性污泥法系统的曝气池和二沉池影响最为明显。当流量增加时，污水在曝气池内的停留时间缩短，有机物的好氧降解和氨氮的硝化就不可能达到预期的效果，使出水水质变差；进水水量的增加，使曝气池的水位也可能增高，若为机械表面曝气，水位的增高会使曝气机运行变得不稳定；水量的增加导致二沉池内水的流速增加，从而对泥水分离产生明显影响。

5.8.2 有机负荷

有机负荷是废水处理设计和运行的重要参数之一。根据经验，当要求的处理效率较高时，设计的污泥负荷一般不宜大于 0.5kgBOD_5/(kgMLSS·d)，以免发生污泥膨胀；如果要求进入消化阶段，一般采用 0.15kgBOD_5/(kgMLSS・d）左右的污泥负荷。有时为了减小曝气池的容积，可以采用高负荷，即污泥负荷在 1.0kgBOD_5/(kgMLSS・d）以上。采用高的污泥负荷虽可减小曝气池的容积，但出水水质要降低，而且使剩余污泥量增多，增加污泥处置的费用和困难。同时，整个处理系统耐冲击性能差，造成运行管理困难；因此，近年来很多研究人员不主张采用高负荷系统，为避免剩余污泥处置上的困难，并确保废水处理系统运行的稳定，拟采用低的污泥负荷，即小于 0.1kgBOD_5/(kgMLSS・d)。采用低负荷，曝气池容积会很大，但系统出水水质好，剩余活性污泥排放少。

在运行过程中随着有机污染物浓度波动，F/M 会偏离设计值，影响处理系统的运行效果。此时应通过调整排泥量，控制曝气池中微生物的量 M，使 F/M 基本保持恒定。

5.8.3 污泥浓度

MLSS 的主体是活性微生物，提高 MLSS 或增加微生物数量，可以降低污泥负荷，提高处理速率，进而缩小曝气池的容积。但采用高的污泥浓度（或 MLSS）并不能加强曝气池的处理效果，甚至可能适得其反。其一，污泥量并不等于微生物的活细胞量。曝气池污泥量的增加意味着污泥龄的增加，污泥龄的增加就使污泥中活细胞的比例减小。其二，过高的污泥浓度导致沉淀池难于泥水分离（发生拥挤沉淀)，影响出水水质。其三，微生物数量多意味着需氧量大，而随着曝气池污泥浓度的增加，混合液的黏滞性会发生变化，从而影响三相传质，使池中氧的传质速率明显降低。其四，各种曝气设备都有其合理的氧传递速率的范围。超出了其合理的氧传递速率范围，其充氧动力效率将明显降低，使能耗增加。因此，采用一定的曝气设备系统，实际上已限定了与之匹配的污泥浓度，MLSS 的提高是有限度的。根据长期的运行经验，采用鼓风曝气设备的传统活性污泥法时，曝气池中 MLSS 在 2000～4000mg/L 是适宜的，并非污泥浓度越高越好。

5.8.4 污泥龄

有机负荷和污泥龄间存在着内在的联系，选择一定的有机负荷和一定的 MLSS，就相应决定了污泥龄。过长的污泥龄易导致微生物老化，使污泥絮凝沉降性能变差，并增加惰性物质引起的浊度。经验表明，通常活性污泥法系统的微生物平均停留时间约为水力停留时间的 20 倍，延时曝气系统的比例为 30∶1，甚至为 40∶1。对于高负荷系统，其比例接近 10∶1。活性污泥法系统的水力停留时间，对城镇污水来讲一般为 4～6h，则相应的微生物停留时间为 3.3～5d。

延时曝气的水力停留时间为 24h，则微生物停留时间为 30d 左右。高负荷系统曝气时间为 2～3h，微生物停留时间约为 1d。

污泥龄使每天的排泥量相对固定，但运行过程中，若遇到二沉池出水悬浮物含量突然增大后，应当缩短污泥龄，增大排放的剩余污泥量。但排泥量过大，又会使污泥龄过短，污泥吸附的有机物来不及氧化分解，二沉池出水有机物含量增大，系统的处理效果恶化。

计算活性污泥法系统的污泥龄是否应包括二沉池中的活性污泥量？二沉池中有可观的活性污泥量，但由于氧的浓度很低，微生物代谢可以忽略。因而大多数活性污泥法系统设计时只根据曝气池的污泥量计算污泥龄。但在吸附再生系统中，因为吸附池和再生池的水力停留时间不同，MLSS 也不同，且二沉池经常用作污泥调蓄池，在这种情况下，根据吸附池和再生池的运行数据计算污泥龄时发生很大变化，而考虑沉淀池污泥量后，则污泥龄比较稳定。

5.8.5 污泥回流比

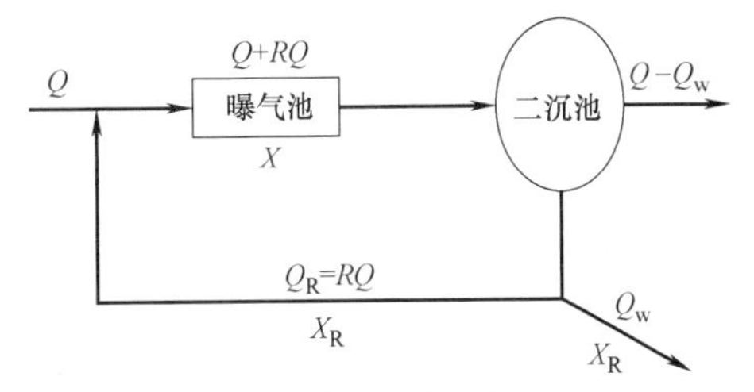

图 5-20 曝气池混合液 MLSS 与回流污泥的关系

曝气池混合液中污泥主要来自回流污泥，根据物料平衡（图 5-20），可得下列关系式：

$$RQX_{R}=(Q+RQ)X \tag{5-29}$$

$$X=\frac{R}{1+R}X_{R} \tag{5-30}$$

式中 R——污泥回流比；

X_{R}——回流污泥浓度，mg/L。

从这个公式可知，曝气池混合液的 MLSS 不可能高于回流污泥浓度，回流比越大，两者越接近。回流污泥来自二沉池，其污泥浓度与活性污泥的沉降浓缩性能和浓缩时间有关。通常情况下，回流污泥浓度为 10000mg/L，对于沉降浓缩性能略差的活性污泥，其回流污泥浓度范围在 5000～8000mg/L。

因为较高的污泥回流比增大进入沉淀池的流量，增加二沉池的负荷，降低沉淀效率，故一般将活性污泥回流比控制在一定的范围中，以降低能耗并保持沉淀池运行稳定。对于小规模污水处理厂，生产运行时回流污泥变数大，为了保持稳定的有机负荷，通常使用变频调节回流泵，根据进水流量调节回流污泥量，但运行过程中会存在一些问题：其一是污水流量的增加并不意味着有机物浓度保持或提高，这时提高污泥回流比有可能降低有机负荷；其二是随着污水流量的增加而加大污泥回流量，这样对二沉池产生集中的水力冲击，使二沉池断面流速增加、沉淀时间缩短，出水水质会明显下降甚至恶化；其三是增加 MLSS 必须同时增加曝气量，使曝气系统处于超负荷状态，如运行管理不当，也可能对出水水质带来影响。

维持一个稳定的污泥回流比，对于保证污水处理厂的正常运行十分重要。因为在入流污水量较低时，沉淀池中有较多的回流污泥进入曝气池，比从曝气池中进入沉淀池的污泥多，曝气池中的 MLSS 增加，这等于为流量和有机负荷的增加做准备，而沉淀池中贮存的污泥体积变得较小。当流量增加和有机负荷增加时，曝气池中较高的 MLSS 已具备适应条件，这时有更多的 MLSS 从曝气池中流向沉淀池，而二沉池已留出空间。MLSS 能自动地响应流量和有机负荷的变化，从而保证出水水质。

5.8.6 氧传递速率

氧传递速率也决定活性污泥法处理系统的能力，氧传递速率要考虑两个过程，即氧溶解到水中以及传递到微生物的膜表面。通常的试验数据只表明氧传递到水中，但这并不意味着同样量的氧已达到微生物表面，而后者则控制微生物活性的发挥。从这个观点看，曝气设备不仅要提供充分的氧，而且要创造足够的紊动条件，以剪切活性污泥絮体，这样可使污泥絮体内部的细菌得到氧。因此，要提高氧的传递速率，除了供应充足的氧外，应使混合液中的悬浮固体保持悬浮状态和紊动条件。另外，曝气设备的选择、布置，以及如何同池型配合，是关系氧传递速率的重要前提。

机械表面曝气机，是把废水粉碎成小的液滴，散布于连续的大气相中，而扩散曝气器则是把空气粉碎成微小气泡，散布于连续的液相。目的都是希望从空气中获得氧，提高液相中的氧浓度。目前两种曝气方法几乎同样广泛使用。并且因为曝气量随进水水量、水质等因素变化较大，目前的曝气设备一般都具有充氧量调节功能。

鼓风曝气过程中气泡上升的阶段，向邻近液体传递氧，因而气泡中的氧浓度降低，相邻液体的氧浓度提高。而细小气泡上升力小，不能促使邻近液体产生紊动，气泡和液体接触界面更新速度缓慢，因而最大的氧传递速率是发生在气泡刚形成时。基于这种认识，要提高氧传递速率，就要尽可能使单位气量分布到尽可能大的面积上。但是当把扩散器布满大部分池底时，在同样的气量下，曝气强度（单位面积上的气体流量）会明显降低，MLSS可能沉下来。此时可把扩散设备布置在池底的一边，从而使曝气池内MLSS保持螺旋运动，促进活性污泥处于悬浮状态。

5.8.7 溶解氧浓度

通常溶解氧浓度不是关键因素，除非DO数值低到接近于零。从理论上讲，只需溶解氧保持约1mg/L就足够发生好氧代谢。有研究认为，对于单个悬浮的好氧细菌，代谢溶解氧浓度只要高于0.1～0.3mg/L，代谢速率就不受溶解氧浓度影响。但是，活性污泥絮体是许多个体集结在一起的絮状物质，要使内部的溶解氧浓度达到0.1～0.3mg/L，絮体周围的溶解氧浓度一定要高出许多，具体数值同絮状体的大小、结构及影响氧扩散性能的混合情况有关。一般认为混合液中溶解氧浓度保持在0.5～2mg/L时，活性污泥法系统可以正常运行。若溶解氧过低，好氧微生物的代谢活动下降，活性污泥会因此发黑发臭，进而处理污水的能力也下降，导致水质恶化。

反之，过量曝气时溶解氧浓度很高，但紊动过分剧烈，易导致菌胶团或污泥絮体破裂，使出水浊度升高。特别是对于耗氧速率不高、污泥龄偏长的系统，强烈混合使破碎的絮体不能很好地再凝聚。这些离散的污泥沉降性能差，污泥中的原生动物也不能去除这些颗粒，因为它缺少原生动物所需的营养。过分地曝气使这些生物残体以泡沫形式积聚在曝气池和沉淀池的表面，形成深褐色的泡沫状浮渣。

5.8.8 pH和碱度

活性污泥混合液本身对pH变化有一定的缓冲作用。在废水生物处理过程中，废水中的蛋白质代谢后产生的碳酸铵碱度和从原水中带来的碱度一般会使混合液的pH维持在6.5～8.5的范围内。如果进水缺少碱度，有机物代谢过程中产生的CO_2和有机酸会使pH降到5.5，甚至低于5.0，从而使系统的硝化作用受到明显抑制。纯氧或富氧曝气系统因为曝气池的密封，非常容易出现这种情况。

工业废水往往缺少蛋白质，因而会产生pH过低的问题，尤其是在糖厂、淀粉厂和某些合成化学厂。糖类、醛类、丙酮和乙醇等被细菌代谢为有机酸，造成pH降低，进而减慢生物代谢速率。为避免这种情况，可将碱或石灰直接添加到曝气池中，以维持所希望的pH。碱或石灰同代谢产生的CO_2作用产生碳酸钠或碳酸钙成为缓冲剂。对于酸性废水，有机酸等酸性物质通常要在进入曝气池前进行中和。当废水pH低于6时，易刺激霉菌和其他真菌生长，改变废水生态系统的平衡，导致丝状菌膨胀。处理强碱性的废水也需要先进行中和，当pH大于9时，微生物代谢速率将受不利影响，菌胶团解体，悬浮物增多，出水恶化。

5.8.9 污泥膨胀及其控制

正常的活性污泥沉降性能良好，其污泥体积指数SVI在50～150之间；当活性污泥不正常时，污泥不易沉淀，SVI明显升高，污泥体积膨胀，这种现象称为活性污泥膨胀。由于膨胀污泥不易沉淀，容易流失，不仅影响二沉池的出水水质，还造成回流污泥量的不足，对这种现象若不及时加以控制，系统中的污泥就会越来越少，从而破坏曝气池的运行。

但是，沉降性能恶化并非都是污泥膨胀现象。例如，在合建型处理工艺中，二沉池里的沉淀污泥反硝化产生氮气而使污泥上浮，部分区域积泥造成厌氧发酵而上浮，也都会出现污泥性能恶化，需要同污泥膨胀现象区分开。膨胀的活性污泥，主要表现在压缩性差，沉淀性能不良，而它的处理功能和净化效果并不差。

评价污泥膨胀的 SVI，目前并不统一。一般认为，SVI 超过 200 就属于污泥膨胀。活性污泥膨胀可分为污泥中丝状菌大量繁殖导致的丝状菌膨胀以及并无大量丝状菌存在的非丝状菌膨胀。

1. 丝状菌膨胀

活性污泥中的微生物是一个以细菌为主的群体，正常的活性污泥以菌胶团形式出现，在不正常的情况下，丝状菌大量出现并过度增长繁殖，这些丝状体相互支撑、交错，严重影响污泥的凝聚、沉降和压缩性能，造成泥水分离困难，如图 5-21 所示。

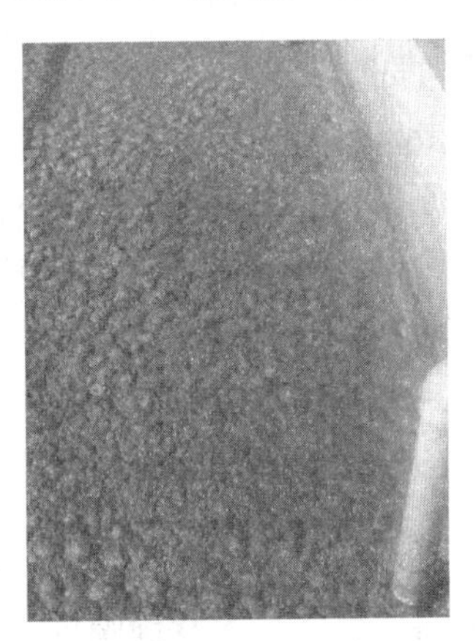

图 5-21　丝状菌膨胀图

研究显示，导致污泥膨胀的主要因素大致为：①废水水质。含溶解性碳水化合物高的废水往往发生由浮游球衣细菌引起的丝状菌膨胀，含硫化物高的废水往往发生由硫细菌引起的丝状菌膨胀；废水中碳、氮、磷的比例对发生丝状菌膨胀影响很大，氮和磷不足都易引发丝状菌膨胀。此外，水温高和 pH 较低时容易产生膨胀。②曝气池的污泥负荷会影响污泥膨胀。污泥负荷在 0.5～1.5kgBOD_5/(kgMLSS・d) 范围内，SVI 较高。但实践表明，导致这种污泥膨胀的最主要因素是水质而不是污泥负荷。对某些污水，不论污泥负荷是高或是低都会发生污泥丝状菌膨胀。③工艺方法。完全混合式比传统推流式较易发生污泥膨胀，而间歇运行的曝气池最不容易发生污泥膨胀；不设初沉池（设有沉砂池）的活性污泥法，SVI 较低，不容易发生污泥膨胀；叶轮式机械曝气与鼓风曝气相比，易发生丝状菌膨胀；射流曝气的供氧方式可以有效地克服浮游球衣细菌引起的污泥膨胀。

2. 非丝状菌膨胀

非丝状菌膨胀与丝状菌膨胀相似，污泥 SVI 高，在沉淀池内很难沉淀，但此时的污水处理效率仍很高，上清液也清澈。如将污泥用显微镜检查，看不到丝状细菌，即使看到也是数量极少的短丝状菌。

非丝状菌膨胀主要发生在废水水温较低而污泥负荷太高的情况。污泥的负荷高使得细菌吸取了大量的营养物，但温度低，微生物代谢速率较慢，大量高黏度的多糖类物质来不及降解而被贮存起来。这些多糖类物质的积贮，使活性污泥的表面附着水大幅增加，污泥SVI升高，形成膨胀污泥，如图5-22所示。

图5-22　非丝状菌膨胀图

在运行中，如发生污泥膨胀，首先应通过微生物镜检确定污泥膨胀类型。结合进水水质及管理状况分析污泥膨胀的原因，针对污泥膨胀成因及类型采取相应配套措施，如：①控制曝气量，使曝气池中保持适量的溶解氧（不低于1.2mg/L，不超过4mg/L）。②调整pH。③投加氮素和磷素，调节氮、磷比例。④投加一些化学药剂（如铁盐混凝剂、有机高分子絮凝剂等）。⑤对现有城镇污水处理厂，废水经过沉砂池后超越初沉池，直接进入曝气池。

二维码7　微生物的种类、形态特征、出现时期

在设计时，为避免发生污泥膨胀，宜采取以下措施：①不设初沉池，增加进入曝气池的废水中悬浮物，改善污泥沉降性能。②在常规曝气池前设置污泥厌氧或缺氧选择池。③对于现有的容易发生污泥膨胀的污水处理厂，可以在曝气池的前端补充设置填料。这样，既降低曝气池的污泥负荷，又改变进入后面曝气池的水质，以有效地克服活性污泥膨胀。

本章关键词（Keywords）

活性污泥	Activated sludge
污泥沉降	Sludge settling
氧转移	Oxygen transfer
颗粒	Particles
曝气设备	Aeration equipment
完全混合活性污泥法	Complete mix activated sludge
推流式活性污泥法	Plug-flow activated sludge
延时曝气法	Extended aeration
氧化沟	Oxidation ditch systems
纯氧活性污泥法	Oxygen activated sludge
水力负荷	Hydraulic load

有机负荷	Organic load
污泥龄	Sludge retention time
污泥膨胀	Sludge bulking
二沉池	Secondary clarifier
澄清	Clarification
沉淀池	Sedimentation tank
絮凝	Flocculent

思考题

1. 简述活性污泥处理的基本流程。

2. 试指出污泥沉降比、污泥浓度、污泥容积指数在活性污泥法运行中的重要意义。

3. 试讨论影响活性污泥法运行的主要环境因素。

4. 衡量曝气设备效能的指标有哪些？什么叫充氧能力？什么叫氧转移效率？

5. 曝气设备按供气方式主要分为几类？各类曝气设备除了要满足充氧要求外，还应满足哪些混合强度要求？

6. 某城镇污水处理厂日处理流量为 $9800m^3/d$，最大时流量为 $Q_{max}=400m^3/h$，污水流量总变化系数 K_z 为 1.3，污水经一级预处理后 BOD_5 为 220mg/L，要求二级生物处理出水水质 $BOD_5<22mg/L$。已知该地区的大气压为 1.013×10^5Pa，水温为 14～25℃，MLVSS/MLSS 为 0.75，污泥龄 $\theta_c=6d$，曝气池悬浮固体浓度 MLSS 为 3000mg/L。求合建式完全混合曝气池有关数据、剩余污泥量和所需空气量。

7. 某城市污水处理厂日处理流量为 $70000m^3/d$，小时最大处理能力 $Q_{max}=3000m^3/h$，污水经一级预处理 BOD_5 为 200mg/L，要求二级生物处理出水 $BOD_5<22mg/L$。已知该地区的大气压为 1.013×10^5Pa，水温为 14～25℃，MLVSS/MLSS＝0.75，污泥龄 $\theta_c=7d$，曝气池污泥浓度 MLSS＝3000mg/L。求普通推流式曝气池有关数据、剩余污泥量和所需空气量。

8. 某城镇污水处理厂最大设计处理废水量 $Q_{max}=3000m^3/h$，设计人口 $N=200000$，污水经曝气池生物处理后进入二沉池，二沉池采用机械刮泥，设计中心进水周边出水辐流式二沉池各部分尺寸。

Chapter 6 Anaerobic Treatment Processes

6. 1 Principle of Anaerobic Treatment

Anaerobic treatment or decomposition involves the breakdown of organic wastes to gas (methane and carbon dioxide) in the absence of oxygen. Although the process kinetics and material balances are similar to those of aerobic systems, certain basic differences require special consideration.

The conversion of organic acids to methane gas yields little energy: hence the rate of growth is slow and the yield of organisms by synthesis is low. The kinetic rate of removal and the sludge yield are both considerably less than in the aerobic activated sludge process. The quantity of organic matter converted to gas will vary from 80% to 90%. Since there is less cell synthesis in the anaerobic treatment process, the nutrient requirements are correspondingly less than in the aerobic system. High process efficiency requires elevated temperatures and the use of heated reaction tanks. The methane gas produced by the reaction can be used to provide this heat. Wastes of low COD or BOD concentration will not provide sufficient methane for heating, and a supplementary source of heat is necessary.

6. 2 Classification of Anaerobic Reactors

The anaerobic reactor is operated in one of several ways.

(1) The anaerobic filter reactor establishes growth of the anaerobic organisms on a packing medium. The filter may be operated upflow or downflow. The packed filter medium, while retaining biological solids, also provides a mechanism for separating the solids and the gas produced in the anaerobic treatment process.

(2) The anaerobic contact process provides for separation and recirculation of seed organisms, thereby allowing process operation at retention periods of 6 to 12h. A degasifier is usually needed to minimize floating solids in the separation step. For high-degree treatment, the solids retention time has been estimated at 10d at 320℃; the estimate doubles for each 110℃ reduction in operating temperature.

(3) In the fluidized-bed reactor (FBR), wastewater is pumped upward through a sand bed on which microbial growth has been developed. Biomass concentrations exceeding 30000mg/L have been reported. Effluent is recycled to mix with the feed in quantities dictated by the strength of the wastewater and the fluidization velocity. Organic removal efficiencies of 80% are

achieved at loadings of 4kgCOD/(m^3 • d) on dilute wastewater.

(4) In the upflow anaerobic sludge blanket (UASB) process, wastewater is directed to the bottom of the reactor, where it must be distributed uniformly. The wastewater flows upward through a blanket of biologically formed granules which consume the waste as it passes through the blanket. Methane and carbon dioxide gas bubbles rise and are captured in the gas dome. Liquid passes into the settling portion of the reactor, where solids-liquid separation takes place. The solids return to the blanket area while the liquid exits over the weirs. Formation of granules and their maintenance is extremely important in the operation of the process.

Some anaerobic reactor classifications are shown in Figure 6-1.

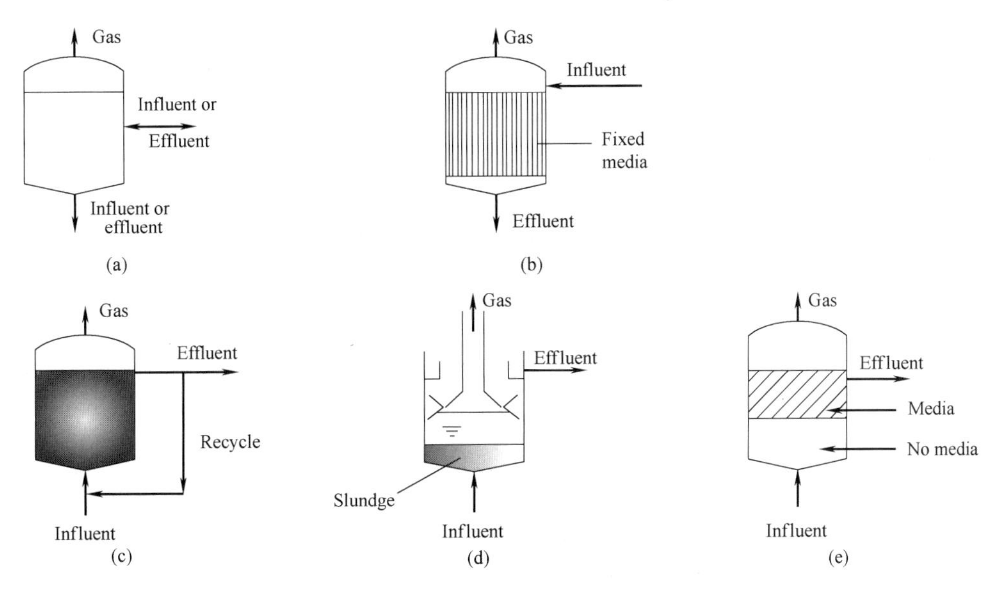

Figure 6-1 Some anaerobic reactor classifications

(a) Upflow/downflow anaerobic fitter; (b) Downflow stationary fixed film; (c) Fluidized bed; (d) Upflow anaerobic sludge blanket (UASB); (e) Hybrid anaerobic sludge reactor

Early applications of anaerobic treatment of industrial wastewater are activated sludge processes (suspended growth treatment processes), which are initially designed in a similar manner as anaerobic sludge digesters. Three types of anaerobic activated sludge are following: (1) the complete-mix suspended growth anaerobic digester, (2) the anaerobic contact process, and (3) the anaerobic sequencing batch reactor, a more recent suspended growth reactor design development.

For the complete-mix anaerobic digester, the hydraulic retention and solids retention times are equal. The hydraulic retention times of reactor may be in the range of 15 to 30d to provide sufficient safety factors for operation and process stability. The complete-mix digester without sludge recycle is more suitable for wastes with high concentrations of solids or extremely high dissolved organic concentrations, where thickening the effluent solids is difficult so that it is more practical to operate with hydraulic retention time equal to the *SRT*.

The anaerobic contact process overcomes the disadvantages of a complete-mix process without recycle. Biomass is separated and returned to the contact reactor so that the process *SRT* is longer than hydraulic retention time. By separating hydraulic retention time and *SRT* values, the anaerobic reactor volume can be reduced. Gravity separation is the most common approach for solids separation and thickening prior to sludge recycle; however, a sludge with poor settling properties is commonly produced and alternative separation processes must be used or other methods must be employed to improve solids capture. Separation of solids using gas flotation by dissolving the process off-gas under pressure has been used in place of gravity separation.

Because the reactor sludge contains gas produced in the anaerobic process and gas production can continue in the separation process, solids-liquid separation can be inefficient and unpredictable. Various methods have been used to minimize the effect of trapped gas bubbles in the sludge-settling step. These include gas stripping by agitation or vacuum degasification, inclined-plate separators, and the use of coagulant chemicals. Clarifier hydraulic application rates range from 0.5 to 1.0m/h. Practical reactor MLVSS concentrations are 4000 to 8000mg/L.

The anaerobic sequencing batch reactor (ASBR) process can be considered a activated sludge or suspended growth process with reaction and solids-liquid separation in the same vessel, much like that for aerobic sequencing batch reactors (SBRs). The success of the ASBR depends on the development of a good settling granulated sludge as found in the upflow anaerobic sludge blanket (UASB) processes. The operation of ASBRs consists of four steps: (1) feed, (2) react, (3) settle, and (4) decant/effluent withdrawal. During the react period, intermittent mixing for a few minutes each hour is done to provide uniform distribution of substrates and solids.

6.3 厌氧处理的影响因素

根据厌氧消化的机理，在大多数情况下产甲烷阶段是厌氧过程的限制性步骤，所以下列影响产甲烷菌活性的因素可看作影响厌氧消化过程的主要因素。

6.3.1 pH

pH 是废水生物厌氧处理最重要的影响因素之一。微生物对 pH 的波动十分敏感，即使在其适宜生长的 pH 范围内，也要比对温度变化的适应慢得多。pH 的变化将直接影响产甲烷菌的生存和活动。在厌氧生物处理中，水解菌和产酸菌对 pH 有较大范围的适应性，这类菌可以在 pH 为 5.0～8.5 范围生长良好，一些产酸菌在 pH 小于 5.0 时仍然能够生长。但对 pH 敏感的甲烷菌适宜 pH 为 6.5～7.8（这也是通常情况下厌氧生物处理所应控制的 pH 范围），最佳 pH 范围应在 6.8～7.2 之间。

厌氧生物处理的 pH 范围指厌氧反应器内反应区的 pH，而不是进水的 pH。事实上废水进入反应器内，生物化学反应和稀释作用可以迅速改变反应器内的 pH，尤其是大量溶解性碳水化合物酸化后能迅速降低反应器的 pH。一般情况下，反应器的出水 pH 接近或等于反应器内的 pH。

废水和泥液中的碱度有缓冲作用，如果有足够的碱度中和有机酸，其 pH 有可能维持在 6.8 之上，酸化和甲烷化两大类细菌就有可能共存，从而消除分阶段现象。此外，消化池池液的充分混合对调整 pH 也是必要的。

6.3.2 温度

有不同的微生物分别适宜在 10～34℃进行低温消化、在 35～38℃进行中温消化，在 52～55℃进行高温消化，三种消化反应能达到的消化速度不同。由于温度制约生物化学的反应速度，即使嗜冷微生物处在最佳的生长温度，它的代谢速率也会低于中温厌氧反应器。在大多数情况下，厌氧消化符合温度每增加 10℃反应速率增加 1 倍的规律，如中温消化的消化时间（产气量达到总量 90％所需时间）约为 20d，高温消化的消化时间约为 10d。

基于温度对厌氧微生物代谢速率的影响，工程技术上不宜选用低温消化（农村小型沼气发酵除外），而常采用中温消化。因中温消化的温度与人体温度接近，故对寄生虫卵及大肠杆菌的杀灭率较低；高温消化对寄生虫卵的杀灭率可达 99％，但高温消化需要的热量比中温消化高很多。有研究表明，高温消化产沼气速率及产沼气量均高出中温消化，但温度升高，加剧除产甲烷菌外其他微生物的活动，因而沼气产量中的甲烷并没有增加，甚至还有所降低，增加的沼气主要为其代谢产物二氧化碳，因此，用沼气产量或产率评价厌氧消化的好坏是值得商榷的，必须进行技术经济性评价证实。若维持高温消化需要耗能时，高温消化显然并非一个好的选择。

对于一个反应器而言，稳定的消化温度是十分必要的。据研究，消化温度波动范围一天不宜超过±2℃，中温和高温消化波动的温度范围为±(1.5～2)℃。当有±3℃的变化时，就会抑制消化作用的进行；当有±5℃的急剧变化时，消化反应器就会停止产气。所以，在选择厌氧消化的温度时，要根据废水本身的温度及环境条件（气温、有无废热可以利用等）确定。

6.3.3 生物固体停留时间（污泥龄）

厌氧消化的效果跟污泥龄（*SRT*）有直接关系。鉴于有机物降解程度是污泥龄的函数，消化池的设计并非类似好氧处理那样按有机负荷设计，而是以污泥龄或水力停留时间（*HRT*）进行设计。

消化池的有效容积可用污泥的投配率表达：

$$V = V'/n \tag{6-1}$$

式中 n——污泥投配率，％，为每日投加新鲜污泥体积所占消化池有效容积的百分数；

V——消化池有效容积，m^3；

V'——新鲜污泥量，m^3。

投配率是消化工艺设计的重要参数。投配率过高，可能影响产甲烷菌的正常生理代谢，导致脂肪酸在反应器中大量积累，pH 下降，污泥消化不完全；投配率过低，污泥消化完全，产气量大，但产气速率会降低、消化池容积加大，使基建投资费用增加。根据实践经验，一般城市污水处理厂污泥中温消化的投配率以 5％～8％为宜，相应的消化时间为 12.5～20d。

6.3.4 搅拌和混合

厌氧消化是由细菌体的内酶和外酶与底物进行的接触反应。在生物反应器中，底物首先需传质到细菌表面，进而被代谢，而传质速率将起到较重要的作用。搅拌是提高传质速率的重要因素之一，是实现两者充分混合的关键所在。影响传质速率的因素主要有厌氧污泥（或生物膜）与介质间的液膜厚度，以及颗粒污泥内部不同细菌菌群间代谢产物的传质速率。液膜厚度大将影响底物的传质速率，可通过搅拌（机械或代谢产气搅拌）降低液膜厚度。

搅拌的方法过去常采用水射器搅拌法、消化气循环搅拌法和混合搅拌法，但现在国内外大型沼气池或消化池已开始采用池内安装搅拌器直接搅拌，如图 6-2 所示。

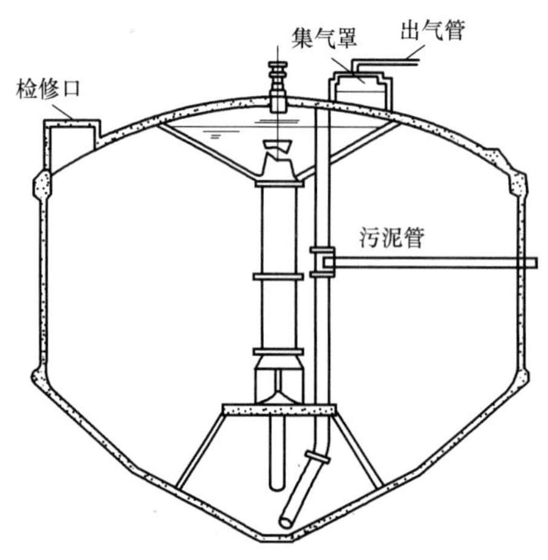

图 6-2　螺旋桨搅拌的消化池

6.3.5 营养与 C/N 比

厌氧细菌由于生长速率低，所以对氮、磷等营养盐需求较少。一般来说，处理含天然有机物的废水时不用投加营养盐。试验表明，COD：N：P 控制在 500：5：1 左右为宜，在厌氧处理装置启动时，可稍微增加氮素，有利于微生物的增殖，并有利于提高反应器的缓冲能力。厌氧消化过程中，如 C/N 比太高，细胞的氮量不足，消化液的缓冲能力低，pH 容易降低；C/N 比太低，氮量过多，pH 可能上升，铵盐容易积累，会抑制消化进程。

6.3.6 氧化还原电位

无氧环境是严格厌氧产甲烷细菌繁殖的最基本条件之一。对厌氧反应器介质中的氧浓度与电位的关系的判断，可用氧化还原电位（ORP）表达。

有资料表明，产甲烷细菌初始繁殖的条件是氧化还原电位不能高于－330mV。按照 Nernst 关系式，氧化还原电位－330mV 相当于 2.36×10^{56}L 水中有 1mol 氧。可见专性厌氧的产甲烷细菌对介质中分子态氧的存在是极为敏感的。对环境严格厌氧的要求是由产甲烷细菌本身的严格厌氧特性决定的。在厌氧发酵全过程中，非产甲烷阶段可在兼氧条件下完成，氧化还原电位在－250～＋100mV；而产甲烷阶段最适氧化还原电位为－500～－300mV。

氧化还原电位还受到 pH 的影响，pH 低，ORP 高；pH 高，ORP 低。因此，在初始富集产甲烷细菌阶段，应尽可能保持介质 pH 接近中性，并应保持反应装置的密封性。

6.3.7 有毒物质

所谓“有毒”是相对的，事实上任何一种物质对微生物的生长都有两方面的作用，既有激活作用又有抑制作用，关键在于它们的浓度界限，即毒阈浓度。

1. 重金属离子的毒害作用

在工业废水中，常含有重金属。微量重金属对厌氧细菌的生长可能起到刺激作用，但当其过量时，却可能抑制微生物生长。众多研究表明，各种重金属离子对厌氧发酵产生抑制的阈限浓度因试验条件、底物成分、厌氧工艺不一以及污泥驯化程度不同，其结果差别较大，不同学者的研究结果并不统一。在大多数情况下，重金属对微生物的毒性大小依次为 Ni＞Cu＞Pb＞Cr＞Cd＞Zn。

重金属离子对甲烷消化的抑制有两个方面：(1) 与酶结合，产生变性物质，使酶的作用消失；(2) 重金属离子及氢氧化物的絮凝作用，使酶沉淀。

2. 氨的毒害作用

氨氮对厌氧微生物的生长亦有刺激浓度和抑制浓度之分。当有机酸积累时，pH 降低，此时 NH_3 转变为 NH_4^+，当 NH_4^+ 浓度超过 150mg/L 时，消化受到抑制。氨氮浓度在 50～200mg/L 时，对厌氧反应器中的微生物有刺激作用，在 1500～3000mg/L 时则有明显的抑制作用。值得注意的是，反应液的 pH 决定了水中氨和铵离子间的分配百分比。当 pH 较高时，对产甲烷细菌有毒性的游离氨的比例也会相应提高。

3. H_2S 的毒害作用

脱硫弧菌（属于硫酸盐还原菌）能将乳酸、丙酮酸和乙醇转化为 H_2，CO_2 和乙酸。但在含硫无机物（SO_4^{2-}、SO_3^{2-}）存在时，它将优先还原 SO_4^{2-} 和 SO_3^{2-}，产生 H_2S，形成与产甲烷菌对基质的竞争，并对产甲烷菌带来毒害作用。因此，当厌氧处理系统中 SO_4^{2-} 和 SO_3^{2-} 浓度过高时，产甲烷过程就会受到抑制，严重时会影响整个系统的正常工作。同时，导致产生的消化气中 CO_2 成分提高，并含有较多的 H_2S。H_2S 的存在降低消化气的质量，在其燃烧时会腐蚀金属设备（管道、锅炉等）。

硫酸盐、硝酸盐和亚硝酸盐的存在将对产甲烷阶段构成一定的竞争抑制，研究表明，厌氧处理有机废水时生物氧化的顺序是：反硝化、反硫化、产酸发酵、产甲烷等。只有在前一种反应条件不具备时才进行后一种反应。因此，必须严格控制厌氧反应器进水中的硫酸盐、硝酸盐和亚硝酸盐的含量，才能使反应器保持有利于产甲烷阶段的运行状态。

4. 难降解有机物的毒害作用

在一些工业废水中，常含有一定浓度的有毒有机物质，其中有天然有机物，也有相当一部分是人工合成的生物异型化合物。有毒有机物的毒性由两种原因引起：①非极性的有机化合物可能损害细胞的膜系统；②通过氢键与菌体蛋白质结合，使酶失活。对有机物来说，分子结构将对微生物的抑制作用有影响。例如，醛基、双键、氯取代基、苯环等结构，可增加对微生物的抑制作用。在脂肪酸中，丙酸、己酸、十二烷酸对厌氧微生物具有抑制作用。此外，几乎所有的苯环化合物对厌氧过程都有一定的抑制性，其中可能是硝基苯毒性最大。

6.4 厌氧处理设备

目前我国将厌氧生物处理工艺的发展历程分为三代，从 1896 年英国出现的第一座厌氧消化池开始，包括化粪池、双层沉淀池等传统设施，称为“第一代厌氧生物反应器”，其共同特点是：(1) 水力停留时间（*HRT*）很长，有时在污泥处理时，污泥消化池的 *HRT* 会长达 90d；(2) 处理效率低，处理效果不好；(3) 有浓臭的气味。

20 世纪 70 年代中后期，相继出现了一批能实现现代高速厌氧消化反应的处理工艺，将其统一称为“第二代厌氧生物反应器”，主要包括：厌氧接触法、滤池（AF）、上流式厌氧污泥床（UASB）反应器、厌氧流化床（AFB）、厌氧生物转盘（AR）和挡板式厌氧反应器等。它们的主要特点有：(1) *HRT* 有所缩短，有机负荷大幅提高，处理效率大幅提高；(2) *HRT* 与 *SRT* 分离，*SRT* 相对很长，*HRT* 则可以较短，反应器内生物量很高。以上优点促使厌氧反应器的实际应用变得越来越广泛。

20 世纪 90 年代以后，随着以颗粒污泥为主要特点的 UASB 反应器的广泛应用，在其基础上又发展起来了同样以颗粒污泥为根本的颗粒污泥膨胀床（EGSB）反应器和厌氧内环（IC）反应器。其中 EGSB 反应器利用外加的出水循环可以使反应器内部形成很高的上升速度，提高反应器内的基质与微生物之间的接触和反应，可以在较低温度下处理较低浓度有机废水，如城市废水等；而 IC 反应器则主要应用于处理高浓度有机废水，依靠厌氧生物过程本身所产生的大量沼气形成内部混合液的充分循环与混合，可以达到更高的有机负荷，这些反应器又被统一称为“第三代厌氧生物反应器”。

6.4.1 厌氧生物滤池

厌氧生物滤池是一种内部填充微生物载体或填料的厌氧生物反应器，结构与一般的好氧生物滤池类似，由池体、滤料、布水设施以及排水、排泥设备组成，不同之处是厌氧生物滤池的池顶是密闭的。滤池按功能可分为布水区、反应区、出水区和集气区 4 个部分。厌氧生物滤池的中心构造是滤料。滤料可采用拳状石质滤料，如碎石、卵石等，粒径在 40mm 左右，也可使用塑料填料。

厌氧生物滤池的工作原理为：污水从池底进入，经过附着大量生物膜的滤料与微生物接触，并被生物膜中的微生物降解转化为沼气，后从池上部排出至后续构筑物。微生物附着生长在滤料上，不随出水流出，而能保持较长的污泥龄。由于填料是固定的，废水进入反应器后逐渐被微生物水解酸化、产氢产乙酸和产甲烷，废水组成在反应器不同高度逐渐变化，其对应的微生物种群分布也呈规律性变化。在进水处发酵菌和产酸菌为主，随反应器高度上升，产氢产乙酸菌和产甲烷菌逐渐增大并占主导地位。

厌氧生物滤池的主要优点是微生物固体停留时间长，去除有机物的能力较强；滤池内可以保持很高的微生物浓度；不需另设泥水分离设备，出水 SS 较低；设备简单、操作方便等。它的主要缺点是：滤料费用较贵；进水分配不易均匀，滤料容易堵塞，池下部生物膜很厚，堵塞后没有简单有效的清洗方法。因此，悬浮固体高的污水不适用此法。

6.4.2 厌氧接触法

厌氧接触法实质上是厌氧活性污泥法，它是在一个完全混合式厌氧反应器基础上增加污泥分离和污泥回流的装置，其流程如图 6-3 所示。经厌氧接触反应器处理的混合液首先在沉淀池固液分离（也可以采用气浮法或膜过滤），沉淀或分离的污泥回流至消化池，保证消化池稳定的高污泥浓度，提高消化池的有机负荷率和处理效率。该工艺为中低负荷工艺，适宜高浓度有机废水和悬浮固体较高的有机污水处理，如酒精糟液、肉联加工废水的处理。

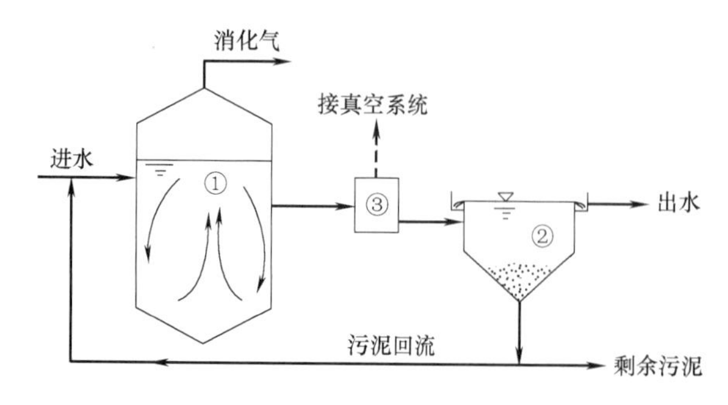

图 6-3 厌氧接触法的流程

①—混合接触池；②—沉淀池；③—真空脱气器

厌氧接触池中的污泥浓度一般在12000～15000mg/L，污泥回流量很大，一般是废水流量的2～3倍。由于污泥量大，厌氧接触池要进行适当搅拌以使污泥保持悬浮状态。搅拌可以用机械方法，也可以用泵循环搅拌。厌氧接触法由于污泥回流，反应器内能够维持较高的污泥浓度，大幅降低水力停留时间，并使反应器具有一定的耐冲击负荷能力。但从厌氧反应器排出的混合液，因污泥附着大量气泡在沉淀池中易于上浮而被出水带走。此外进入沉淀池的污泥仍有产甲烷菌在活动，并产生沼气，使已沉淀的污泥上翻，影响固液分离效果，使回流污泥浓度降低，影响反应器内污泥浓度。对此需采取下列技术措施加以防治：

(1) 在反应器与沉淀池之间设脱气器，尽可能将混合液中的沼气脱除。但这种措施不能抑制产甲烷菌在沉淀池内继续产气；

(2) 投加混凝剂，提高沉淀效果；

(3) 用膜过滤代替沉淀池。

6.4.3 两级厌氧消化和两相厌氧消化

两级厌氧消化是根据消化过程沼气产生的规律进行设计的，目的是节省污泥加温与搅拌所需的能量。根据中温消化的产气率与消化时间的关系，消化前8d的产气量约占全部产气量的80%，如把消化池设计成两级，第一级消化池有加温、搅拌设备，并有集气罩收集沼气；第二级不设加热和搅拌设备，依靠第一级的余热继续消化，产气量约占20%，可收集或不收集。由于温度低、消化时间长，加之不搅拌，所以二级消化池具有浓缩功能。

两相厌氧消化（two phase anaerobic digestion）是根据消化机理进行设计的，目的是使各相消化池具有更适合于消化过程三阶段各自微生物种群生长繁殖的环境。1971年戈什（Ghosh）和波兰特（Pohland）首次提出了两相发酵的概念，即使产酸和产甲烷两个阶段的反应分别在两个独立的反应器内进行，以创造各自最佳的环境条件，并将这两个反应器串联起来，形成两相厌氧发酵系统。目前城市污水处理厂污泥消化处理一般采用两相厌氧消化。由于水解酸化和产甲烷在两个独立的反应器内分别进行，从而使本工艺具有下列特点：

(1) 为产酸菌、产甲烷菌分别提供各自最佳的生长繁殖条件，在各自反应器内得到最高的反应速率，提高处理效率；

(2) 酸化反应器有一定的缓冲作用，缓解冲击负荷对后续的产甲烷反应器的影响。当废水含有硫酸盐等抑制性物质时，其可以减轻对甲烷菌的影响；

(3) 酸化反应器反应进程快，水力停留时间短，COD可去除20%～25%，能够大幅减轻产甲烷反应器的负荷。

两相厌氧消化第一相消化池采用100%的投配率，水力停留时间为1d；第二相消化池采用15%～17%的投配率，水力停留时间为6～6.5d。为节省能源，一般只对第二相设加温、搅拌和集气装置，产气量约为1.0～1.3m^3/m^3污泥，即每去除1kg有机物的产气量约为0.9～1.1m^3。

6.4.4 厌氧膨胀床和厌氧流化床

厌氧膨胀床和厌氧流化床是20世纪70年代开发的一种高效厌氧反应器，如图6-4所示。床体内填充细小的固体颗粒填料，如石英砂、无烟煤、活性炭、陶粒和沸石等，填料

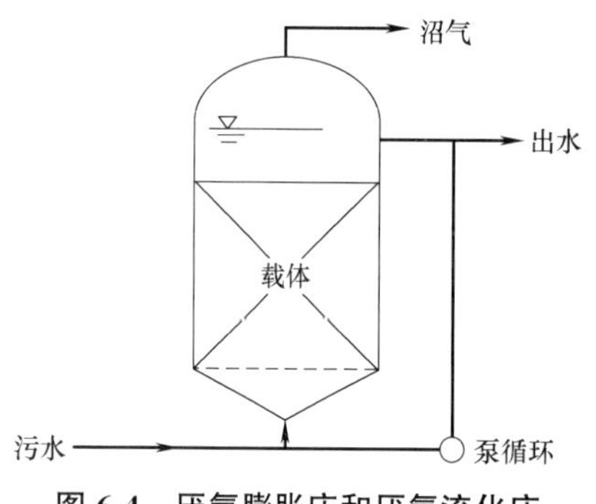

图 6-4 厌氧膨胀床和厌氧流化床

粒径一般为 0.2～1mm。废水从床底部流入，在浮力和摩擦力的作用下使生物颗粒处于悬浮状态，并与废水充分接触而完成厌氧生物降解过程净化后的水从上部溢出，产生的气体也从上部溢出。为使填料层膨胀，需将部分出水用循环泵回流，提高床内水流的上升流速。一般认为膨胀率为 10%～20%的为厌氧膨胀床，膨胀床的颗粒保持相互接触。当采用部分出水回流，提高床内水流上升流速，使膨胀率控制在 30%以上、载体颗粒在反应器内处于流化状态，即为厌氧流化床。

这类反应器的主要特征是，有机物容积负荷较高，水力停留时间短，耐冲击负荷能力强，运行稳定，载体不易堵塞。

但是，厌氧流化床反应器存在几个尚未解决的问题。为实现反应器处于流化态并使污泥与载体不致从反应器中流失，必须使生物膜颗粒保持均匀的大小、形状和密度，这是难以实现的；为取得较高上升流速以保持流化态，流化床需要大量回流水，导致能耗增加；对处理系统的设计与运行管理要求高，从而限制了该技术的推广和应用。

6.4.5 上流式厌氧污泥床反应器

上流式厌氧污泥床反应器（UASB）是目前应用最成功的厌氧生物处理工艺，如图 6-5 所示，其广泛用于各种有机废水的处理。1977 年由荷兰的 Lettinga 教授研发的 UASB 反应器，主体为无填料的空容器，其中含有大量厌氧污泥。由于废水以一定流速自下而上流动以及厌氧过程产生大量沼气的搅拌作用，废水与污泥充分混合，有机物被吸附分解。所产生的沼气经反应器上部的三相分离器的集气室排出，含有悬浮污泥的废水进入三相分离器的沉降区进行泥水分离，沉淀污泥返回主体反应器，澄清污水从出水口排放。

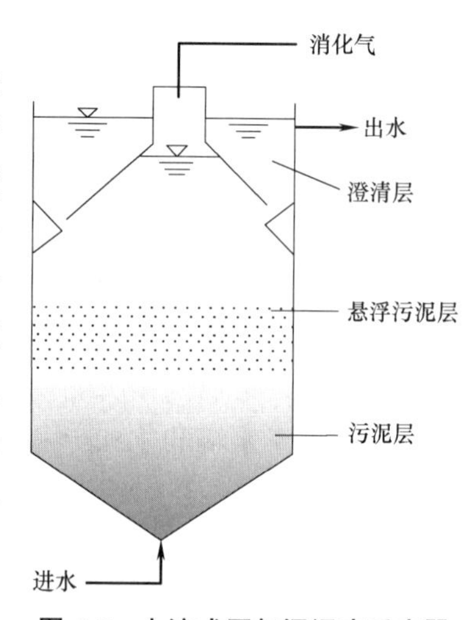

图 6-5 上流式厌氧污泥床反应器

由于废水流动和沼气的搅拌、黏附作用，细小的污泥絮体会随废水流出或洗出，经过一段时间运行后，会在反应器底部形成一个高浓度（可达 60～80g/L）、高活性和良好沉降性能的颗粒污泥层，使反应器能够承受较大的上升流速和很高的容积负荷，能适应负荷冲击和温度、pH 的变化。

UASB 反应器的三相分离器结构与反应器的进水系统是该工艺的特色。由于需分离的混合物由气体、液体和固体（污泥）组成，所以这一系统要具有气、液、固三相分离的功能：(1) 在水和污泥的混合物进入沉淀区前，必须首先将气泡分离出来；(2) 为避免在沉淀区内产气，污泥在沉淀区内的滞留时间必须是短的；(3) 由于厌氧污泥形成积聚的特征，沉淀器内存在的污泥层对液体通过它向上流动影响不大。

污泥颗粒化是 UASB 的主要特点，所以影响污泥颗粒化的因素就是影响 UASB 运行效果的因素。

颗粒污泥的形成受污泥接种物性质、底物成分，反应器工艺条件，微生物性质以及微生物菌种间、微生物与底物间的相互作用等影响，是生物、化学及物理因素等多种作用的结果。

1. 接种污泥

Lettinga 提出稠密型厌氧污泥（约为 60kg/m^3）比稀薄型的污泥好。前者的单位生物量产甲烷能力（比产甲烷活性）虽然低于后者，但前者的沉淀性能好，不易因产生过度膨胀而流失。

2. 废水性质

废水的性质包括有机组分、浓度、悬浮物含量及可生物降解性能等，这些对污泥结团（颗粒化）都有影响。

底物种类对污泥颗粒化影响较大，含碳水化合物和易降解废水易形成颗粒污泥。对于生物降解性差的化工等废水，在启动时适当加入淀粉等易生物降解物质是有利的。

COD 对污泥的颗粒化有一定影响，在低浓度的废水里结团会更快，其原因尚不清楚。启动时，COD 以 4000～5000mg/L 为宜，对浓度过高的废水最好采用稀释的方法。

进水的营养宜保持 COD∶N∶P＝200∶5∶1，当氮、磷缺乏时，应加以补充，并要求 pH 控制在 6.5～7.5 为宜。例如，适当补充钙和铁，会利于颗粒污泥的形成。

进水悬浮物的含量应控制在一定范围内，一般要求在 2g/L 以下。一般来说，高浓度的惰性分散固体不利于颗粒污泥的形成。

进水 pH 一般在 6.0～8.0 之间。

3. 工艺条件

在培养颗粒污泥的过程中，各种条件都应控制在有利于细菌生长的范围内，主要控制参数有温度、挥发酸、污泥龄（*SRT*）以及容积负荷等。

反应器的温度以中温为宜（33～41℃），高温下污泥结团过程与中温类似，但颗粒较小，易流失。

反应器内需要高浓度的活性污泥，一般平均污泥浓度为 30～40g/L，所以 UASB 反应器污泥龄（*SRT*）很长，一般为 30d 以上，甚至在启动和运行最初的 100～200d 几乎不需排泥。

污泥颗粒化使反应器具有很高的容积负荷，中温消化的 COD 容积负荷一般为 10～20kgCOD/(m^3·d)，对应的有机负荷较高，反应器的水力停留时间较短。

颗粒污泥的直径随有机负荷提高而增大，实际上，颗粒污泥的大小受底物传质过程中所能进入颗粒内部的深度所支配。当颗粒大小与传质之间不相适应时，颗粒内即会因营养不足发生细胞自溶，最终导致颗粒破碎。高的负荷或高的进水有机物浓度可使更多底物进入颗粒内部，从而允许大的颗粒存在和生长，其较大的上升流速与高的负荷产生大量生物气也有助于洗出细小污泥（使细小的絮状污泥随水流一起冲出反应器），这是高负荷下颗粒污泥平均直径较大的又一原因。

正常运行状态下，颗粒污泥形状不规则，一般呈球形或者椭球形，直径为 0.1～2mm，最大可达 3mm；颜色呈灰黑色或褐黑色，相对密度一般为 1.01～1.05；污泥容积指数（SVI）与颗粒大小有关，细小颗粒一般为 20mL/gSS；而沉淀性能较好的絮状污泥约为 40mL/gSS；颗粒污泥在反应器中的沉降速率一般为 0.3～0.8m/h；成熟的颗粒污泥，VSS/SS一般为0.7左右，其与颗粒污泥在反应器中的分布位置有关，一般越往反应器顶部VSS/SS越高，最高可达0.8；反之，反应器最底部最低，可达0.5甚至更低。

厌氧生物处理系统的比较见表6-1，不同废水厌氧处理出水指标范围见表6-2，新型厌氧设备的原理及实物照片见表6-3。

几种厌氧生物处理系统的比较　　表6-1

方法名称	工艺特点	优点	缺点
厌氧接触系统	由完全混合式消化池及沉淀池组成，中间设脱气装置	能处理SS含量高的废水，负荷率较高，出水SS较少	负荷高时污泥会流失，操作要求高
厌氧生物滤池	池内填充粒径为20～50mm的滤料或装置软性滤料，污泥龄很长	负荷率较高，能承受冲击负荷，运行稳定，能耗低，出水悬浮物较少	易堵塞，不适应处理SS含量高的废水
上流式厌氧污泥床	池内底部为污泥床，其上有一层悬浮污泥层，顶部设三相分离器	负荷率较高，容积小，抗冲击负荷能力强	如设计不当，污泥会大量消失，池的构造复杂
厌氧膨胀床/流化床	池内填充粒径为0.5mm左右的挂膜介质，全部悬浮于上升水流中，废水常需回流	负荷率较高，容积小，抗冲击负荷能力强	管理较复杂
两段厌氧消化系统	由酸发酵池及甲烷发酵池两池组成，两段可采用不同的组合方式	抗冲击负荷能力强，运行稳定，酸化采用较低的温度	两池工作条件不同，运行管理复杂

不同废水厌氧处理出水指标范围　　表6-2

废水类型	BOD_5(mg/L)	COD(mg/L)
糖类	50～500	250～1500
乳品业	150～500	250～1200
玉米淀粉	—	500～1500
马铃薯	200～300	250～1500
蔬菜	100	700
葡萄酒	3500	—
纸浆	350～900	1400～8000
纤维板工业	2500～5500	8800～14900
造纸厂	100～200	280～300
垃圾渗滤液	—	500～4000
消化上清液	400	800～1400
啤酒厂	—	200～350
酿酒厂	—	320～400

新型厌氧设备的原理及实物照片图 **表 6-3**

名称	工作原理及形成	现场图
UASB 上流式 厌氧污配床	1. 反应器内污泥浓度高，反应器内设三相分离器，无配合搅拌设备，污泥床内不填载体；被分离的污泥则自动滑落到悬浮污泥层。出水则从澄清区流出； 2. 总容积小，能耗低，不需搅拌，污泥颗粒化后使反应器对不利条件的抗性增强；在一定的水力负荷下，可以靠反应器内产生的气体实现污泥与基质的充分接触； 3. 若设计不善，污泥会大量流失；对水质和负荷突然变化较为敏感	
EGSB 厌氧颗粒 污配膨胀床	1. 通过设计较大的高径比，同时采用出水循环，提高反应器的上升流速，使颗粒污泥膨胀流化； 2. 活性高，沉降性能好，抗冲击负荷能力强。改善泥水接触效果及传质效果。处理低温低浓度的有机废水，适用范围广，可用于 SS 含量高和对微生物有抑制性的废水处理； 3. 不耐高悬浮物，技术有待研究，运行条件和控制要求较高	
UBF 上流式 污配床过滤器	1. 由 UASB 和 AF 构成复合式反应器，其下部是污泥床，上部是填料及其附着的生物膜组成的滤料层； 2. 增加生物总量，防止生物量的突然流出（填料表面生长微生物，空隙载量悬浮微生物）；由于填料的存在，夹带污泥的气泡在上升过程中与之碰撞，加快污泥与气泡的分离，降低污泥的流失； 3. 不适合连续含 SS 较多的有机废水，否则填料层容易堵塞	

6.5 厌氧反应器的设计计算

6.5.1 UASB 反应器的设计计算

1. 设计规定

（1）UASB 反应器设计流量应按最高日平均时废水量设计，如厂区内设置调节池且停留时间大于 8h，UASB 反应器设计流量可按平均日平均时设计。

（2）UASB 反应器进水应符合表 6-4 的规定。

UASB 反应器进水条件 **表 6-4**

项目		单位	参数值
温度	常温厌氧	℃	20～25
	中温厌氧		35～40
	高温厌氧		50～55
pH		—	6.0～8.0
营养比（COD：NH_4-N：TP）		—	（100～500）：5：1
BOD_5/COD		—	≥0.3

续表

项目	单位	参数值
悬浮物(SS)浓度	mg/L	≤1500
氨氮浓度	mg/L	≤2000
磷酸盐浓度	mg/L	≤1000
化学需氧量	mg/L	≥1500
重金属、酚类、氰化物等有毒有害物	mg/L	严格控制

(3) 反应器的最大单体体积应小于 3000m³。反应器的有效水深为 5～8m，废水在反应器内的上升流速应小于 0.8m/h。

2. UASB 反应器设计

(1) UASB 反应器对常规污染物的去除效果参考表 6-5。

UASB 反应器对常规污染物的去除效果　　表 6-5

悬浮物(SS)	化学需氧量(COD)	5d 生化需氧量(BOD_5)
30%～50%	80%～90%	70%～80%

(2) UASB 反应器容积宜采用容积负荷法计算，反应器的容积负荷应通过试验或参照类似工程确定，缺少相关资料时可参考表 6-6，处理高浓度复杂废水时参考表 6-7。

国内不同类型废水 UASB 反应器设计负荷统计表　　表 6-6

废水类型	容积负荷[kgCOD/(m³·d)]		
	最高	平均	最低
啤酒废水	10.0	7.3	5
酒精废水	20	6.5	2
造酒厂	10	6.4	4
淀粉	8.0	5.4	2.7
土豆加工	10	6.8	6
柠檬酸生产	20	14.8	6.5
味精	4.0	2.3	2.0
食品加工	4.0	3.5	3.0
制药厂	8.0	5.0	0.8
屠宰场	4.0	3.1	2.3

高浓度复杂废水 UASB 反应器设计负荷参考表　　表 6-7

废水 COD(mg/L)	35℃时反应器容积负荷[kgCOD/(m³·d)]	
	絮状污泥	颗粒污泥
2000～6000	3～5	4～6
6000～9000	4～6	5～8
>9000	5～8	6～10

注：常温厌氧环境下反应器容积负荷在本表基础上适当降低，高温厌氧环境反应器容积负荷适当提高。

(3) UASB反应器计算

$$V=\frac{QS_0}{L_v} \tag{6-2}$$

式中 V——UASB反应器有效容积，m^3；

Q——污水设计流量，m^3/d；

S_0——反应器进水COD，g/L；

L_v——反应器COD容积负荷，kgCOD/(m^3·d)。

(4) 沼气产量计算

$$Q_G=\frac{Q(S_0-S_e)\eta}{1000} \tag{6-3}$$

式中 Q_G——沼气产量，Nm^3/d；

Q——污水设计流量，m^3/d；

S_0——反应器进水COD，mg/L；

S_e——反应器出水COD，mg/L；

η——沼气产率，一般为0.45～0.5Nm^3/kgCOD。

【例6-1】 某啤酒厂废水有机物浓度高，设计污水最大流量 $Q=2000m^3/d$，进水COD为3500mg/L，BOD_5 为2100mg/L，悬浮物浓度为460mg/L，磷酸盐（PO_4^-）浓度为240mg/L，拟采用厌氧加好氧生物处理工艺，其中厌氧工艺采用UASB反应器进行处理，COD去除率为85%，采用中温消化，设计UASB反应器。

【解】 根据《升流式厌氧污泥床反应器污水处理工程技术规范》HJ 2013—2012规定，进水COD宜大于1500mg/L，悬浮物浓度小于1500mg/L，BOD_5/COD=2100/3500=0.6，比值大于0.3，硫酸盐浓度小于1000mg/L，有机负荷参考已建UASB反应器中温条件下处理类似污水的运行数据，当COD去除率为85%时，取容积负荷8kgCOD/(m^3·d)。

(1) 确定反应器容积

取容积负荷 L_v=8kgCOD/(m^3·d)，则反应器容积为：

$$V=\frac{QS_0}{L_v}=\frac{2000\times3.5}{8.0}=875m^3$$

取有效容积系数 $E=0.85$，反应器实际容积为：

$$V_T=\frac{V}{E}=\frac{875}{0.85}=1029.4m^3$$

(2) UASB反应器主要构造尺寸

UASB采用圆形池，布水均匀，处理效果好。

取水力负荷 $q=0.6m^3/(m^2\cdot h)$，则反应器表面积为：

$$A=\frac{Q}{q}=\frac{2000}{0.6\times24}=138.9m^2$$

确定反应区液相高度为：

$$H_L=\frac{V_L}{A}=\frac{875}{138.9}=6.3m$$

UASB 反应区上部为储气空间，取储气空间高度 $H_G=2.5m$，则反应器总高度为：

$$H_T=H_L+H_G=6.3+2.5=8.8m$$

采用 2 座尺寸相同的 UASB 反应器，每个 UASB 反应器面积为：

$$A_1=\frac{A}{2}=\frac{138.9}{2}=69.45m^2$$

UASB 反应器池体为圆形，则池体直径为：

$$D=\sqrt{\frac{4A_1}{\pi}}=\sqrt{\frac{4\times69.45}{\pi}}=9.4m$$

（3）UASB 反应器水力停留时间

$$HRT=\frac{V}{Q}=\frac{875\times24}{2000}=10.5h$$

（4）产气量

UASB 反应器产气率 η 取 0.5Nm^3/kgCOD，反应器 COD 去除率为 85%，出水 COD 取 520mg/L，则沼气产量为：

$$Q_G=\frac{Q(S_0-S_e)\eta}{1000}=\frac{2000\times(3500-520)\times0.5}{1000}=2980Nm^3/d$$

6.5.2 水解酸化反应器的设计计算

1. 设计规定

（1）水解酸化反应器设计水质应根据工程实测水质确定或参考同行业同规模的废水排放资料类比确定。

（2）水解酸化反应器进水水质要求：

pH 宜为 5.0～9.0；营养物 COD：N：P 宜为（100～500）：5：1；若污水可生化性较好，化学需氧量 COD 宜低于 1500mg/L，污水可生化性较差时，化学需氧量 COD 可适当放宽。

（3）水解酸化反应器污染物去除率

处理城镇污水宜采用升流式水解酸化反应器。处理工业废水时，根据废水水质、水量选用不同形式的水解酸化反应器，若反应器中微生物增长缓慢可采用复合式水解酸化反应器。水解酸化反应器对不同类型污（废）水污染物的去除率见表 6-8。

水解酸化反应器对不同污（废）水污染物的去除率　　表 6-8

污（废）水类型	进水水质要求	污染物去除率		
		SS*	COD	BOD_5
城镇污水	可生化性较好或一般	50%～80%	30%～50%	20%～40%
啤酒废水、屠宰废水、食品废水、制糖废水等	可生化性较好，非溶解性 COD 比例大于 60%	50%～80%	30%～50%	20%～40%
造纸废水、焦化废水、煤化工废水、石化废水、制革废水、含油废水、纺织染整废水等，包括工业园区废水	可生化性一般，非溶解性 COD 比例为 30%～60%	30%～50%	10%～30%	20%～40%

续表

污(废)水类型	进水水质要求	污染物去除率		
		SS*	COD	BOD_5
其他难降解有机废水	可生化性较差，非溶解性COD比例小于30%	30%～50%	10%以下	10%以下

注：SS* 此值为升流式水解酸化反应器参考值。

（4）升流式水解酸化反应器宜为圆形或矩形，矩形反应器的长宽比宜为（1～5）∶1。

（5）升流式水解酸化反应器可采用钢筋混凝土结构或不锈钢、碳钢加防腐涂层等材料；有效水深宜为4.0～8.0m，超高宜为0.5～1.0m；升流式水解酸化反应器污水上升流速宜为0.5～2.0m/h，对于难降解污水应适当降低污水流速或增加出水回流。

2. 升流式水解酸化反应器设计计算

升流式水解酸化反应器有效容积采用水力负荷或水力停留时间法，按下式计算

$$V = K_z \cdot Q \cdot HRT \text{ 或 } V = \frac{K_z QS}{L_v} \tag{6-4}$$

式中 V——水解酸化反应器有效容积，m^3；

K_z——污水流量变化系数；

Q——设计流量，m^3/h；

HRT——水力停留时间，h；

S——水解酸化反应器进水COD浓度，kg/m^3；

L_v——反应器容积负荷，需要通过试验或参考类似工程实际运行经验确定，kgCOD/(m^3·d)。

升流式水解酸化反应器有效容积的水力停留时间应通过实验或参照类似工程确定，在缺少相关资料时可参考表6-9取值。

升流式水解酸化反应器水力停留时间参考值 **表6-9**

污(废)水类型	进水水质要求	水力停留时间(h)
城镇污水	可生化性较好或一般	2～4
啤酒废水、屠宰废水、食品废水、制糖废水等	可生化性较好，非溶解性COD比例大于60%	2～6
造纸废水、焦化废水、煤化工废水、石化废水、制革废水、含油废水、纺织染整废水等，包括工业园区废水	可生化性一般，非溶解性COD比例为30%～60%	4～12
其他难降解有机废水	可生化性较差，非溶解性COD比例小于30%	10以上

【例6-2】 某石化园区污水处理厂设计处理水量为8000m^3/d，污水流量总变化系数K_z=1.4。设计进水水质BOD_5为380mg/L，COD为1050mg/L，悬浮物（SS）浓度为350mg/L，pH为6～9，设计升流式水解酸化反应器。

【解】（1）设计流量

$$Q = 8000m^3/d = 333.3m^3/h$$

（2）水解酸化反应器的容积

污水在水解酸化反应器的水力停留时间取 8h，则水解酸化反应池容积为：

$$V=K_zQ\cdot HRT=1.4\times333.3\times8=3733\text{m}^3$$

水解酸化反应器采用矩形结构，水解酸化反应池分为 4 格，每格容积为：

$$V_{单}=\frac{3733}{4}=933.3\text{m}^3$$

（3）水解酸化反应器尺寸

水解酸化反应器有效水深 H 取 6m，则每格水解酸化反应器面积为：

$$F=\frac{V_{单}}{H}=\frac{933.3}{6}=155.6\text{m}^2$$

水解酸化反应器按长宽比 2∶1 设计，每格水解酸化反应器宽度为：

$$B=\sqrt{\frac{F}{2}}=\sqrt{\frac{155.6}{2}}=8.8\text{m}$$

每格水解酸化反应器长度为：

$$L=2B=2\times8.8=17.6\text{m}$$

（4）水解酸化反应器内上升流速核算

$$v=\frac{Q}{A}=\frac{V}{HRT\cdot A}=\frac{H}{HRT}=\frac{6}{8}=0.75\text{m/h}$$

上升流速在 0.5～2.0m/h 之间，符合要求。

（5）排泥系统设计

采用静水压排泥装置，沿矩形池纵向多点排泥，排泥点设在反应器污泥区中下部，污泥层与水面高度保持在 1.0～1.5m，排泥管干管管径大于 150mm。污泥排放采用定时排泥，每日 1～2 次。

本章关键词（Keywords）

厌氧处理（分解）	Anaerobic treatment or decomposition
甲烷	Methane
过滤介质	Filter medium
厌氧消化器	Anaerobic digester
厌氧生物滤池	Anaerobic biological filtration
厌氧接触法	Anaerobic contact process
厌氧流化床	Anaerobic fluidized bed
上流式厌氧污泥床反应器	Upflow anaerobic sludge blanket reactor（UASB）
水解酸化反应器	Hydrolysis acidification reactor
水力停留时间	Hydraulic retention time
厌氧序批式反应器	Anaerobic sequencing batch reactor

思考题

1. 与好氧处理相比，厌氧处理的优缺点是什么？
2. 请分析影响厌氧处理的主要因素有哪些。
3. 存在有机物的厌氧过程中，主要经历哪两个阶段？其作用机理是什么？

4. 培养厌氧活性污泥过程中有哪些注意事项?

5. 请简述厌氧处理流程各组成部分的工作原理和作用。

6. 废水厌氧生物处理基本原理是什么?

7. 厌氧消化产生的甲烷应该如何处理?

8. 简述 UASB 系统中三相分离器的组成部分及作用。

9. UASB 运行的前提是什么?

10. 请分析影响 UASB 运行效果的因素主要有哪些。

11. 某啤酒厂废水有机物浓度高，设计污水流量 $Q=2000m^3/d$，进水 COD 为 3500mg/L，BOD_5 为 2100mg/L，悬浮物浓度为 460mg/L，磷酸盐（PO_4^-）浓度为 240mg/L，拟采用厌氧加好氧生物处理工艺，其中厌氧工艺采用 UASB 反应器进行处理，COD 去除率为 85%，采用中温消化，设计 UASB 反应器。

12. 某石化园区污水处理厂设计处理水量为 $10000m^3/d$，污水流量总变化系数 $K_z=1.3$。设计进水水质 BOD_5 为 400mg/L，COD 为 1000mg/L，悬浮物 SS 为 360mg/L，pH 为 6～9，设计升流式水解酸化反应器。

Chapter 7 Combined Anoxic/Aerobic Biological Treatment Processes

7.1 Principle of Nitrogen Removal

In the anoxic/aerobic process, nitrate is fed to the anoxic reactor from nitrate in the return activated-sludge flow and by pumping mixed liquor from the aerobic zone. In step-feed anoxic/aerobic processes, nitrate will be fed to the anoxic zone by flow of mixed liquor from a previous nitrification step. The electron donor is provided by the influent wastewater fed to these preanoxic zones.

A number of alternative treatment systems are available to achieve nitrification and denitrification, in which some form of aerobic-anoxic sequencing is provided. The systems differ in whether they utilize a single sludge or two sludges in separate nitrification and denitrification reactors. The single-sludge system uses one tank or basin and clarifier and the raw wastewater or endogenous reserves as the carbon and energy sources for denitrification.

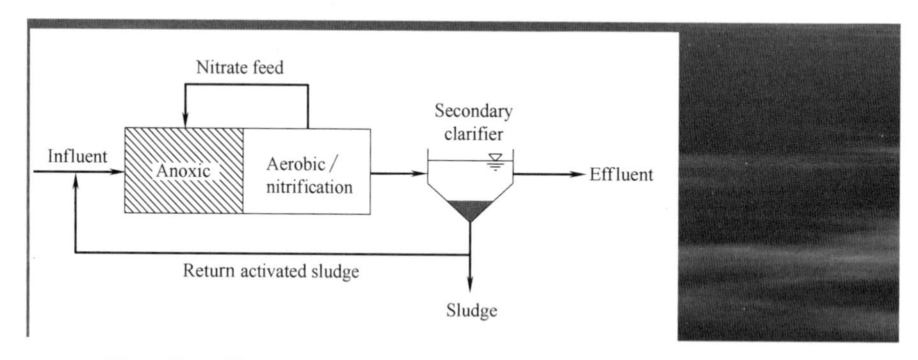

Figure 7-1 Single-sludge system to achieve nitrification and denitrification

The term "single-sludge" means only one solid separation device (normally a secondary clarifier) which is used in the process (see Figure 7-1). The activated-sludge tank may be divided into different zones of anoxic and aerobic conditions and mixed liquor may be pumped from one zone to another (internal recycle), but the liquid-solid separation occurs only once. In the two-sludge system, the most common system consists of an aerobic process (for nitrification) followed by an anoxic process (for denitrification), each with its own clarifier, thus producing two sludges. For postanoxic denitrification, an organic substrate, usually methanol, must be added to create a biological demand for the nitrate. Because single-sludge systems are used more often, they are discussed in more detail in the following paragraphs. The two-sludge system is discussed briefly later in this chapter.

7.1.1 The Principle of Nitrification

Nitrification is the biological oxidation of ammonia to nitrate with nitrite formation as an intermediate step. The microorganisms involved are the autotrophic species, nitrosomonas and nitrobacter, which carry out the reaction in two steps:

$$2NH_4^+ + 3O_2 \rightarrow 2NO_2^- + 4H^+ + 2H_2O \quad \text{(Nitrosomonas)} \tag{7-1}$$

$$2NO_2 + O_2 \rightarrow 2NO_3^- \quad \text{(Nitrobacter)} \tag{7-2}$$

The cell yield for nitrosomonas has been reported as 0.05 to 0.29mgVSS/mgNH_3-N and for nitrobacter 0.02 to 0.08mgVSS/mgNH_3-N. A value of 0.15mgVSS/mgNH_3-N is usually used for design purposes. It is generally accepted that the biochemical reaction rate of nitrobacter is faster than the reaction rate of nitrosomonas and hence there is no accumulation of nitrite in the process and the reaction rate of nitrosomonas will control the overall reaction.

1. Nitrification of High-Strength Wastewater

Wastewater containing high ammonia concentration and negligible BOD can be treated by biological nitrification. For example, wastewater from a fertilizer manufacturing complex was treated by the activated sludge process. The NH_3-N content of the influent wastewater varied from 339 to 420mg/L and inorganic suspended solids varied from 313 to 598mg/L. The TDS was 6300mg/L. Because of the high inert suspended solids, the mixed liquor was only 20% volatile with a sludge volume index (SVI) of 30~40 mL/g. A small fragile floc was generated, which provided an effluent TSS of 55mg/L. Alkalinity was supplied to the system in the form of sodium bicarbonate. The temperature correction coefficient is significantly higher than for typical domestic wastewater, indicating that the nitrification rate was more sensitive to the operating mixed liquor temperature.

2. Inhibition of Nitrification

In treating industrial wastewater, nitrification is frequently inhibited. Or in some instances prevented, by the presence of toxic organic or inorganic compounds.

In cases where nitrification is significantly reduced or totally inhibited, the application of powdered activated carbon (PAC) to adsorb the toxic agents may enhance nitrification. However, in some cases, excessive quantities of PAC are required to achieve single-stage nitrification. In some cases, a second-stage nitrification step can be successfully employed after a first-stage biological process for removal of carbonaceous material and reduction of toxicity.

Metals have been found to be toxic to growing Nitrosomonas culture with complete inhibition for the following metals and concentrations; Ni, 0.25mg/L; Cr, 0.25mg/L; and Cu, 0.1 to 0.5mg/L.

Un-ionized ammonia (NH_3) inhibits both nitrosomonas and nitrobacter. Since the unionized fraction increases with pH, a high pH combined with a high total ammonia concentration will severed inhibit or prevent complete biological nitrification. Since nitrosomonas is less sensitive to ammonia toxicity than nitrobacter, the nitrification process may only be partially complete and result unaccumulation of nitrite ion. This can have severe consequences since NO_2^- is strongly toxic to many aquatic organisms whereas NO_3-N is not. Ammonia toxicity to activated sludge biomass is rarely a problem in treating municipal wastewaters, since the concentration of total ammonia is low and the mixed liquor pH is near neutral. Industrial wastewater with high ammonia concentrations or levels and the potential for high pH, however, may cause biotoxicity and loss of the nitrification process. Under these conditions, it is necessary to control the mixed liquor pH to avoid biotoxicity due to an ammonia spill or shock load. In extreme cases, two stages operated at different pH values may be required to separate the nitrosomonas and nitrobacter and allow complete nitrification.

7.1.2 The Principle of Denitrification

Some industrial wastewater such as those from fertilizer, explosive/propellant manufacture, and the synthetic fibers industry contain high concentrations of nitrates, while others generate nitrates by nitrification. Since biological denitrification generates one hydroxyl ion while nitrification generates two hydrogen ions, it may be advantageous to couple the nitrification and denitrification processes to provide "internal" buffering capacity. While many organics inhibit biological nitrification, this is not generally true for denitrification. Denitrification uses BOD as a carbon source for synthesis and energy and nitrate as an oxygen source.

The denitrification process consumes approximately 3.7gCOD per gNO_3-N reduced and produces 0.45gVSS and 3.57 g alkalinity per gNO_3-N reduced. This amounts to one-half the alkalinity that is consumed during nitrification. Some of this alkalinity, however, is lost by reaction with the CO_2 generated by microbial respiration.

The denitrification rate will depend on the biodegradability of the organics in the wastewater and the concentration of active biomass aeration is similar to the aerobic process.

In cases where a carbon source is not available in the wastewater, methanol has been used as a carbon source. Various industrial effluent can also be used as a carbon source.

Denitrification in secondary clarifiers (final clarifiers) causes floating sludge and increased effluent suspended solids. The nitrogen gas rate production depends on the carbon source available for denitrification, the solid retention time (*SRT*), temperature and sludge concentration.

Biological nitrate removal (i. e. denitrification) is also achievable through the use of extractive membrane biological reactor (MBR) hybrid processes.

In all biotreatment processes, the treated water must be separated from the sludge or biomass. Fixed film process effluent is notionally low in biological material since the latter forms a biofilm on the growth media, although biofilms can slough off into the product water, whereas in the conventional activated sludge process (CASP) separation is normally by sedimentation. This means that CASPs rely on the solids (which are flocculated particles and referred to as flocs) growing to a size where they can be settled out, which means that they must be retained in the bioreactor for an appropriate length of time. The solid retention time (*SRT*) is thus coupled with the hydraulic retention time (*HRT*), the retention time being the time taken for the solids and water respectively to pass through the reactor. For commercial MBR technologies, separation is by membrane filtration, eliminating the requirement for substantial floc growth and the associated long *HRT*s. The key advantage offered by the MBR process, with specific respect to biotreatment, is thus the uncoupling of the *HRT* and *SRT*; However, MBRs can also be configured as fixed film processes, using the membrane to support a biofilm.

In 2008, the MBR market in China exceeded 1.6 billion CNY (over $US230 million). Today, China has become one of the most MBR-active countries. MBR applications include treatment/reuse of municipal wastewater, industrial wastewater, landfill leachate, bathing wastewater, hospital wastewater and polluted river water. Municipal wastewater applications account for about 60% of installed capacity and industrial wastewater plants about 30%, and the rest are for polluted river water treatment, of which the Wenyu River plant is an example, and other applications.

7.2 Principle of Phosphorus Removal

The removal of phosphorus from wastewater involves the incorporation of phosphate into TSS and the subsequent removal of those solids. Phosphorus can be incorporated into either biological solids (e.g. microörganisms) or chemical precipitates. The fundamentals of biological phosphorus removal are considered in Chapter 3 and this section. The removal of phosphorus in chemical precipitates is introduced in this section. The topics to be considered include (1) the chemistry of phosphate precipitation, (2) strategies for phosphorus removal, (3) phosphorus removal using metal salts and polymers, and (4) phosphorus removal using lime.

7.2.1 Principle of Chemical Phosphorus Removal

1. Chemistry of Phosphate Precipitation

The chemical precipitation of phosphorus is brought about by the addition of the salts of multivalent metal ions that form precipitates of sparingly soluble phosphate. The multivalent metal ions used most commonly are calcium [Ca (Ⅱ)], aluminum [Al (Ⅲ)], and iron [Fe (Ⅲ)]. Polymers have been used effectively in conjunction with alum and lime as flocculant aids. Because the chemistry of phosphate precipitation with calcium is quite different than with aluminum and iron, the two different types of precipitation are considered separately in the following discussion.

(1) Phosphate precipitation with calcium. Calcium is usually added in the form of lime $Ca(OH)_2$. The reaction is as following:

$$10Ca^{2+} + 6PO_4^{3-} + 2OH^- \leftrightarrow Ca_{10}(PO_4)_6(OH)_2 \tag{7-3}$$

Because of the reaction of lime with the alkalinity of the wastewater, the quantity of lime required will, in general, be independent of the amount of phosphate present and will depend primarily on the alkalinity of the wastewater. The quantity of lime required to precipitate the phosphorus in wastewater is typically about 1.4 to 1.5 times the total alkalinity expressed as $CaCO_3$. Because a high pH value is required to precipitate phosphate, coprecipitation is usually not feasible.

(2) Phosphate precipitation with aluminum and iron. The basic reactions involved in the precipitation of phosphorus with aluminum and iron are as follows.

Phosphate precipitation with aluminum:

$$Al^{3+} + H_nPO_4^{3-n} \leftrightarrow AlPO_4 + nH^+ \tag{7-4}$$

Phosphate precipitation with iron:

$$Fe^{3+} + H_nPO_4^{3-n} \leftrightarrow FePO_4 + nH^+ \tag{7-5}$$

Because of the many competing reactions, Equations (7-4) and (7-5) cannot be used to estimate the required chemical dosages directly. Therefore, dosages are generally established on the basis of bench-scale tests and occasionally by full-scale tests, especially if polymers are used.

2. The Addition Place of Phosphate Precipitation

(1) Metal salt addition to primary sedimentation tanks. When aluminum or iron salts are added to untreated wastewater, they react with the soluble orthophosphate to produce a precipitate. Organic phosphorus and polyphosphate are removed by more complex reactions and by adsorption onto floc particles. The insolubilized phosphorous, as well as considerable quantities of BOD and TSS, are removed from the system as primary sludge. Adequate initial mixing and flocculation are necessary upstream of primary facilities, whether separate basins are provided or existing facilities are modified to provide these functions. Polymer addition may be required to aid in settling. The exact application rate is determined by onsite testing, and varies with the characteristics of the wastewater and the desired phosphorus removal.

Lime has been used customarily either as a precipitant in the primary sedimentation tanks or following secondary treatment clarification.

Both low and high lime treatment can be used to precipitate a portion of the phosphorus (usually about 65% to 80%). When lime is used, both the calcium and the hydroxide react with the orthophosphorus to form an insoluble hydroxyapatite [$Ca_5(OH)(PO_4)_3$]. A residual phosphorus level of 1.0mg/L can be achieved with the addition of effluent filtration facilities to which chemicals can be added. In the high lime system, sufficient lime is added to raise the pH to about 11. After precipitation, the effluent must be recarbonated before biological treatment. In activated-sludge systems, the pH of the primary effluent should not exceed 9. 5 or 10. 0; higher pH values can result in biological process upsets. In the trickling filter process, the carbon dioxide generated during treatment is usually sufficient to lower the pH without recarbonation. The dosage for low lime treatment is usually in the range of 75 to 250mg/L as $Ca(OH)_2$ at pH values of 8. 5 to 9. 5.

(2) Metal salt addition to secondary treatment. Metal salts can be added to the untreated wastewater, in the activated-sludge aeration tank, or the final clarifier influent channel. In trickling filter systems, the salts are added to the untreated wastewater or to the filter effluent. Multipoint additions have also been used. Phosphorus is removed from the liquid phase through a combination of precipitation, adsorption, exchange, and agglomeration, and removed from the process with either the primary or secondary sludges, or both. Theoretically, the minimum solubility of $AlPO_4$ occurs at about pH 6. 3, and that of $FePO_4$ occurs at about pH 5. 3; however, practical applications have yielded good phosphorus removal anywhere in the range of pH 6. 5 to 7. 0, which is compatible with most biological treatment processes.

The use of ferrous salts is limited because they produce low phosphorus concentrations or levels only at high pH values. In low-alkalinity water, either sodium aluminate and alum or ferric plus lime, or both can be used to maintain the pH higher than 5. 5 improved settling and lower effluent BOD result from chemical addition, particularly if polymer is also added to the final clarifier.

Lime can be added to the waste stream after biological treatment to reduce the level of phosphorus and TSS. When lime is used, the principal variables controlling the dosage are the degree of removal required and the alkalinity of the wastewater. The operating dosage must usually be determined by onsite testing.

The use of lime for phosphorus removal is declining because of (1) the substantial increase in the mass of sludge to be handled compared to metal salts, and (2) the operation and maintenance problems associated with the handling, storage, and feeding of lime.

(3) Metal salt and polymer addition to secondary clarifiers. In certain cases, such as trickling filtration and extended aeration activated-sludge processes, solids may not flocculate and settle well in the secondary clarifier. This settling problem may become acute in plants that are over loaded. The addition of aluminum or iron salts will cause the precipitation of metallic hydroxides or phosphates, or both. Aluminum and iron salts, along with certain organic polymers, can also be used to coagulate colloidal particles and to improve removals on filters, the resultant coagulated colloids and precipitates will settle readily in the secondary clarifier, reducing the TSS in the effluent and effecting phosphorus removal. Dosages of aluminum and iron salts usually fall in the range of 1 to 3 metal ion/phosphorus on a molar ration basis if the residual phosphorus in the secondary effluent is greater than 0. 5mg/L. To achieve phosphorus concentrations or levels below 0. 5mg/L, significantly higher metal salt dosages and filtration will be required.

Polymers should not be subjected to insufficient or excessive mixing, because the process efficiency will diminish, resulting in poor settling and thickening characteristics.

7.2.2 The Principle of Biological Phosphorous Removal

At present, people have the following consensus on the physiological characteristics of phosphorus accumulating microorganisms:

(1) Biological phosphorus removal is mainly completed by a class of microorganisms collectively referred to as phosphorus accumulating bacteria, which belong to heterotrophic bacteria, such as Acinetobacter, Aeromonas, Pseudomonas and Corynebacterium. The generation period of phosphorus accumulating bacteria is short, and phosphorus removal is realized by removing excess sludge. Therefore, in order to ensure good phosphorus removal effect, the system must operate with short sludge age.

(2) Under anaerobic conditions, phosphorus accumulating bacteria hydrolyze polyphosphate in cells to orthophosphate, release extracellular and obtain energy; The energy released by polyphosphate hydrolysis is used to absorb easily degradable COD in sewage, such as volatile fatty acids; Synthetic energy storage material β-Hydroxybutyrate (PHB), etc.

(3) Under aerobic conditions, phosphorus accumulating bacteria oxidize PHB stored in cells with free oxygen as electron acceptor, use the energy generated by this reaction to absorb phosphate excessively from sewage, synthesize high-energy substances ATP and phosphorus accumulating, and store phosphorus accumulating in cells as storage. The amount of aerobic phosphorus absorption is greater than that of anaerobic phosphorus release. The purpose of efficient phosphorus removal can be achieved through the removal of excess sludge.

(4) Under aerobic conditions, the content of PHB in cells decreases exponentially with time; Under anaerobic conditions, it increases linearly, and the increase of PHB is linearly related to the decrease of intracellular phosphorus accumulation. Under certain conditions, the more complete the anaerobic effective phosphorus release of phosphorus accumulating bacteria, the greater the phosphorus uptake under aerobic conditions.

(5) Some phosphorus accumulating bacteria have nitrogen removal function. Under the condition of no free oxygen, nitrate can be used as electron acceptor to reduce nitrate to N_2 or N_xO_y, and can also absorb a large amount of phosphorus. When nitrate is mixed into the anaerobic section, part of the easily degradable carbon source is used by denitrification, which has an adverse impact on the phosphorus release of phosphorus accumulating bacteria.

(6) The degree of anaerobic phosphorus released by phosphorus accumulating bacteria is closely related to the type of substrate. Generally, the substrate that can be directly used is mainly short chain volatile fatty acids, and other substrates need to be transformed into short chain volatile fatty acids before they can be used.

Over the long time, several biological suspended growth processes have been used to accomplish biological phosphorus removal, and they all include the basic steps of an anaerobic zone followed by an aerobic zone. The modifications of the basic process include (1) combining the anaerobic/aerobic sequence with various biological nitrogen-removal designs, (2) recycling mixed liquor to the anaerobic zone from a downstream anoxic zone instead of only from the secondary clarifier underflow, (3) adding volatile fatty acids to the anaerobic zone as either acetate or a liquid stream from a fermentation reactor processing primary clarifier sludge, and (4) using multiple-staged anaerobic and aerobic reactors.

The alternating exposure to anaerobic conditions can be accomplished in the main biological treatment process, or "mainstream", or in the return sludge stream, or "sidestream". Several mainstream biological phosphorus-removal processes and one sidestream process, Phostrip, are described in next section. The Phostrip process combines biological and chemical processes for phosphorus removal. Also included in next section is process design considerations, process control, analysis of biological phosphorus-removal performance, design parameters, and process selection considerations. The first "mainstream" biological phosphorus-removal process that was included with biological nitrogen removal is the Bardenpho process in next section.

7.3 脱氮工艺

由上可知，生物脱氮过程中，废水中的有机氮及氨氮经过氨化作用、硝化反应、反硝化反应，最后可转化为氮气排放，所以工程上应设置相应的好氧硝化段和缺氧反硝化段达到生物脱氮的目的。

硝化作用由两类不同的硝化细菌分工进行：亚硝酸菌（又称氨氧化菌）负责将氨氧化为亚硝酸，硝酸菌（又称亚硝酸氧化菌）负责将亚硝酸氧化为硝酸。这两类细菌都是革兰氏染色阴性，不生芽孢的球状或短杆状的细菌，有强烈的好氧性；适宜于中性或碱性环境，不能在强酸性条件下生长。大多数为专性化能自养型，不能在有机培养基上生长；少数为兼性自养型，也能在某些有机培养基上生长。到目前为止，水族上的硝化细菌已经发展到第五代。

1. 第一代硝化细菌

第一代硝化细菌主要由亚硝化单胞菌和硝化杆菌等自养菌组成，生长周期长，其平均代时（即细菌繁殖一代所需要的时间）在10h以上。产品为液体，杂菌较多，有恶臭味。目前市场上已不常见。

2. 第二代硝化细菌

第二代硝化细菌实际上是由能降低水体中氨氮的光合细菌组成的，因为是自然水体的土著菌种，适应性强，其平均代时在3h以上。产品为红色液体，杂菌较多，有腥臭味。

3. 第三代硝化细菌

第三代硝化细菌指由芽孢杆菌纯种发酵后的芽孢休眠体组成的淡乳白色液体，有一定的降氨氮和清水功能，芽孢的萌发需要24h以上，其平均代时跟大多数异养菌一样，在30min左右。产品为淡乳白色液体，无味或有淡腥味，杂菌很少。

4. 第四代硝化细菌

第四代硝化细菌指由芽孢杆菌和乳酸菌混合发酵后冷冻干燥的粉剂，也称 EM 菌，白色的是精制品，菌含量较高，杂质少，棕褐色为直接干燥的产物，含有培养基等杂质，菌含量相对较低，有一些产品复合了酶制剂。产品为粉剂（或胶囊）和片剂，无杂菌，无味或淡腥味。

5. 第五代硝化细菌

第五代硝化细菌也称产酶硝化细菌，是由产酶异养硝化菌、产酶芽孢杆菌、好氧反硝化菌、乳酸菌、放线菌等分别发酵，经微胶囊化工艺进行包被后冷冻干燥的粉剂经科学配比而成的。微胶囊技术包被过的菌种活力强，保质期长，可以抵抗低浓度药物和自来水中氯的损伤。其平均代时在 25min 左右。产品有白色粉剂（或胶囊）和片剂，无杂菌，无味或淡甜味，可食用。

生物脱氮技术的开发是在 20 世纪 30 年代发现生物滤床中的硝化、反硝化反应开始的。但其应用还是在 1969 年美国的 Barth 提出三段生物脱氮工艺之后。下面是几种典型的生物脱氮工艺介绍。

7.3.1 三段生物脱氮工艺

该工艺是将有机物氧化、硝化及反硝化段独立开来，每一部分都有其自己的沉淀池和各自独立的污泥回流系统，并分别控制在适宜的条件下运行，处理效率高。其流程如图 7-2 所示。

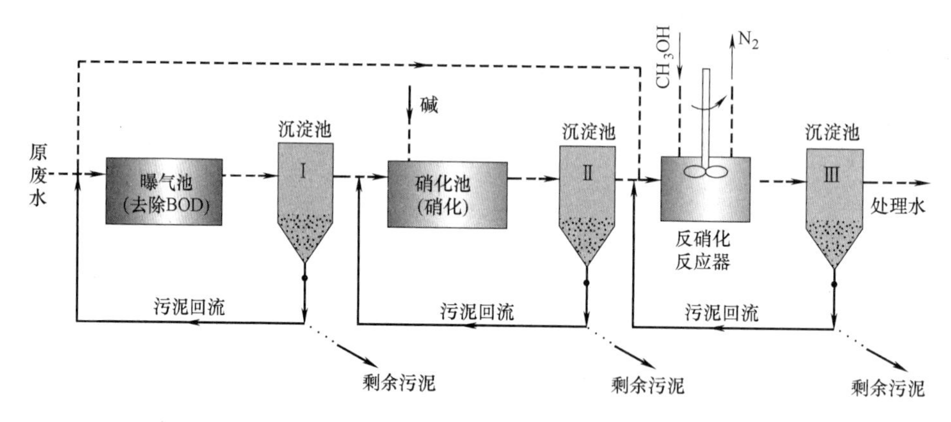

图 7-2 三段生物脱氮工艺

由于反硝化段设置在有机物氧化和硝化段之后，主要靠内源呼吸利用碳源进行反硝化，效率很低，所以必须在反硝化段投加碳源保证高效稳定的反硝化反应。随着对硝化反应机理认识的加深，将有机物氧化和硝化合并成一个系统以简化工艺，从而形成两段生物脱氮工艺（图 7-3）。各段同样有其自己的沉淀及污泥回流系统。除碳和硝化作用在一个反应器中进行时，设计的污泥负荷率要低，水力停留时间和污泥龄要长，否则，硝化作用不完全。在反硝化段仍需要外加碳源维持反硝化的顺利进行。

7.3.2 前置缺氧-好氧 A_N/O 生物脱氮工艺

该工艺于 20 世纪 80 年代初开发，其工艺流程、运行工况如图 7-4、图 7-5 所示。该工艺将反硝化段设置在系统的前面，因此，又称为前置式反硝化生物脱氮系统，是目前较

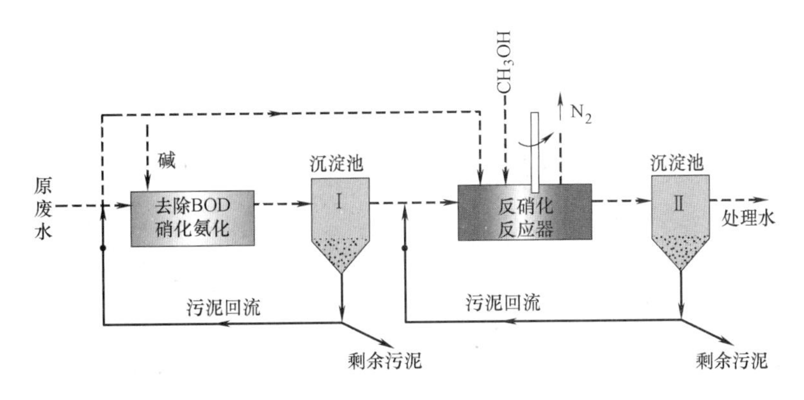

图 7-3 补充外碳源的两段硝化反硝化工艺

为广泛采用的一种脱氮工艺。反硝化反应以污水中的有机物为碳源，曝气池混合液中含有大量硝酸盐，通过内循环回流到缺氧池中，在缺氧池内进行反硝化脱氮。

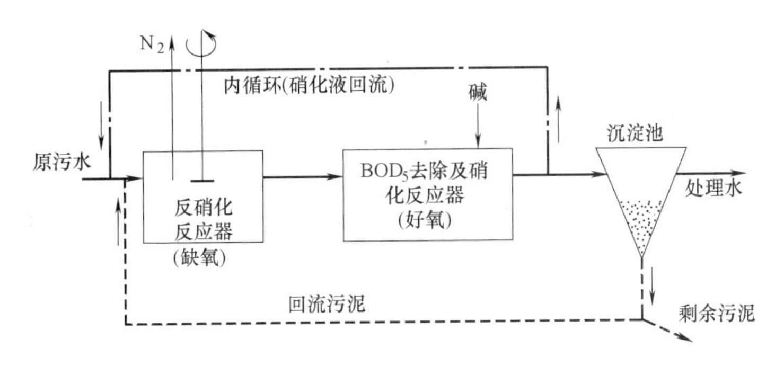

图 7-4 前置缺氧—好氧生物脱氮工艺

前置缺氧反硝化具有以下特点：反硝化产生碱度补充硝化反应之需，约可补偿硝化反应中所消耗的碱度的 50%；利用原污水中有机物，无需外加碳源；利用硝酸盐作为电子受体处理进水中有机污染物，这不仅可以节省后续曝气量，而且反硝化细菌对碳源的利用更广泛，甚至包括难降解有机物；前置缺氧池可以有效控制系统的污泥膨胀。该工艺流程简单，因而基建费用及运行费用较低，对现有设施的改造比较容易，脱氮效率在 70%左右，但由于出水中仍有一定浓度的硝酸盐，在二沉池中，有可能进行反硝化反应，造成污泥上浮，影响出水水质。典型的二沉池参数如图 7-6 所示，二沉池污泥上浮如图 7-7 所示。

图 7-5 A_N/O 生物脱氮工艺的运行工况

7.3.3 Bardenpho 脱氮工艺

它是在 A_N/O 脱氮工艺的基础上又增设缺氧池Ⅱ和好氧池Ⅱ，缺氧池Ⅱ可对好氧池

图 7-6　典型的二沉池参数

图 7-7　二沉池污泥上浮

Ⅰ流入的混合中的硝态氮进行反硝化反应，经第一段处理脱氮大体完成，进一步提高脱氮效率，废水进入第二段反硝化反应器利用内源呼吸进行反硝化，好氧池Ⅱ可吹脱水中氮气，提高污泥沉降性能，防止在二沉池中发生污泥上浮现象。Bardenpho 脱氮工艺（图 7-8）具有两次反硝化过程，脱氮效率在 90%～95%之间。

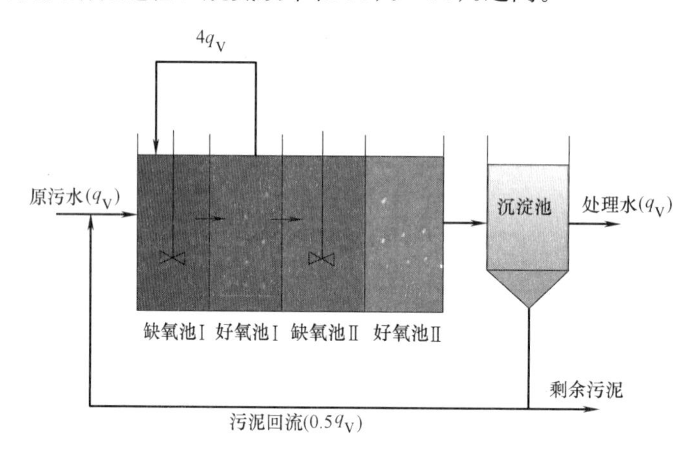

图 7-8　Bardenpho 脱氮工艺

7.3.4　同步硝化反硝化过程

同步硝化反硝化过程指在没有明显独立设置缺氧区的活性污泥法处理系统内总氮被大量去除的过程。对同步硝化反硝化过程的机理解释主要有以下 3 个方面：

（1）反应器溶解氧分布不均理论：在反应器的内部，由于充氧不均衡，混合不均匀，形成反应器内部不同部分的缺氧区和好氧区，分别为反硝化细菌和硝化细菌的作用提供了优势环境，造成硝化和反硝化作用的同时进行。除了反应器不同空间内的溶解氧不均外，反应器在不同时间点的溶解氧变化也可认为是同步硝化反硝化过程。

图 7-9 为一种氧化沟处理系统原理图，由一系列同心的圆形或椭圆形廊道组成，污水和回流污泥由最外圈沟渠进入，然后依次进入内圈沟渠，最后由位于中心的沟渠进入二沉池进行泥水分离。三个廊道的溶解氧分别控制为 0～0.3mg/L、0.5～1.5mg/L、2～

3mg/L，通过控制曝气强度，外圈廊道的供氧速率与渠道内耗氧速率相近，保证混合液的硝化反应，同时因为溶解氧浓度低，反硝化细菌可以利用硝酸盐作为电子受体进行反硝化反应。氮素在外圈廊道的反应过程是一个同步硝化反硝化过程。常用的有转碟曝气器和转刷曝气器。

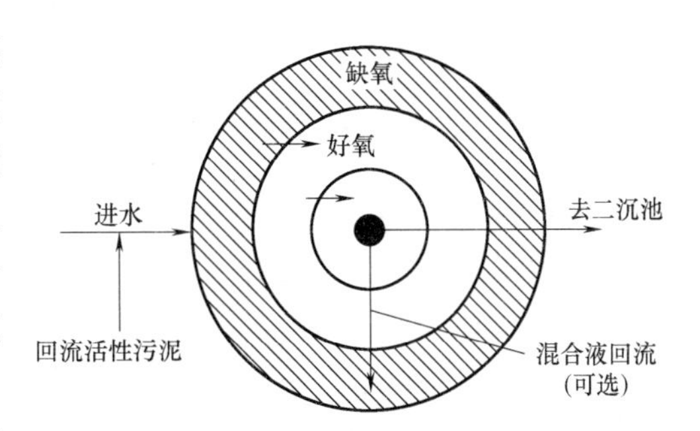

图 7-9 控制溶解氧浓度的同步硝化反硝化（Orbal 氧化沟）

（2）缺氧微环境理论：缺氧微环境理论被认为是同步硝化反硝化发生的主要原因之一。其基本观点认为：在活性污泥的絮体中，从絮体表面至其内核的不同层次上，由于氧传递的限制原因，氧的浓度分布是不均匀的，微生物絮体外表面氧的浓度较高，内层浓度较低。在生物絮体颗粒尺寸足够大的情况下，可以在菌胶团内部形成缺氧区，在这种情况下，絮体外层好氧硝化细菌占优势，主要进行硝化反应，内层反硝化细菌占优势，主要进行反硝化反应（图 7-10）。除了活性污泥絮体外，一定厚度的生物膜中同样可存在溶解氧梯度，使得生物膜内层形成缺氧微环境。

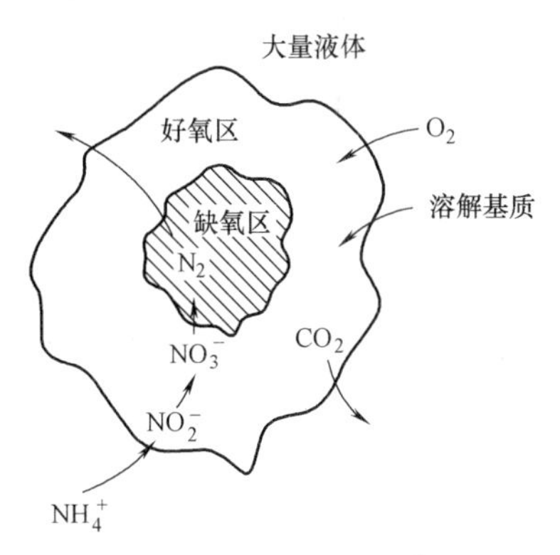

图 7-10 活性污泥内部的缺氧微环境

（3）微生物学解释：传统理论认为硝化反应只能由自养菌完成，反硝化只能在缺氧条件下进行，有研究已经证实存在好氧反硝化细菌和异养硝化细菌。在好氧条件下很多反硝化细菌可以进行氨氮硝化作用。在低浓度氧状态下，硝化细菌和亚硝酸细菌可以进行反硝化作用。

如果刻意追求同步硝化反硝化的效果，则要求更大的反应池体积，且对运行操作管理的要求更高。

在诸多的生物脱氮工艺中，目前前置缺氧反硝化使用较为普遍。随着生物脱氮技术的发展，新的工艺不断被研究开发，同时，人们将生物脱氮与除磷工艺相结合形成了许多新的生物脱氮除磷处理工艺。

7.3.5 新型生物脱氮工艺

传统生物脱氮工艺需要消耗大量的溶解氧、碳源，造成较高的运行成本，同时硝化与反硝化反应的进行存在相互制约的关系：在有机物大量存在的情况下，自养硝化菌对氧气和营养物的竞争力不如好氧异养菌，无法占据主导地位；反硝化需要有机物作为电子供体，但是硝化过程去除了大量的有机物，导致反硝化过程中碳源缺乏，所以为平衡两单元的不同需求，发展出多种生物脱氮方法相结合的工艺。

传统的生物脱氮工艺主要依靠调整工艺流程缓解硝化菌反应环境和反硝化菌反应环境之间存在的矛盾。如果硝化反应阶段在前，则需要外加电子供体，例如甲醇等物质，提高运行费用；如果硝化反应阶段在后，则需要将硝化废水回流，容易产生污泥上浮并且需要提高回流比以获得更高的去除率。这个矛盾在处理氨氮浓度较低的市政废水中尚不明显，但在处理垃圾渗滤液、畜牧废水等高浓度氨氮废水时，极大地限制了系统脱氮效率。

近年来通过理论研究和实践创新，国内外开发了一些超越传统生物脱氮理论的生物脱氮方法，如SND工艺、由亚硝酸型硝化反硝化技术发展的SHARON工艺、由厌氧氨氧化技术发展的ANAMMOX工艺、SHARON-ANAMMOX组合工艺等。

1. 同步硝化反硝化（SND）脱氮工艺

根据传统生物脱氮理论，脱氮途径一般包括硝化和反硝化两个阶段，硝化和反硝化两个过程需要在两个隔离的反应器中进行，或者在时间或空间上造成交替缺氧和好氧环境的同一个反应器中；实际上，较早时期，在一些没有明显的缺氧及厌氧段的活性污泥工艺中，人们就曾多次观察到氮的非同化损失现象，在曝气系统中也曾多次观察到氮的消失。在这些处理系统中，硝化和反硝化反应往往发生在同样的处理条件及同一处理空间内，因此，这些现象被称为同步硝化/反硝化（SND）。

对于各种处理工艺中出现的SND现象已有大量的报道，包括生物转盘、连续流反应器以及序批式SBR反应器等。与传统硝化-反硝化处理工艺比较，SND能有效地保持反应器中pH稳定，减少或取消碱度的投加；减少传统反应器的容积，节省基建费用；对于仅由一个反应池组成的序批式反应器来讲，SND能够降低实现硝化-反硝化所需的时间；曝气量的节省，能够进一步降低能耗。

因此，SND系统提供了今后降低投资并简化生物除氮技术的可能性。

2. 短程硝化脱氮SHARON工艺

SHARON（single reactor for high ammonia removal over nitrite）即亚硝化脱氮工艺，是荷兰Delft技术大学1997年提出并开发的一种新型生物脱氮技术。其基本原理是在同一个反应器内，在有氧的条件下，自养型亚硝酸菌将NH_4^+转化为NO_2^-，然后在缺氧的条件下，异养型反硝化菌以有机物为电子供体，以NO_2^-为电子受体，将NO_2^-转化为N_2。

其理论基础是亚硝酸型硝化反硝化技术，生化反应式可用式（7-6）表示：

$$NH_4^+ + 0.75O_2 + HCO_3^- \rightarrow 0.5NH_4^+ + 0.5NO_2^- + CO_2 + 1.5H_2O \quad (7\text{-}6)$$

该工艺的关键是如何将氨氧化控制在亚硝酸阶段，并持久维持较高浓度的亚硝酸盐积累。由于硝化过程中的两类细菌（亚硝酸菌和硝酸菌）的生长特性不同，对环境的要求也不同，这为将硝化控制在亚硝化阶段提供了条件。

SHARON工艺使用单个无需污泥停留的完全混合反应器（CSTR）来实现，在较短的水力停留时间和30～35℃的条件下，利用高温下硝酸菌的活性比亚硝酸菌的活性低，同时利用硝酸菌的水力停留时间大于亚硝酸菌的水力停留时间，使水力停留时间介于两者之间，从而通过“洗泥”的方式进行种群筛选，产生大量的亚硝酸菌，淘汰硝酸菌。经过小试、中试，第一个运用SHARON工艺的Dokhaven污水处理场于1998年初在荷兰鹿特丹建成并投入运行。该SHARON的进水氨氮质量浓度为1g/L，进水氨氮总量为1200kg/d，氨氮的去除率为85%。SHARON工艺适用于高浓度氨（500mg/L）废水的处理，尤其适用于具有脱氨要求的预处理或旁路处理。

SHARON工艺与传统的脱氮工艺相比，具有能够节省25%的氧气、节省40%的碳源、污泥产量少、反应器容积减少、反应时间短等优点。同时，它也存在一些问题，如反应时较高的温度不适合城市污水的处理，仅比较适合处理污泥消化上清液和垃圾渗滤液等高氨高温废水，适合 C/N 较低的废水，亚硝化产物 NO_2^- 是致癌、致畸、致突变物质，对受纳水体和人体健康有害。

3. ANAMMDX 工艺

ANAMMOX（anaerobic ammonium oxidation）即厌氧氨氧化工艺，是由荷兰 Delft 大学 1990 年提出的一种新型脱氮工艺。在厌氧的条件下，微生物以 NH_4^+ 为电子供体，NO_2^- 为电子受体，把 NH_4^+、NO_2^- 转化为 N_2。其生化反应式可用式（7-7）表示：

$$NH_4^+ + NO_2^- \rightarrow N_2 + 2H_2O \tag{7-7}$$

Graaf 等通过同位素 15N 示踪研究，提出厌氧氨氧化可能的代谢途径，如图 7-11 所示。他认为 ANAMMOX 是通过生物氧化的途径实现的，过程中最可能的电子受体是羟胺（NH_2OH），而羟胺本身是由亚硝酸盐产生的。

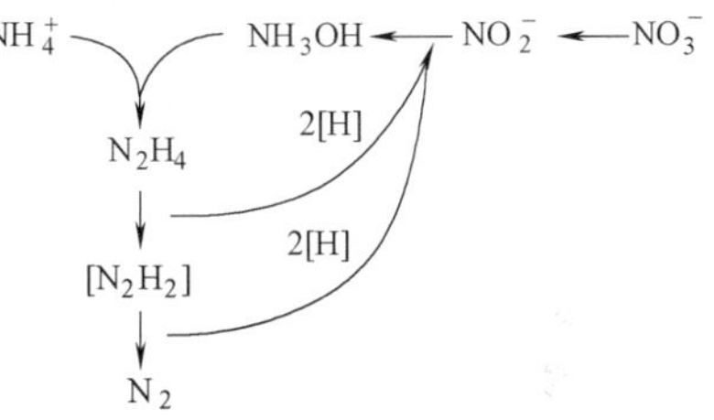

图 7-11　Graaf 提出的 ANAMMOX 工艺的可能途径

ANAMMOX 工艺的影响因素主要集中在系统环境对 Anammox 菌的抑制。主要的影响因素包括反应器的生物量、基质浓度、pH、温度、水力停留时间和污泥龄等。ANAMMOX 工艺具有不少突出的优点：相对传统的脱氮过程，耗氧下降 62.5%；不需外加碳源，节约成本；不需调节 pH 降低运行费用。但该工艺还存在以下几个方面的问题：工艺还没有实现实用化和长期稳定运行；Anammox 细菌生长缓慢，启动时间长，为保持反应器内足够多的生物量，需要有效地截留污泥等。荷兰的研究者们于 2002 年通过数学模型模拟设计出世界上第一个具有生产性规模的 ANMAMOX 反应器，该反应器建在荷兰鹿特丹 Dokhaven 污水处理厂内，主要用于污泥消化液的脱氮处理。

4. SHARON-ANAMMOX 工艺

SHARON-ANAMMOX 工艺即为 SHARON 和 ANAMMOX 的组合工艺。SHARON 作为硝化反应器，在此反应器内含 NH_4^+ 的污水中约 50%的 NH_4^+ 氧化成 NO_2^-；ANAMMOX 作为反硝化反应器，含 NH_4^+ 和 NO_2^- 的 SHARON 反应器的出水作为此反应器的进水，在此反应器内，厌氧条件下 NH_4^+ 和 NO_2^- 被转化为 N_2 和 H_2O。生化反应式如式（7-8）：

$$NH_4^+ + 0.75O_2 + HCO_3^- \rightarrow 0.5N_2 + CO_2 + 2.5H_2O \tag{7-8}$$

典型的 SHARON-ANAMMOX 工艺流程如图 7-12 所示。

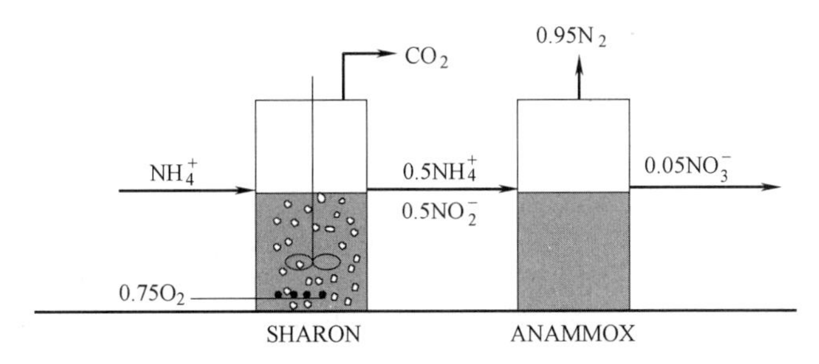

图 7-12　SHARON-ANAMMOX 工艺流程

SHARON-ANAMMOX工艺中反应的主要控制条件为温度、碱度和水力停留时间；同时，ANAMMOX反应器中不得有溶解氧的存在。SHARON-ANAMMOX工艺中发挥作用的细菌主要为氨氧化菌和Anammox菌，两者均为自养型细菌，因此，该工艺无需外加碳源；同时还可以节约氧气约50%，污泥产量低，可以节约90%以上的运行成本，具有很好的应用前景。SHARON-ANAMMOX工艺主要适用于处理污泥上清液和高氨氮、低碳源工业废水。对污泥上清液而言，应用此工艺时并不需要调节pII，因为污泥上清液中含有HCO_3^-，当一半的NH_4^+被转化后，污水中的碱度也几乎被耗光，导致反应器中pH下降，从而抑制硝化反应的进行，使SHARON反应器的出水中NH_4^+/NO_2^-保持在1.0左右，为ANAMMOX工艺中反应的发生创造条件。

世界上第一个生产性SHARON-ANAMMOX工艺已于2002年6月在荷兰鹿特丹Dokhaven污水处理厂正式运行，主要用于处理污泥消化上清液。

新的生物脱氮工艺相对于传统脱氮工艺，具有明显的优势，如降低供氧能耗、无需外加碳源、减少反应器容积、节省运行费用等。新工艺中反应的发生往往需要特定的条件，如较高的温度，一定的pH，低碳源、高氨氮的进水等，这通常不适于处理常规的生活污水，而对特殊的废水如污泥消化上清液和垃圾渗滤液等，则具有良好的处理效果。

7.4 生物脱氮过程的影响因素

7.4.1 硝化过程影响因素

（1）溶解氧浓度

硝化细菌为了获得足够的能量用于生长，必须氧化大量的NH_4^+和NO_2^-，氧是硝化反应过程的电子受体，反应器内溶解氧含量的高低，必将影响硝化反应的进程，在硝化反应的曝气池内，溶解氧含量不得低于1mg/L，多数学者建议溶解氧应保持在1.2～2.0mg/L。

（2）碱度

硝化反应过程释放H^+，使pH下降，为保持适宜的pH，应当在污水中保持足够的碱度，以调节pH的变化，1g氨态氮（以N计）完全硝化，需碱度（以$CaCO_3$计）7.14g。

$$NH_4^+ + 2HCO_3^- + 2O_2 \rightarrow NO_3^- + 2CO_2 + 3H_2O \tag{7-9}$$

（3）pH

硝化细菌对pH的变化十分敏感，最佳pH为8.0～8.4，在最佳pH条件下，硝化细菌的最大比增长速率可以达到最大值。

（4）反应温度

硝化反应的适宜温度是20～30℃，在15℃以下时，硝化反应速度下降，5℃时完全停止。

（5）混合液中有机物含量

硝化细菌是自养菌，有机基质浓度并不是它的增殖限制因素，但它们需要与普通异养菌竞争电子受体，若BOD过高，将使增殖速度较快的异养型细菌迅速增殖，从而使硝化细菌在利用溶解氧作为电子受体方面处于劣势而不能成为优势种属。

(6) 污泥龄

为了使硝化菌群能够在反应器内存活并繁殖，微生物在反应器内的固体平均停留时间（污泥龄）SRT_n，必须大于其最小的世代时间，否则将使硝化细菌从系统中流失殆尽，一般认为硝化细菌最小世代时间在适宜的温度条件下为3d。SRT_n 值与温度密切相关，温度低，SRT_n 取值应相应明显提高。

(7) 重金属及有害物质

除有毒有害物质及重金属外，对硝化反应产生抑制作用的物质还有：高浓度的 NH_4^+-N、高浓度的 NO_x-N、高浓度的有机基质以及络合阳离子等。

7.4.2 反硝化过程影响因素

(1) 碳源

反硝化细菌为兼性异养菌，必须提供有机物作为电子供体，能为反硝化细菌所利用的碳源较多，从污（废）水生物脱氮考虑，可有下列三类：一是原污（废）水中所含碳源，对于城市污水，当原污水 BOD_5/TKN＞3～5时，即可认为碳源充足；二是外加碳源，如市售的甲醇、醋酸钠等，工程中多采用甲醇（CH_3OH），因为甲醇作为电子供体反硝化速率高，被分解后的产物为 CO_2 和 H_2O，不留任何难降解的中间产物；三是利用微生物组织进行内源反硝化。在反硝化反应中，目前面临最大的问题是碳源的浓度，就是污（废）水中可用于反硝化的有机碳源的多少及其可生化程度。

实际上用乙酸钠作为外源性碳源的较多，加入后系统脱氮效率比其他明显要高，并且加入乙酸钠后反硝化还原菌的脱氮效率比加入其他外源性碳源要高。几种碳源的对比见表7-1。

几种碳源的对比　　表7-1

外加碳源种类	优点	缺点
甲醇	1. 运行费用低； 2. 污泥产量小； 3. 氮去除率高	1. 成本相对较高； 2. 响应时间较慢； 3. 有毒害作用
乙醇	1. 无毒； 2. 污泥产量小； 3. 氮去除率较高，与甲醇相似	成本较高
葡萄糖	1. 无毒； 2. 脱氮效率高于甲醇	1. 易引起细菌的大量繁殖，导致污泥膨胀； 2. 增加水中的COD，影响出水水质； 3. 与醇相比更容易导致亚硝态氮积累
乙酸钠	1. 立刻响应反硝化过程（适用于应急处理）； 2. 脱氮效果好	1. 成本相对较高； 2. 污泥产量高

(2) pH

反硝化反应最适宜的pH是6.5～7.5，pH高于8或低于6，反硝化速率将大幅下降。

(3) 溶解氧浓度

反硝化细菌在无分子氧的同时存在硝酸根或亚硝酸根离子的条件下，能够利用这些离子作为电子受体进行呼吸，使硝酸盐还原，如果溶解氧浓度过高，则反硝化细菌将把电子供体提供的电子转交溶解氧以获得更多能量，这时硝酸盐无法得到电子而被还原完成脱氮过程。另一方面，反硝化细菌体内的某些酶系统组分，只有在有氧条件下，才能够合成。这样，反硝化反应宜在缺氧、好氧交替的条件下进行，反硝化时溶解氧浓度应控制在0.5mg/L以下。

（4）温度

反硝化反应的最适宜温度是20～40℃，低于15℃反硝化反应速率降低。为了保持一定的反硝化速率，在低温季节，可采用如下措施：提高生物固体平均停留时间；降低负荷率；提高污水的水力停留时间。

7.5 除磷工艺

生物除磷效率受整个活性污泥工艺以及进水特性的影响。对于污泥龄（*SRT*）较长、向厌氧区输入更多硝酸盐或氧气以及较难生物降解的COD的系统，除磷效率较低。

比较常用的两种具有厌氧/好氧条件的生物除磷装置是A/O（仅厌氧/好氧）和A^2O（厌氧/缺氧/好氧）工艺。交替暴露在厌氧条件下可在主要生物处理工艺（主流）或回流污泥流（侧流）中完成。

较低的运行*SRT*用于防止硝化作用的开始。理想的*SRT*在20℃时为2～3d，在10℃时为4～5d，以便在不进行硝化作用的情况下进行生物除磷。A^2O和UCT（University of Cape Town）工艺是通过生物除磷装置去除硝酸盐的两种基本主流系统。在A^2O工艺中，含有硝酸盐的回流活性污泥循环被导流至厌氧区。在UCT工艺中，回流污泥循环被导流至缺氧区，混合液循环至厌氧区在硝酸盐浓度最低的缺氧区之后进行。UCT和其他类似工艺通常用于硝酸盐浓度相对较低的废水，其中添加硝酸盐会对生物除磷装置性能产生显著影响。

由上可知，生物除磷工艺中，污泥必须交替经过厌氧和好氧过程，所以目前最基本流程为A_p/O工艺，而Phostrip工艺为生物除磷与化学除磷的结合。

7.5.1 A_p/O工艺

A_p/O工艺是由厌氧区和好氧区组成的同时去除污（废）水中有机污染物及磷的处理系统，其流程如图7-13所示。

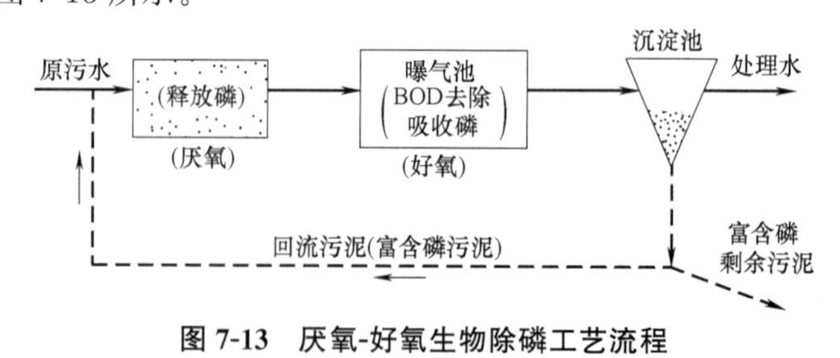

图7-13 厌氧-好氧生物除磷工艺流程

为了使微生物在好氧池中易于吸收磷，溶解氧应维持在2mg/L以上，pH应控制在7～8之间。磷的去除率还取决于进水中的易降解COD含量，一般用其与磷浓度之比表示。据报道，如果比值大于10：1，出水中磷的浓度可降至1mg/L左右。由于微生物吸收磷是可逆的过程，过长的曝气时间及污泥在沉淀池中长时间停留都有可能造成磷的释放。

厌氧-好氧（A_p/O）生物除磷工艺特点主要有以下五个优点：

（1）去除有机物的同时可以生物除磷；

（2）污泥沉降性能好；

（3）用于大型污水处理厂费用较低；

（4）污泥经过厌氧消化达到稳定；

（5）沼气可以回收利用。

厌氧-好氧（A_p/O）生物除磷工艺特点主要有以下三个缺点：

（1）生物脱氮效果差；

（2）用于中、小型污水处理厂费用偏高；

（3）污泥渗出液需要化学除磷。

基于以上优缺点，厌氧-好氧生物除磷工艺一般用于要求除磷但不要求硝化脱氮的大型和较大型污水处理厂。

采用生物除磷处理污水时，剩余污泥宜采用机械浓缩。

7.5.2 Phostrip 工艺

Phostrip 工艺过程将生物除磷和化学除磷结合在一起，在回流污泥过程中增设厌氧释磷池和上清液的化学沉淀处理系统，称为旁路（图 7-14）。一部分富含磷的回流污泥送至厌氧释磷池，释磷后的污泥再回到曝气池进行有机物降解和磷的吸收，用石灰或其他化学药剂对释磷上清液进行沉淀处理。Phostrip 除磷效率不像其他生物除磷系统那样受进水的易降解 COD 的影响，处理效果稳定。

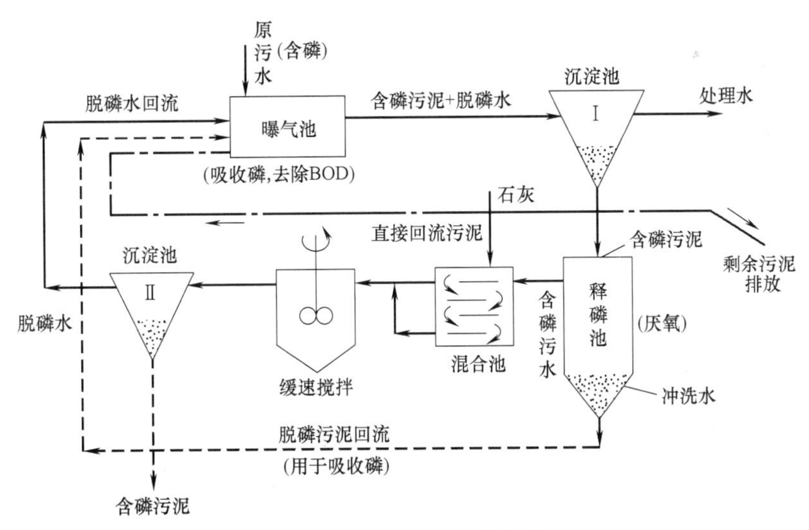

图 7-14 Phostrip 工艺流程

该工艺将在常规的好氧活性污泥法工艺中增设厌氧释磷池和化学沉淀池。工艺流程为：部分回流污泥（约为进水量的 10%～20%）通过旁路进入厌氧池，在厌氧池中的停留时间为 8～12h，使磷由固相中释放，并转移到水中；脱磷后的污泥回流到好氧池中继续吸磷，厌氧池上清液含有高浓度磷（可超过 100mg/L），将此上清液排入石灰混凝沉淀池进行化学处理生成磷酸钙沉淀，该含磷污泥可作为农业肥料，而混凝沉淀池出水应流入初沉池再进行处理。Phostrip 工艺不仅通过高磷剩余污泥除磷，而且还通过化学沉淀除磷。该工艺具有生物除磷和化学除磷双重作用，所以 Phostrip 工艺具有高效脱氮除磷功能。

Phostrip 工艺比较适合于对现有工艺的改造，只需在污泥回流管线上增设少量小规模的处理单元即可，且在改造过程中不必中断处理系统的正常运行。总之，Phostrip 工艺受外界条件影响小，工艺操作灵活，脱氮除磷效果好且稳定，具有以下特点：

（1）生物除磷和化学除磷双重除磷工艺，除磷效果好，出水磷小于 1mg/L；

（2）因为有化学除磷，产泥量高；

（3）SVI<100，污泥易沉淀、浓缩、脱水，肥分高；

（4）可根据 BOD/P 灵活调节污泥回流量及混凝污泥量比例。

（5）工艺流程复杂、运行管理麻烦、处理成本较高。

7.5.3 SBR 工艺

SBR 工艺是将除磷脱氮的各种反应，通过时间顺序上的控制，在同一反应器中完成。其基本运行模式由进水、反应、沉淀、出水和闲置 5 个基本过程组成，从污（废）水流入到闲置结束构成一个周期，在每个周期里上述过程都是在一个设有曝气或搅拌装置的反应器内依次进行的，如图 7-15 所示。

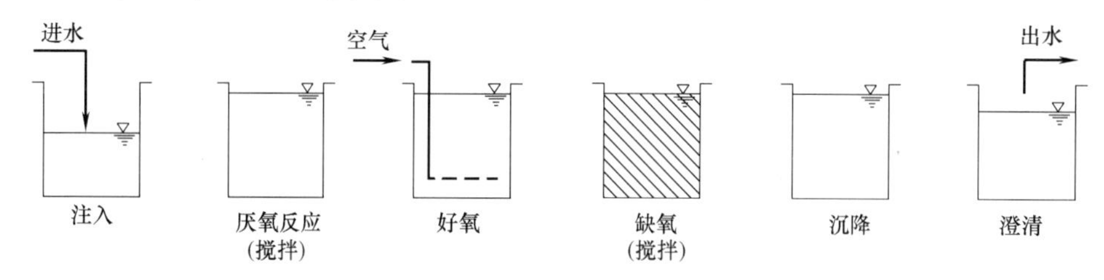

图 7-15　SBR 除磷工艺示意图

如果在序批式反应器（SBR 工艺）运行期间去除了足够的硝酸盐，则可在 SBR 注入期间和之后形成厌氧反应阶段。在经过足够的好氧硝化和硝酸盐产生时间后，开始缺氧运行。或者，可在反应期间使用循环好氧和缺氧周期。因此，在沉淀阶段之前，硝酸盐浓度降至最低，在注入和初始反应期间，几乎没有硝酸盐可用于竞争 COD。因此，厌氧条件发生在注入和初始反应期间，所以聚磷细菌可以吸收和储存 COD，而不是硝酸盐还原细菌消耗 COD。

工艺的一个关键之处是在排出废水之前的沉降期内污泥的沉降速度。使用的沉降时间约为 30min。在沉降之前，在反应期结束时，有机物去除率和产气率较低，从而为固体沉降提供更好的条件。在足够的操作时间后，形成密度较大的颗粒污泥，提高了固液分离率。在试验条件范围内，废水 TSS 范围为 50～100mg/L。温度较低时，出水 TSS 较高。*SRT* 范围为 50～200d。

SBR 工艺与连续流活性污泥工艺相比的优点如下：

（1）工艺系统组成简单，不设二沉池，曝气池兼具二沉池的功能，无污泥回流设备；

（2）耐冲击负荷，在一般情况下［包括工业污（废）水处理］无须设置调节池；

（3）反应推动力大，易于得到优于连续流系统的出水水质；

（4）运行操作灵活，通过适当调节各单元操作的状态可达到脱氮除磷的效果；

（5）污泥沉淀性能好，SVI较低，能有效防止丝状菌膨胀；

（6）该工艺的各操作阶段及各项运行指标可通过计算机加以控制，便于自控运行，易于维护管理。

但是SBR工艺还存在着以下缺点：

（1）更复杂的脱氮除磷操作；

（2）比仅脱氮工艺需要更大的占地面积；

（3）出水水质取决于澄清设施；

（4）设计更复杂；

（5）需要熟练地维护，更适合较小的流量。

7.6 生物除磷过程的影响因素

1. 厌氧环境条件

厌氧释磷要控制厌氧反应器的溶解氧、硝酸盐等电子受体浓度，保证厌氧反应条件。（1）氧化还原电位：Barnard，Shapiro等人研究发现，在批式试验中，反硝化完成后，ORP（氧化还原电位，oxidation-reduction potential）突然下降，随后开始释放磷，释磷时ORP一般小于－150mV；（2）溶解氧浓度：厌氧区如存在溶解氧，兼性厌氧菌就不会启动其发酵代谢，不会产生脂肪酸，也不会诱导放磷，好氧呼吸会消耗易降解有机质；（3）NO_x^-浓度：产酸菌利用NO_x^-作为电子受体，消耗易生物降解有机质，抑制厌氧发酵过程。

所以，如果厌氧池存在溶解氧、硝酸盐等电子受体，在聚磷菌厌氧释磷前，异养菌或反硝化细菌将会利用相当的时间和一定的空间完成溶解氧消耗和反硝化过程，同时必定会损失易生物降解有机物浓度。

2. 有机物浓度及可利用性

碳源的性质对磷释放及其速率影响很大，传统水质指标COD或BOD中的挥发性脂肪酸（VFA）、其他结构简单的易降解的有机物等是聚磷菌最理想的碳源。

3. 污泥龄

因为聚磷菌的内源呼吸衰减速率仅为普通异养菌的1/15～1/10，所以在厌氧释磷条件良好的厌氧-好氧生物除磷处理系统中，延长污泥龄可以增加系统中的聚磷菌含量，从而提高剩余污泥中的含磷量，进水中易降解有机物浓度越高，污泥龄越长，聚磷菌含量越大，这时通过排放剩余污泥达到降磷的效果越显著，因此，脱氮除磷系统应处理好污泥龄的矛盾。

同时还需考虑F/M与SRT、回流比R、水力停留时间HRT、溶解氧DO、BOD_5/TP、pH、温度对除磷效果的影响等。

（1）F/M与SRT

A-O生物除磷工艺是一种高F/M低SRT系统。这是因为磷的去除是通过排放剩余污泥完成的。F/M较高时，SRT较小，剩余污泥排放量也就较多，因而在污泥含磷量一定的条件下，除磷量也就越多。但SRT不能太低，必须以保证BOD_5的有效去除为前提。另外，SRT对污泥的含磷量也有影响，一般认为SRT在7～10d时，污泥中的含磷量最高，但并不意味着必须在这个范围内运行，因为总体还应着眼于总除磷量。有的处理厂发现，当 *SRT* 大于 15d 时，除磷效率在 50%以下，而当 *SRT* 降至 6d 以下时，除磷效率升至 80%以上。

（2）回流比 *R*

总起来看，A-O 除磷系统的 *R* 不宜太低，应保持足够的回流比，尽快将二沉池内的污泥排出，防止聚磷菌在二沉池内遇到厌氧环境发生的释放。在保证快速排泥的前提下，应尽量降低 *R*，以免缩短污泥在厌氧段的实际停留时间，影响磷的释放。已经证明，A-O 除磷系统的污泥沉降性能一般都良好，*R* 在 50%～70%范围内，即可保证快速排泥。而有的处理厂将 *R* 降至 25%，也未发现磷在二沉池大量释放。

（3）水力停留时间 *HRT*

污（废）水在厌氧段的水力停留时间一般在 1.5～2.0h 的范围内。停留时间太短，一是不能保证磷的有效释放，二是污泥中的兼性酸化菌不能充分地将污水中的大分子有机物（如葡萄糖）分解成低级脂肪酸（如乙酸），以供聚磷菌摄取，从而也影响磷的释放。停留时间太长，不但没有必要，还可能产生一些副作用。污（废）水在好氧段的停留时间一般在 4～6h，这样即可保证磷的充分吸收。

（4）溶解氧 DO

厌氧池应尽量保持严格的厌氧状态，实际运行中应控制 DO 在 0.2mg/L 以下。因为聚磷菌只有在严格厌氧状态下，才进行磷的释放，如果存在 DO，则聚磷菌将首先利用 DO 吸收磷或进行好氧代谢，这样就会大幅影响其在好氧段对磷的吸收。大量实践证明，只有保证聚磷菌在厌氧段有效地释放磷，才能使之在好氧段充分地吸收磷，从而保证应有的除磷效果。放磷越多，则吸收越多，放磷量与吸磷量成正比。厌氧状态下，聚磷菌每多释放 1mg 磷，进入好氧状态后就可多吸收 2.0～2.4mg 磷。

好氧段的 DO 应保持在 2.0mg/L 之上，一般控制在 2.0～3.0mg/L 之间。这是因为聚磷菌只有在绝对好氧的环境中才能大量吸收磷。另外，保持好氧段的高氧环境，还可以防止聚磷菌进入二沉池后，由于厌氧而产生磷的释放。

（5）BOD_5/TP

一般认为，要保证除磷效果，应控制进入厌氧段的污（废）水中 BOD_5/TP 大于 20，以保证聚磷菌对磷的有效释放。聚磷菌大多为不动菌属，其生理活动较弱，只能摄取有机物中已分解的部分，只能吃“极可口”的食物，例如乙酸等挥发性脂肪酸。对于 BOD_5 中的大部分有机物，例如固态的 BOD_5 部分、较大的 BOD_5，部分聚磷菌是不能吸收的，甚至对已溶解的葡萄糖，聚磷菌也“懒”得摄取。因而在运行控制中，如能测得 BOD_5 中极易分解的那部分有机物量，将是非常有用的，但是很难办得到。国外一些处理厂运行控制中，常将 $SBOD_5$/TP 作为控制指标，$SBOD_5$ 是溶解性 BOD_5 或过滤性 BOD_5。根据以上分析，采用 $SBOD_5$/TP 控制运行要比单纯采用 BOD_5/TP 准确得多。有些处理厂运行发现，要使出水 TP＜1mg/L，应控制 $SBOD_5$/TP＞10，而要使出水 TP＜0.5mg/L，应控制 $SBOD_5$/TP＞20。

（6）pH

pH 对磷的释放和吸收有不同的影响。在 pH＝4.0 时，磷的释放速率最快，当 pH＞4.0 时，释放速率降低，pH＞8.0 时，释放速率将非常缓慢。在厌氧段，其他兼性菌将部分有机物分解为脂肪酸，会使污水的 pH 降低，从这一点看，对磷释放也是有利的。在 pH 为 6.5～8.5 的范围内，聚磷菌能在好氧状态下有效地吸收磷，且 pH=7.3 左右吸收速率最快。

综上所述，低 pH 有利于磷的释放，而高 pH 有利于磷的吸收，而除磷效果是磷释放和吸收的综合。所以在生物除磷系统中，宜将混合液的 pH 控制在 6.5～8.0 的范围内。当 pH<6.5 时，应向污水中投加石灰，调节 pH。

（7）温度

温度对除磷效果的影响较复杂，目前尚不太清楚。各种研究和不同处理厂的运行结果相差较大，有的甚至得出完全相反的结论。例如，有的处理厂发现除磷效果随温度降低而提高，而有的处理厂则发现随温度降低而降低。一般认为，在 5～35℃的范围内，均能进行正常除磷，因而一般温度的变化不会影响除磷工艺的正常运行。

（8）其他

影响系统除磷效果的还有污泥沉降性能和剩余污泥处理方法等，二沉池溢流带出的悬浮固体几乎与剩余污泥含有相同的磷酸盐含量，出水悬浮固体浓度越高，带出的磷酸盐浓度越高，根据运行数据曲线，如果污泥中磷含量按 5%计算，磷排放标准小于 1.0mg/L，在出水 TSS 为 20mg/L 时就很难达到排放要求。生物除磷系统的剩余污泥如果采用重力浓缩等处理方式，会导致污泥在浓缩池内进行厌氧磷释放，上清液进入污水处理厂内的排水系统而导致磷酸盐在处理系统中进行循环处理。建议尽可能减少贮泥池的容积，并采用带充氧的搅拌设备，或者运用气浮浓缩，或者采用机械浓缩脱水一体化设备，尽量减少污泥处理过程中的磷释放量。对污泥处理过程中的回流上清液进行单独加药沉淀处理，也是减少磷再次进入污（废）水处理系统的一个有效方法。

7.7 联合生物处理工艺

组合好氧、缺氧和厌氧处理工艺包括活性污泥（单级或多级工艺，各种专有工艺）和混合工艺（单级或多级工艺，带有用于生物膜附着生长的填料），用于去除碳质 BOD、硝化、反硝化和除磷（图 7-16）。

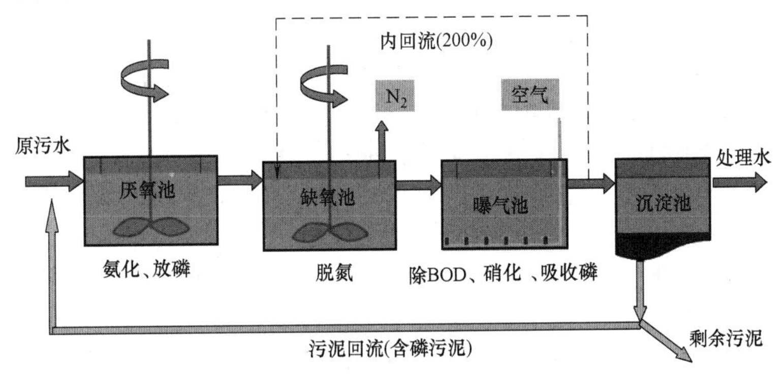

图 7-16　典型的联合生物脱氮除磷工艺流程

传统活性污泥法以去除有机污染物和悬浮物为主要目标，并不考虑对污（废）水中氮和磷的去除，随着水体富营养化的不断加剧和再生水回用率要求的提高，对污水处理厂氮、磷排放浓度和排放总量提出了更高的要求，如何高效地降低污（废）水中的氮和磷成为选择污（废）水处理工艺的一个重要影响因素。

为经济有效地降低污（废）水中的氮和磷，利用生物脱氮除磷的原理，科学家开发了多种生物脱氮除磷的工艺，主要有缺氧 A_N/O（缺氧-好氧）生物脱氮、A_p/O（厌氧-好氧）生物除磷、厌氧-缺氧-好氧（A^2/O）工艺、Bardenpho 工艺、UCT 工艺、Phoredox 工艺以及 SBR 工艺等。

城镇污水处理厂通常需要在一个流程中同时完成脱氮、除磷功能，依据生物脱氮除磷的理论而产生的最基本的工艺是由美国气体产品与化学公司在 20 世纪 70 年代发明的 A^2/O 工艺。近年来，随着对生物脱氮除磷的机理研究不断深入，以及各种新技术、新设备的不断运用，衍生出许多新的生物脱氮除磷工艺，本节主要介绍 A^2/O 工艺与 UCT 工艺两种典型的生物脱氮除磷工艺。

7.7.1 A^2/O 工艺

A^2/O 工艺也称 AAO 工艺，在一个处理系统中同时具有厌氧区、缺氧区、好氧区，能够同时做到脱氮、除磷和有机物的降解。其工艺流程如图 7-17 所示。

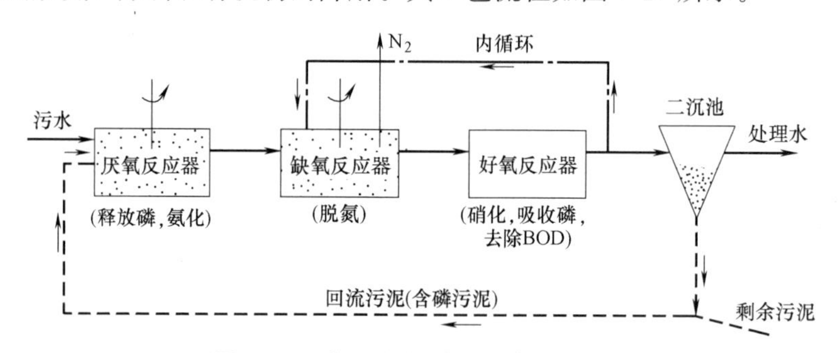

图 7-17　A^2/O 生物脱氮除磷工艺流程

污（废）水进入厌氧反应区，同时进入的还有从二沉池回流的活性污泥，聚磷菌在厌氧环境条件下释磷，同时把易降解的 COD、VFA 转化为 PHB，部分含氮有机物进行氨化。

污（废）水经过厌氧反应器后进入缺氧反应器，该反应器的首要功能是进行脱氮。硝态氮通过混合液内循环由好氧反应器转输过来，通常内回流量为 2～4 倍原污（废）水流量，部分有机物在反硝化细菌的作用下利用硝酸盐作为电子受体而得到降解去除。

混合液从缺氧反应区进入好氧反应区，如果反硝化反应进行基本完全，混合液中的 COD 已基本接近排放标准，在好氧反应区除进一步降解有机物外，主要进行氨氮的硝化和磷的吸收，混合液中硝态氮回流至缺氧反应区，污泥中过量吸收的磷通过剩余污泥排除。

如果污（废）水中可生物降解的有机物很少，则聚磷菌无法正常进行磷的释放，导致好氧段也不能大量地吸收污（废）水中的磷，从而影响除磷的效果。试验证明：进水中溶解性磷与溶解性 BOD_5 之比应小于 0.06，才会有较好的除磷效果。

缺氧段：C/N 较高时，NO_x-N 反硝化速率大，则 HRT＝0.5～1.0h；C/N 较低时，NO_x-N 反硝化速率小，则 HRT＝2.0～3.0h。

对于低 BOD_5 的城市污水，C/N 较低，脱氮率不高。一般来说，污（废）水中

COD/TKN>8，N的总去除率可达80%。

1. A^2/O 的工艺影响因素

（1）污泥龄 *SRT*

污泥龄受硝化和除磷两个方面的影响：一方面硝化反应要求污泥龄 *SRT* 比普通活性污泥工艺时间长；另一方面由于除磷的要求，污泥龄不能过长，A^2/O 工艺中的 *SRT* 一般为15～20d。

（2）溶解氧DO

好氧段DO过高，DO会随污泥回流和混合液回流带至厌氧段与缺氧段，造成厌氧段的厌氧不完全而影响聚磷菌释放磷。而缺氧段DO升高则影响 NO_x-N 的反硝化。相反，好氧段DO下降，则氨氮的硝化速度下降，即氧化速度下降。因此，在好氧段DO以2mg/L左右为好，缺氧段DO≤0.5mg/L，厌氧段DO<0.5mg/L。

（3）有机物负荷率 N_s

好氧段：N_s≤0.18kgBOD_5/(kgMLVSS·d)，否则异氧菌会大幅超过硝化菌，使硝化反应受到抑制；

厌氧段：N_s>0.1kgBOD_5/(kgMLVSS·d)，否则除磷效果会下降。

（4）TKN/MLSS负荷率

过高浓度的氨氮对硝化菌会产生抑制作用，影响其硝化，一般控制TKN/MLSS<0.05kgTKN/(kgMLSS·d)。

（5）污泥回流比 *R* 与混合液回流比 *RN*

R 为25%～100%为宜。*R* 太高，污泥将DO和 NO_x-N 带入厌氧段太多，影响其厌氧状态，使释磷不利；如果 *R* 太低，可能维持不了反应池内污泥正常浓度在2500～3500mg/L之间，影响生化反应速率。缺氧段的脱氮效果对混合液回流比 *RN* 有较大的影响，一般采用 *RN*≥200%。

2. A^2/O 的工艺特点和注意事项

A^2/O 法是一种常用的工艺，优点是可以同时达到去除有机物、脱氧和除磷的多重目的，总水力停留时间少于分别完成各项任务的工艺，A^2/O 工艺流程简单，运行费用较低，且好氧、厌氧交替运行的条件使丝状菌不易生长繁殖，避免了常规活性污泥法的污泥膨胀问题，所以常用于二级污水处理或三级污水处理，以及中水回用。

但传统的 A^2/O 单泥系统，高效脱氮与高效除磷两个过程之间存在着多种矛盾冲突，包括混合生长在同一系统里的聚磷菌、反硝化菌、硝化菌等各自的适宜污泥龄矛盾，碳源竞争的矛盾，硝酸盐及溶解氧的残余干扰等。例如，A^2/O 法的回流污泥全部进入厌氧段，为了维持较低的污泥负荷，要求有较大的回流比才能获得较好的硝化效果，但回流污泥也会将大量的硝酸盐带入厌氧段。当厌氧段存在大量硝酸盐时，反硝化菌会以有机物为碳源进行反硝化，等脱氮完全后才开始磷的厌氧释放，进而使得厌氧段进行磷释放的有效容积大为减少，从而使在脱氧效果较好时，除磷效果较差。反之，如果好氧段硝化作用不好，则回流污泥进入厌氧段的硝酸盐减少，改善了厌氧段的厌氧环境，使磷能充分地进行厌氧释放，所以除磷的效果较好，但由于硝化不完全，故脱氧效果不佳。即 A^2/O 法不可能同时取得脱氮和除磷都较好的双重效果。

3. A^2/O 工艺的运行管理

一般规定进入系统的污（废）水应符合下列要求：

（1）脱氮时，污（废）水中的5d生化需氧量（BOD_5）与总凯氏氮（TKN）之比宜大于4；

（2）除磷时，污（废）水中的BOD_5与总磷（TP）之比宜大于17；

（3）同时脱氮、除磷时，宜同时满足前两项的要求；

（4）好氧池（区）剩余碱度宜大于70mg/L（以碳酸钙$CaCO_3$计）；

（5）当工业废水进水COD超过1000mg/L时，前处理可采用升流式厌氧污泥床反应器（UASB）等厌氧处理措施；

（6）当工业废水进水的BOD_5/COD小于0.3时，前处理需采用水解酸化等预处理措施。

运行管理时需要注意以下问题：

1）外加碳源

当进入反应池废水的BOD_5/TKN小于4时，应在缺氧池中投加碳源。投加碳源量可按式（7-10）确定。

$$\Delta C = 2.86 \Delta N \cdot Q \qquad (7\text{-}10)$$

式中　ΔC——投加的碳源对应的BOD_5量，g/d；

ΔN——硝态氮的脱除量，mg/L；

Q——设计污水流量，m^3/d。

2）化学除磷

当出水总磷不能达到排放标准要求时，宜采用化学除磷作为辅助手段；

最佳药剂种类、投加量和投加点宜通过试验或参照类似工程确定。化学药剂储存罐容量应为理论加药量的4～7d投加量，加药系统应不少于2套，应采用计量泵投加。

化学除磷时应考虑产生的污泥量，污泥增量可参照表7-2设计。

絮凝剂种类及对应的污泥增量　　　　**表7-2**

絮凝剂	投加位置	污泥增量
铝盐或铁盐作絮凝剂	前置投加	40%～75%
铝盐或铁盐作絮凝剂	后置投加	20%～35%
铝盐或铁盐作絮凝剂	同步投加	15%～50%

接触铝盐和铁盐等腐蚀性物质的设备和管道应采取防腐措施。

3）硝化液回流系统

污泥回流设施应采用不易产生复氧的离心泵、混流泵、潜水泵等设施；

回流设施宜分别按生物处理工艺系统中的最大污泥回流比和最大混合液回流比计算确定；

回流设备不应少于2台，并设备用，回流设备宜有调节流量的措施。

4）减少加入厌氧段的回流污泥量，将回流污泥分两点加入，从而减少进入厌氧段的硝酸盐和溶解氧。在保证总的回流比不变（60%～100%）的情况下，加入厌氧段的回流污泥比为10%，这样既可以满足磷的需要，而其余的回流污泥则回流到缺氧段以保证氮的需要。

5）A^2/O法工艺系统中剩余污泥含磷量较高，在其消化过程中磷会重新释放和溶出。同时，由于剩余污泥沉淀性能较好，所以可以取消消化池避免磷的释放，直接浓缩脱水后作为堆肥使用。

6）在硝化好氧段，污泥负荷率应小于0.18kgBOD_5/（kgMLSS·d），而在除磷厌氧段，污泥的负荷率应在0.1kgBOD_5/（kgMLSS·d）以上。

7.7.2 倒置A^2/O工艺与UCT生物脱氮除磷工艺

A^2/O工艺发展至今，为了进一步提高脱氮除磷效果和节约能耗，又有了多种变形和改进的工艺流程。近年来，同济大学研究开发的改进型A^2/O工艺（又称倒置A^2/O工艺，如图7-18所示），由于具有明显的节能和提高除磷效果等优点，在我国一些大、中型城镇污水处理厂的建设和改造工程中得到较为广泛的应用。

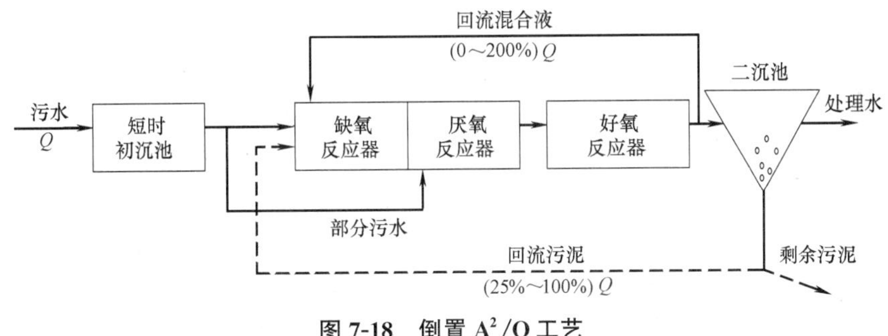

图7-18 倒置A^2/O工艺

该工艺的特点是：采用较短停留时间的初沉池，使一部分进水中的细小有机悬浮固体进入生物反应器，以满足反硝化细菌和聚磷菌对碳源的需要，并使生物反应器中的污泥能达到较高的浓度；整个系统中的活性污泥都完整地经历过厌氧和好氧的过程，因此，排放的剩余污泥都能充分地吸收磷；避免了回流污泥中的硝酸盐对厌氧释磷的影响；由于反应器中活性污泥浓度较高，从而促进好氧反应器中的同步硝化、反硝化，因此，可以用较少的总回流量（污泥回流和混合液回流）达到较好的总氮去除效果。

又如，UCT工艺（图7-19）为南非开普敦大学研究开发的一种工艺，其基本思想是减少回流污泥中的硝酸盐对厌氧区的影响，所以与A^2/O不同的是，UCT工艺的回流污泥回到缺氧区而不是厌氧区，从缺氧区出来的混合液硝酸盐含量较低，回流到厌氧区后为污泥的释磷反应提供了最佳的条件。由于混合液悬浮固体浓度较低，厌氧区停留时间较长。

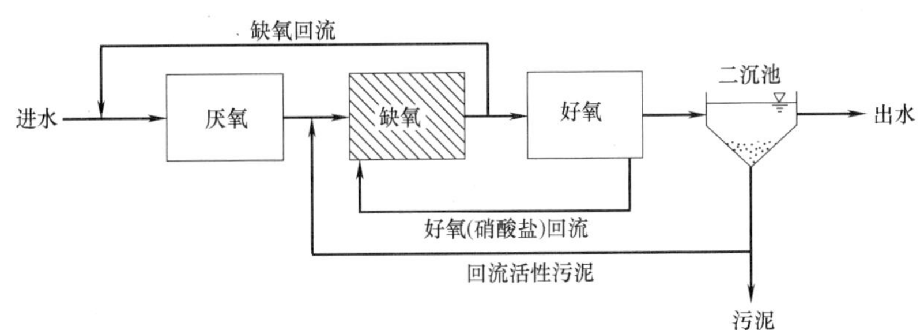

图7-19 UCT生物脱氮除磷工艺流程

在 UCT 基础上出现了改良 UCT 工艺，此时污泥回流到分隔的第一缺氧区，不与混合液回流到第二缺氧区，第一缺氧区主要是回流污泥中的硝酸盐反硝化，第二缺氧区是系统的主要反硝化区（图 7-20）。UCT 工艺和改良 UCT 工艺比 A^2/O 工艺多了一套混合液回流系统，流程较为复杂。

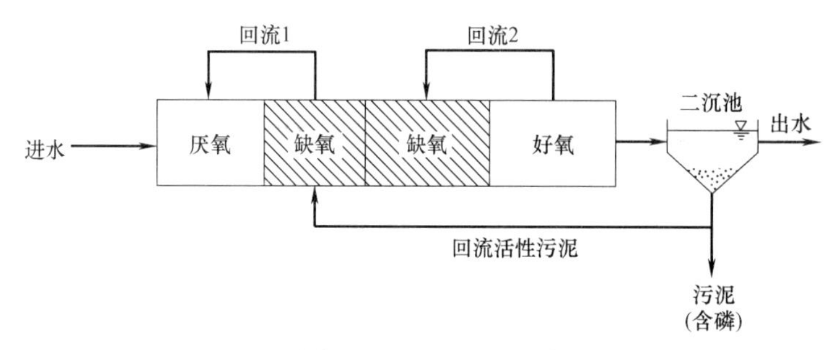

图 7-20　改良 UCT 生物脱氮除磷工艺流程

7.8　联合生物处理工艺的设计计算

7.8.1　A_N/O 生物脱氮设计

1. 设计要点

设计时所采用的硝化菌和反硝化菌的反应速度常数应取冬季水温时的数值。

（1）硝化工况

好氧池出口溶解氧在 1～2mg/L 以上；污（废）水适宜温度为 20～30℃，最低水温应不低于 13℃，低于 13℃时硝化速度明显降低；总氮负荷应小于 0.05kgTN/(kgMLSS・d)；最佳 pH 为 6.0～8.4。

（2）反硝化工况

溶解氧浓度趋近于零；生化反应池进水溶解性 BOD_5 与硝态氮浓度之比在 4 以上（S-BOD：NO_x-N≥4：1）；最佳 pH 为 6.5～8.0。

（3）缺氧/好氧法（A_N/O 法）生物脱氮的主要设计参数，宜根据试验资料确定或按表 7-3 的规定取值。

缺氧/好氧法（A_N/O）生物脱氮主要设计参数　　表 7-3

项目	单位	参数值
BOD_5 污泥负荷	$kgBOD_5$/(kgMLSS・d)	0.05～0.10
总氮负荷率	kgTN/(kgMLSS・d)	≤0.05
污泥浓度 MLSS	g/L	2.5～4.0
污泥龄 θ_c	d	11～23
污泥产率系数	kgVSS/($kgBOD_5$)	0.3～0.6
需氧量 O_2	kgO_2/($kgBOD_5$)	1.1～2.0
水力停留时间 *HRT*	h	9～22
		其中缺氧段 2～10
污泥回流比	%	50～100

续表

项目		单位	参数值
混合液回流比		%	100～400
总处理效率	BOD_5	%	90～95
	TN	%	60～85

2. 设计计算

（1）生化反应池容积

$$V=\frac{Q(S_0-S_e)}{N_s X} \tag{7-11}$$

式中 V——生化反应池容积，m^3；

Q——生化反应池设计进水流量，m^3/d；

S_0——生化反应池进水 BOD_5，mg/L；

S_e——生化反应池出水 BOD_5，mg/L；

X——生化反应池混合液 MLSS，mg/L；

N_s——生化反应池 BOD_5 污泥负荷，$kgBOD_5/(kgMLSS \cdot d)$。

（2）混合液回流比

混合液回流比越大，生物脱氮效率越高，混合液回流比和生物脱氮效率可按式（7-12）计算。

$$R_i=\frac{\eta_{TN}}{1-\eta_{TN}} \tag{7-12}$$

式中 R_i——混合液回流比，%；

η_{TN}——系统的脱氮效率，%。

（3）剩余污泥量

按污泥龄计算剩余污泥量见式（7-13）。

$$\Delta X=\frac{V \cdot X}{\theta_c} \tag{7-13}$$

式中 ΔX——剩余污泥量，kgSS/d。

按污泥产率系数、衰减系数和不可生物降解及惰性悬浮物计算剩余污泥量见式（7-14）。

$$\Delta X_v=YQ(S_0-S_e)-K_d V X_v+fQ(SS_0-SS_e) \tag{7-14}$$

式中 ΔX_v——剩余污泥量，kgVSS/d；

Y——污泥产率系数，$kgVSS/kgBOD_5$；

K_d——衰减系数，d^{-1}，20℃时取 0.040～0.075；

X_v——生化反应池混合液挥发性悬浮固体（MLVSS）浓度，kg/m^3；

f——SS 的污泥转化率，宜根据试验资料确定，无试验资料时可取 0.5～0.7；

SS_0——生物反应池进水悬浮物浓度，kg/m^3；

SS_e——生物反应池出水悬浮物浓度，kg/m^3。

（4）需氧量计算

生物反应池好氧区的需氧量，根据微生物去除的 5d 生化需氧量、氨氮的硝化和除氮等要求，按式（7-15）计算。

$$m_{O_2}=aQ(S_0-S_e)-c\Delta X_v+b[Q(N_k-N_{ke})-0.12\Delta X_v]$$
$$-0.62b[Q(N_t-N_{ke}-N_{0e})-0.12\Delta X_v] \tag{7-15}$$

式中 m_{O_2}——生化反应所需氧量，kg/d；

a——碳的氧当量，当含碳物质以 BOD_5 计时，应取 1.47；

b——常数，氧化每千克氨氮所需的氧量，应取 4.57；

c——常数，细菌细胞的氧当量，应取 1.42；

ΔX_V——生化反应池系统排出微生物量，kg/d；

N_k——生化反应池进水总凯氏氮浓度，kg/m^3；

N_{ke}——生化反应池出水总凯氏氮浓度，kg/m^3；

N_t——生化反应池进水总氮浓度，kg/m^3；

N_{0e}——生化反应池出水硝态氮浓度，kg/m^3；

$0.12\Delta X_v$——排出生物反应池系统的微生物中含氮量，kg/d。

式中其他各项参数意义同前。

【例 7-1】 某城市污水处理厂日最大处理流量为 72000m^3/d，水温为 15～25℃，污水经一级预处理后 BOD_5≤150mg/L，TN≤20mg/L，TKN≤19mg/L，SS≤130mg/L，采用 A_N/O 生物脱氮处理，要求生物处理后出水 BOD_5≤20mg/L，SS≤15mg/L，NH_3-N≈0，TKN≈0，NO_3-N≤3mg/L，TN≤5mg/L。设计缺氧池及好氧池有关数据，计算需氧量及剩余污泥量。

【解】（1）相关设计参数选取

1）生化反应池 BOD_5 污泥负荷：$N_s=0.08$kgBOD_5/(kgMLSS·d)；

2）污泥体积指数：SVI=130mL/g；

3）污泥回流比：$R=80\%$；

4）生物池内混合液悬浮固体浓度：MLSS=3500mg/L；

5）生化池内混合液 MLVSS/MLSS=0.75；

6）TN 去除率：$\eta_{TN}=\dfrac{N_t-N_e}{N_t}=\dfrac{20-5}{20}\times100\%=75\%$

7）混合液回流比：$R_i=\dfrac{\eta_{TN}}{1-\eta_{TN}}=\dfrac{0.75}{1-0.75}\times100\%=300\%$

（2）缺氧池/好氧池（A/O）主要尺寸

1）生化反应池总有效容积：$V=\dfrac{Q(S_0-S_e)}{N_sX}=\dfrac{72000\times(150-20)}{0.08\times3500}=33428.6\text{m}^3$

2）缺氧池和好氧池有效面积

取生化池有效水深 $h=5.0$m，生化池总有效面积为：

$$S_{总}=\frac{V}{h}=\frac{33428.6}{5.0}=6686\text{m}^2$$

缺氧池和好氧池分为 4 组，每组有效面积为：

$$S=\frac{S_{总}}{2}=\frac{6686}{4}=1672\text{m}^2$$

3）缺氧池和好氧池尺寸

设采用 4 廊道式缺氧池和好氧池，单廊道宽 $b=8$m，总宽为 32m，单组缺氧池/好氧池长度为：$l=\frac{S}{4b}=\frac{1672}{4\times8}=52.3$m

生化池长度介于 50～70m 之间，符合规范要求。

廊道长宽比 $\frac{l}{b}=\frac{52.3}{8}=6.5$ 满足 $l/b=5\sim10$ 要求。

生物池宽 $b=8$m，则宽深比：$b/h=8/5=1.6$，介于 1～2 之间，符合规范要求。

取生物池超高 $h_1=0.5$ m，则生物池总高度 H 为：

$H=h_1+h=5+0.5=5.5$m。

4）水力停留时间

$$t=\frac{V}{Q}=\frac{33428.6\times24}{72000}=11\text{h}$$

设缺氧池的水力停留时间为 2h，则好氧段的水力停留时间为 9 h。

(3) 剩余污泥量

污泥产率系数 Y 取 0.45，衰减系数 K_d 取 0.05d^{-1}，转换系数 f 取 0.6，计算剩余污泥量：

$$\begin{aligned}\Delta X_v&=YQ(S_0-S_e)-K_dVX_v+fQ(SS_0-SS_e)\\&=0.45\times72000\times(0.15-0.02)-0.05\times33428.6\times3.5\times0.75+0.6\times72000\times\\&\quad(0.13-0.015)\\&=4212-4387.5+4968\\&=4792.5\text{kgVSS/d}\end{aligned}$$

计算总排泥量：$\Delta X=\frac{4792.5}{0.75}=6390\text{kgSS/d}$

(4)剩余湿污泥量

设污泥含水率为 99.3%，则每日剩余湿污泥量为：

$$\frac{6390}{1000}=6.39\text{t(干泥)/d}$$

$$\frac{6.39}{100\%-99.3\%}=913\text{m}^3/\text{d}$$

(5) 需氧量

$$\begin{aligned}m_{O_2}&=aQ(S_0-S_e)-c\Delta X_v+b[Q(N_k-N_{ke})-0.12\Delta X_v]\\&\quad-0.62b[Q(N_t-N_{ke}-N_{0e})-0.12\Delta X_v]\\&=1.47\times72000\times(0.15-0.02)-1.42\times4792.5+4.57\times[72000\times(0.019-0)-\\&\quad0.12\times4792.5]-0.62\times4.57\times[72000\times(0.02-0-0.003)-0.12\times4792.5]\\&=13759.2-6805.4+3623.6-1838.6\\&=8738.8\text{kg/d}\end{aligned}$$

7.8.2 A²/O 生物脱氮除磷设计

(1) 污水中 5d 生化需氧量与总氮之比应不小于 4，5d 生化需氧量与总磷之比应不小于 17。

(2) 好氧区剩余总碱度应大于 70mg/L（以 $CaCO_3$ 计），当进水碱度不能满足要求时，应采取增加碱度的措施。

（3）常规生物脱氮除磷工艺有 A^2/O、SBR、氧化沟、改良 UCT 工艺等。A^2/O 工艺的主要设计参数，宜通过试验确定，无试验资料时，可参考类似规模、相同水质的污水处理厂数据或按表 7-4 取值。

A^2/O 生物脱氮除磷主要设计参数 **表 7-4**

项目		单位	设计参数
BOD_5 活性污泥负荷		$kgBOD_5/(kgMLSS \cdot d)$	0.05～0.10
反应池混合液 MLSS		g/L	2.5～4.5
污泥龄		d	10～22
水力停留时间		h	10～23
			其中厌氧段 1～2
			缺氧段 2～10
污泥回流比		%	20～100
需氧量		$KgO_2/kgBOD_5$	1.1～1.8
混合液回流比		%	≥200
污泥产率		$kgVSS/kgBOD_5$	0.3～0.6
总处理效率	BOD_5	%	85～95
	TP	%	60～85
	TN	%	60～85

（4）缺氧池及厌氧池应采用机械搅拌，机械设备功率应采用 $2\sim8W/m^2$。

【例 7-2】 某城市污水处理厂日最大处理流量 $Q=30000m^3/d$，污水经一级预处理后出水 COD≤360mg/L，BOD_5≤150mg/L，SS≤130mg/L，TN≤25mg/L，TKN≤24mg/L，TP≤4.5mg/L，水温为 12～25℃；要求二级生物处理出水 COD≤70mg/L，BOD_5≤20mg/L，SS≤20mg/L，NH_3-N≈0，TKN≈0，NO_3-N≤3mg/L，TN≤5mg/L，TP≤1mg/L，设计厌氧/缺氧/好氧（A^2/O）反应池。

【解】（1）判断是否可采用 A^2/O 工艺

$$\frac{BOD_5}{TP}=\frac{150}{4.5}=33.33>17$$

$$\frac{BOD_5}{TN}=\frac{150}{25}=6>4$$

根据《室外排水设计标准》GB 50014—2021 规定，满足生物除磷脱氮要求。

（2）相关设计参数选取

污泥负荷率：$N_s=0.065kgBOD_5/(kgMLSS \cdot d)$；

衰减系数：$K_d=0.04d^{-1}$；

污泥产率系数：$Y=0.45kgVSS/kgBOD_5$；

污泥龄：$\theta_c=15d$；

生化反应池内混合液 MLVSS/MLSS=0.75；

污泥回流比：$R=50\%$；

混合液悬浮固体（MLSS）浓度：$X=3000mg/L$。

(3) 回流污泥浓度

$$X_R=\frac{(1+R)X}{R}=\frac{(1+50\%)\times3000}{50\%}=9000\text{mg/L}$$

(4) 混合液回流比

生物脱氮效率：$\eta_{TN}=\frac{N_t-N_e}{N_t}=\frac{25-5}{25}\times100\%=80\%$

混合液回流比：$R_i=\frac{\eta_{TN}}{1-\eta_{TN}}=\frac{0.8}{1-0.8}\times100\%=400\%$

(5) 反应池总容积

$$V=\frac{Q(S_0-S_e)}{N_s\cdot X}=\frac{30000\times(150-20)}{0.065\times3000}=20000\text{m}^3$$

(6) 反应池总停留时间

废水在生物反应池内总停留时间为：$t=\frac{V}{Q}=\frac{20000\times24}{30000}=16\text{h}$

设厌氧池水力停留时间为 1.5h，缺氧池水力停留时间为 4.5h，则好氧池水力停留时间为 10h，生物池各段容积计算如下：

厌氧池容积：$V_{厌}=\frac{1.5}{16}\times20000=1875\text{m}^3$

缺氧池容积：$V_{缺}=\frac{4.5}{16}\times20000=5625\text{m}^3$

好氧池容积：$V_{好}=\frac{10}{16}\times20000=12500\text{m}^3$

(7) 确定生物池各部分尺寸

设采用 4 组生物池，单组生物池容积为：$V_{单}=\frac{20000}{4}=5000\text{m}^3$

取生物池有效水深 $h_1=4.0\text{m}$，单组生物池面积为：$F_{单}=\frac{V_{单}}{h_1}=\frac{5000}{4}=1250\text{m}^2$

取生物池宽 $b=6\text{m}$，则宽深比：$b/h_1=6/4=1.5$，介于 1～2 之间，符合规定。

采用 4 廊道推流式生物池，单组生物池总宽度为 24m，生物池长度为：

$$l=\frac{F}{4b}=\frac{1250}{4\times6}=52.1\text{m}$$

生物反应池长度介于 50～70m 之间，符合规定。

廊道长宽比：$\frac{l}{b}=\frac{52.1}{6}=8.7$，满足 $l/b=5\sim10$ 要求。

取生物池超高 $h_2=0.5\text{m}$，则生物池总高度为：

$H=h_1+h_2=4+0.5=4.5\text{m}$。

(8) 剩余污泥量

污泥产率系数 Y 取 0.45，衰减系数 K_d 取 0.04d^{-1}，转换系数 f 取 0.6，计算剩余污泥量：

$$\Delta X_v=YQ(S_0-S_e)-K_dVX_v+fQ(SS_0-SS_e)$$

$=0.45\times30000\times(0.15-0.02)-0.04\times20000\times3\times0.75+0.6\times30000\times(0.13-0.02)$

$=1755-1800+1980$

$=1935\text{kgVSS/d}$

计算总排泥量：$\Delta X=\frac{1935}{0.75}=2580\text{kgSS/d}$

（9）剩余湿污泥量

设污泥含水率为 99.2%，则每日剩余湿污泥量为：

$$\frac{2580}{1000}=2.58\text{t(干泥)/d}$$

$$\frac{2.58}{100\%-99.2\%}=322.5\text{m}^3\text{/d}$$

（10）需氧量

$m_{O_2}=aQ(S_0-S_e)-c\Delta X_v+b[Q(N_k-N_{ke})-0.12\Delta X_v]-0.62b[Q(N_t-N_{ke}-N_{0e})-0.12\Delta X_v]$

$=1.47\times30000\times(0.15-0.02)-1.42\times1935+4.57\times[30000\times(0.024-0)-0.12\times1935]-0.62\times4.57\times[30000\times(0.025-0-0.003)-0.12\times1935]$

$=5733-2747.7+2229.2-1212.1$

$=4002.4\text{kg/d}$

7.8.3 SBR 生物脱氮除磷设计

1. 设计要点

（1）SBR 反应池的数量不宜少于 2 个，每天的周期数宜为正整数；

（2）沉淀时间宜为 1.0h，排水时间宜为 1.0～1.5h；

（3）反应池宜采用矩形池，水深宜为 4.0～6.0m；反应池长度和宽度之比：间歇进水时宜为 1∶1～2∶1，连续进水时宜为 2.5∶1～4∶1；

（4）利用 SBR 工艺进行生物脱氮除磷处理城镇污水或水质与城镇污水相似的工业废水时，主要设计参数见表 7-5。如工业废水的水质与城镇污水水质差异较大，设计参数应通过实验确定。

SBR 生物脱氮除磷主要设计参数 **表 7-5**

项目名称		符号	单位	参考值
反应池 BOD_5 污泥负荷	BOD_5/MLVSS	L_s	kg/(kg·d)	0.15～0.25
	BOD_5/MLSS		kg/(kg·d)	0.07～0.15
反应池混合液悬浮固体(MLSS)浓度		X	kg/m^3	2.5～4.5
总氮负荷率(TN/MLSS)			kg/(kg·d)	≤0.06
污泥产率系数（VSS/BOD_5）	设初沉池	Y	kg/kg	0.3～0.6
	不设初沉池		kg/kg	0.6～0.8
厌氧水力停留时间占反应时间比例			%	5～10
缺氧水力停留时间占反应时间比例			%	10～15
好氧水力停留时间占反应时间比例			%	75～80
总水力停留时间		HRT	h	20～30

续表

项目名称	符号	单位	参考值
污泥回流比（仅适用于 CASS 或 CAST）	R	%	20～100
混合液回流比（仅适用于 CASS 或 CAST）	R_i	%	≥200
需氧量	m_{O_2}	kg/kg	1.5～2.0
活性污泥容积指数	SVI	mL/g	70～140
充水比	m		0.3～0.35
BOD_5 总处理率	η	%	85～95
TP 总处理率	η	%	50～75
TN 总处理率	η	%	55～80

2. 主要计算公式

（1）反应时间 t_R：指一个运行周期内进水时间和曝气工序中停止所需要的时间，反应时间按下式计算：

$$t_R=\frac{24S_0m}{1000L_sX} \tag{7-16}$$

式中 t_R——每个周期反应时间，h；

m——充水比；

L_s——反应池 BOD_5 污泥负荷，$kgBOD_5/(kgMLSS \cdot d)$。

（2）SBR 反应池容积为：

$$V=\frac{24Q'S_0}{1000L_sXt_R} \tag{7-17}$$

式中 V——SBR 反应池容积，m^3；

Q'——每周期处理的水量，m^3/d。

式中其他各项参数意义同前。

（3）剩余污泥量和需氧量计算同 7.8.1 节 A_N/O 生物脱氮设计。

【例 7-3】 某城市污水处理厂海拔高度为 600m，设计日最大处理水量 $Q=13000m^3/d$，设计进水水质为 COD≤380mg/L，BOD_5≤160mg/L，SS≤150mg/L，TN≤30mg/L，TKN≤28mg/L，TP≤4.0mg/L，水温为 12～25℃；要求采用生物处理后出水 COD≤60mg/L，BOD_5≤20mg/L，SS≤20mg/L，NH_3-N≈0，TKN≈0，NO_3-N≤5mg/L，TN≤10mg/L，TP≤1.5mg/L，进行 SBR 生物脱氮除磷设计。

【解】（1）判断废水水质是否满足 SBR 生物脱氮除磷要求

$$\frac{BOD_5}{TP}=\frac{160}{4.0}=40>17$$

$$\frac{BOD_5}{TN}=\frac{160}{30}=5.33>4$$

根据《序批式活性污泥法污水处理工程技术规范》HJ 577—2010 规定，满足生物脱氮除磷要求。

（2）反应时间

根据表 7-5，充水比（m）取 0.35；反应池 BOD_5 污泥负荷，取 0.1 $kgBOD_5$/

(kgMLSS·d)；反应池 MLSS，取 3.0gMLSS/L，反应时间为：

$$t_R=\frac{24S_0m}{1000L_sX}=\frac{24\times160\times0.35}{1000\times0.1\times3}=4.5\text{h}$$

根据《序批式活性污泥法污水处理工程技术规范》HJ 577—2010 规定，SBR 每天运行周期宜为整数，如 2d、3d、4d、5d、6d 等，本设计每格运行 3 个周期，每周期运行时间 t 为 24/3=8h，沉淀时间 t_s 取 1.0h，排水时间 t_D 宜为 1.0～1.5h，取 1.3h，则闲置时间为：

$$t_b=t-t_R-t_s-t_D=8-4.5-1-1.3=1.2\text{h}$$

闲置时间可以延长反应时间和沉淀时间，更好地适应水质水量的变化，提高出水水质。根据表 7-5，厌氧反应时间占总时间的 10%，为 0.45h，缺氧反应时间占总时间的 15%，为 0.68h，好氧反应时间占总时间的 75%，为 3.37h。

(3) 反应池容积

本设计 SBR 反应池分为 4 格，每周期运行时间为 8h，每格每天运行 3 个周期，每周期处理的水量为：

$$Q'=\frac{Q}{4\times3}=\frac{13000}{4\times3}=1083.3\text{m}^3/\text{d}$$

因此，SBR 反应池容积为：

$$V=\frac{24Q'S_0}{1000L_sXt_R}=\frac{24\times1083.3\times160}{1000\times0.1\times3\times4.5}=3081.4\text{m}^3$$

(4) 剩余污泥量

污泥产率系数 Y 取 0.4，衰减系数 K_d 取 0.05d^{-1}，系数 f 取 0.6，因此，剩余污泥量为：

$$\begin{aligned}\Delta X_v&=YQ(S_0-S_e)-K_dVX_v+fQ(SS_0-SS_e)\\&=0.4\times13000\times(0.16-0.02)-0.05\times3081.4\times0.75\times3+0.6\times13000\times(0.15-0.02)\\&=728-346.7+1014\\&=1395.3\text{kgVSS/d}\end{aligned}$$

设 MLVSS 与 MLSS 的比值为 0.75，则总排泥量为：

$$\frac{1395.3}{0.75}=1860.4\text{kgSS/d}$$

(5) 剩余湿污泥量

设污泥含水率为 99.2%，每天排放的湿污泥量为：

$$\frac{1860.4}{1000}\approx1.86\text{t/d}$$

$$\frac{1.86}{1-99.2\%}=232.5\text{m}^3/\text{d}$$

(6) 生化池所需氧量

$$\begin{aligned}m_{O_2}&=aQ(S_0-S_e)-c\Delta X_v+b[Q(N_k-N_{ke})-0.12\Delta X_v]\\&\quad-0.62b[Q(N_t-N_{ke}-N_{0e})-0.12\Delta X_v]\\&=1.47\times13000\times(0.16-0.02)-1.42\times1395.3+4.57\times[13000\times(0.028-0)-0.12\times1395.3]-0.62\times4.57\times[13000\times(0.03-0-0.005)-0.12\times1395.3]\end{aligned}$$

$=2675.4-1981.3+898.3-446.4$

$=1146kg/d$

本章关键词（Keywords）

硝化作用	Nitrification
硝化细菌	Nitrobacteria
芽孢	Spore
化能自养	Chemoautotrophic
反硝化作用	Denitrification
可生物降解性	Biodegradability
碳源	Carbon source
聚磷菌	Phosphorus accumulating bacteria
厌氧释磷	Anaerobic phosphorus release
世代期	Generation period
水解	Hydrolyze
挥发性脂肪酸	Volatile fatty acids
磷酸盐	Phosphate
粉末活性炭	Powdered activated carbon
常规活性污泥法	Conventional activated sludge process（CASP）
膜生物反应器	Membrane biological reactor（MBR）
助凝剂	Flocculant aids

思考题

1. 简述城镇污水生物脱氮的基本步骤，如何计算反硝化池的容积？若出水氮含量仍超标，如何进一步脱氮？

2. 生物脱氮除磷的环境条件要求有哪些？说明主要生物脱氮除磷工艺的特点。

3. A^2/O 工艺的运行管理人员称只能在脱氮和除磷两个功能中选择一个为主要任务，否则两者效率都会不高。请分析原因，并提出应对或改进措施。

4. Phostrip 工艺除磷效率不像其他生物除磷系统那样受进水 BOD 的影响，分析其原因。

5. 试比较 A/O 污水处理技术与 A^2/O 污水处理技术的差异。

Chapter 8 Advanced Treatment of Wastewater

Many effluent standards require tertiary or advanced wastewater treatment to remove particular contaminants or to prepare the water for reuse. Some common tertiary operations are removal of phosphorus compounds by coagulation with chemicals, removal of nitrogen compounds by ammonia stripping with air or by nitrification-denitrification in biological reactors, which are discussed in Chapter 7, removal of residual organic and color compounds by adsorption on activated carbon and removal of dissolved solids by membrane processes (reverse osmosis and electrodialysis). The effluent water is often treated with chlorine or ozone to destroy pathogenic organisms before discharge into the receiving water, which is discussed in this Chapter.

8.1 Introduction of Adsorption

A solid surface in contact with a solution tends to accumulate a surface layer of solute molecules because of the unbalance of surface forces. Chemical adsorption or chemisorption results in the formation of a monomolecular layer of the adsorbate on the surface through forces of residual valence of the surface molecules. Physical adsorption results from molecular condensation the capillaries of the solid. In general, substances of the highest molecular weight are most easily adsorbed. There is a rapid formation of an equilibrium interfacial concentration, followed by slow diffusion into the carbon particles. The overall rate of adsorption is controlled by the rate of diffusion of the solute molecules within the capillary pores of the carbon particles. The rate varies reciprocally with the square of the particle diameter, increases within creasing concentration of solute, increases with increasing temperature, and decreases with increasing molecular weight of the solute.

The adsorptive capacity of a carbon for a solute will likewise be dependent on both the carbon and the solute. Most wastewater is highly complex and varies widely in the adsorbability of the compounds present.

Many industrial wastes contain organics which are refractory and which are difficult or impossible to remove by conventional biological treatment processes. Examples are ABS and some of the heterocyclic organics. These materials can frequently be removed by adsorption on an active-solid surface. The most commonly used adsorbent is activated carbon.

Activated carbon columns are employed for the treatment of toxic or non-biodegradable wastewater and for tertiary treatment following biological oxidation. In industrial application, contact times of less than 1h are usually used. Equilibrium is probably closely realized when high carbon dosages are employed, since the rate of adsorption increases with carbon dosage.

PAC (powdered activated carbon) can be integrated into existing biological treatment facilities at minimum capital cost. Since the addition of PAC enhances sludge settleability, conventional secondary clarifiers will usually be adequate, even with high carbon dosages. In some industrial waste applications, nitrification is inhibited by the presence of toxic organics. The application of PAC has been shown to reduce or eliminate this inhibition.

8.2 Introduction of Chemical Oxidation

Chemical oxidation generally refers to the use of oxidizing agents such as ozone (O_3), hydrogen peroxide (H_2O_2), permanganate (MnO_4^-), chloride dioxide (ClO_2), chlorine (Cl) or HOCl, or even oxygen, without the need for microorganisms for the reactions to proceed. These reactions frequently require one or more catalysts in order to increase the rate of reaction to acceptable levels. Catalysts include simple pH adjustment, transition metal cations, enzymes, and a variety of proprietary catalysts of unreported composition.

Chemical oxidation is typically applied to situations where organic compounds are nonbiodegradable (refractory), toxic, or inhibitory to microbial growth. However, chemical oxidation is also effective for the destruction of many organic compounds and the elimination of odorous compounds such as oxidation of sulfides.

While oxygen serves as a readily available and extremely economical oxidant for biological treatment processes, other chemical oxidants are relatively expensive and cannot compete economically with aerobic biological treatment. However, it is not necessary to early chemical oxidation to the fullest extent of reaction conversion of organic carbon to CO_2. Partial oxidation of compounds may be sufficient to render specific compounds, such as priority pollutants, more amenable to subsequent biological treatment. On a general basis, the oxidation of specific compounds may be characterized by the extent of degradation of the final oxidation products:

(1) Primary degradation. A structural change in the parent compound.

(2) Acceptable degradation (defusing). A structural change in the parent compound to the extent that toxicity is reduced.

(3) Ultimate degradation (mineralization). Conversion of organic carbon to inorganic CO_2.

(4) Unacceptable degradation. A structural change in the parent compound resulting in an increase in toxicity.

8.2.1 Ozone Oxidation

Ozone is a powerful oxidant that is commonly used for disinfection and wastewater treatment and is a metastable gas at normal temperatures and pressures and must be generated on site. At high pressures, decomposition is rapid. Therefore, generation and mass transfer operation are carried out at low pressure.

Ozone generation is the key to economical operation. Generators operate on electrical current and produce ozone from either air or pure oxygen gas streams. Moisture retards the process and air must be dehumidified prior to ozone generation (dew point<-60℃).

Ozone decomposes in water, especially at high pH values, to produce free radicals. The mechanisms of ozone oxidation of organics are:

(1) Oxidation of alcohols to aldehydes and then to organic acids;

(2) Substitution of all oxygen atom onto an aromatic ring;

(3) Cleavage of carbon double bonds.

Ozonation can be employed for the removal of color and residual refractory organics in effluents. Oxidation of unsaturated aliphatic or aromatic compounds causes a reaction with water and oxygen to form acids, ketones, and alcohols. At a pH greater than 9.0 in the presence of redox salts such Fe, Mn, and Cu, aromatics may form some hydroxyaromatic structures (phenolic) which may be toxic. Many of the by-products of ozonation are readily biodegradable.

Phenol can be oxidized with ozone, producing as many as 22 intermediate products between phenol and CO_2 and H_2O. The reaction is first order with respect to phenol and proceeds optimally over a pH of 8 to 11. The ozone consumption is 4 to 6 mole O_3, consumed/mole phenol oxidized. This requires in the order of 25 mole O_3/mole phenol to be generated in the gas phase.

Ozone may be used in combination with UV radiation to catalyze the oxidation of non-reactive organics, such as saturated hydrocarbons and highly chlorinated organics. Ozone reacts with UV light (253.7nm) in aqueous systems to produce hydrogen peroxide.

In addition, UV light may react directly with some organics, further promoting reaction of the first by-products with O_3.

8.2.2 Hydrogen Peroxide Oxidation

Hydrogen peroxide is commercially available in a variety of grades, with 30% or 50% (by weight) solutions being most common for wastewater applications. Inhibitors, typically phosphates, are added to prolong storage times. Hydrogen peroxide has a long history of use for oxidation of sulfides in sewer lines and wastewater treatment plants, and more recently has been widely applied to toxic and refractory organics. Sulfide oxidation with H_2O_2, is summarized in (Table 8-1).

Sulfide oxidation with hydrogen peroxide **Table 8-1**

Acidic or neutral pH	Reaction time: seconds
$H_2O_2+H_2S \rightarrow 2H_2O+S$	Basin pH
Reaction time: 15~45min	$4H_2O_2+S^{2-} \rightarrow 4H_2O+SO_4^{2-}$
Catalyst: Fe^{2+}	Reaction time: 15min
pH: 6.0~7.5	

Alkaline peroxidation (pH 10 to 12) is an effective means of providing total cyanide destruction. But reactions with H_2O_2 alone are slow, and a catalyst is generally required. A wide variety of reaction schemes are possible, using high pH (alkaline catalysis); metals such as ferrous sulfate (Fenton's reagent), complexed Fe (Fe-EDTA or Heme), Cu, or Mn; or natural enzymes such as horseradish peroxidase.

Other catalysts include UV radiation (UV-H_2O_2). It is hypothesized that the H_2O_2 molecule can be split directly into hydroxyl radicals by UV light.

Ultraviolet light is derived from lamps that have been developed for a high quantum yield in the appropriate wavelengths for hydroxyl radical generation, based on decomposition of ferrioxalate as an indicator. Additionally, heat generated by the lamps can increase the rates of reaction significantly. The UV-H_2O_2, process is generally applied to aqueous wastes of lower color, turbidity, and concentration, such as contaminated ground waters. However, proprietary additives are available for higher-strength wastes. Many compounds are effectively treated by UV-H_2O_2, including benzene, toluene, xylene, trichloroethylene, and perchloroethylene, but some are refractory to treatment, including chloroform, acetone, trinitrobenzene, and n-octane.

Hydrogen peroxide has been shown to be useful in many cases for removal of specific undesirable pollutants such as priority pollutants, reduction of toxicity, and/or improvements in biodegradability (both late and extent of degradation).

8.2.3 Potassium Permanganate Oxidation

Permanganate is a powerful oxidizing agent that is reactive over a wide pH range for a variety of organic and inorganic compounds. This is supplied in stable form as a solid or in concentrated aqueous form. Applications of $KMnO_4$ traditionally include odor control (oxidation of inorganic and organic sulfides) and textile, tannery, steel processing, metal finishing, pulp and paper, and oil refinery wastes.

Unlike other oxidants, a solid reaction by-product, MnO_2(s), is formed and must be disposed of as a waste sludge, along with other precipitates/solids already under-stood as part of the waste itself. This sludge can be considerable for concentrated wastewater. Data shows that $KMnO_4$ can be effective in the destruction of specific compounds and in toxicity reduction for phenolics and other aromatic compounds, and even certain chlorinated aliphatics, such as trichloroethylene, perchloroethylene, and trichloroethane.

8.2.4 Chlorine Oxidation

Chlorine has a long history of use as an oxidant in water and wastewater treatment, and has been especially successful for color removal where organic dyes are present. However, recent concerns regarding the formation of chlorinated by-products, such as chlorination, have greatly reduced the applications of chlorine in wastewater.

Chlorine is still frequently applied for the oxidation of cyanides in metal finishing operation. The oxidation of cyanide by chlorine proceeds through several reactions that are highly pH-dependent.

In practice，the dosages of chlorine depend heavily on the identity of the metals complexed with CN^- and the presence of other background constituents in the wastewater. Since these concentrations change continually，the system is usually operated on the basis of oxidation reduction potential（ORP）.

8.3 Hydrothermal Processes

Hydrothermal processes refer to aqueous treatment of wastewater at elevated temperatures and pressures. There are three basic operating regimes that are in use or have been investigated in the laboratory：

（1）Wet air oxidation（WAO）. A commonly used process for more concentrated waste，especially those that are toxic and/or biologically refractory. These processes use an oxidant，primarily O_2 from air，to partially oxidize organics，yielding a variety of low-molecular-weight organic acids（readily biodegradable）.

（2）Hydrothermal hydrolysis. Hydrolysis of organic compounds can occur at elevated temperatures and pressures. Temperatures proposed for hydrothermal hydrolysis range from 200 to 374℃.

（3）Supercritical water oxidation（SCWO）. Aqueous oxidation of organics takes place to completion，at even higher temperature and pressures than wet air oxidation. SCWO takes place beyond the critical point of water（about 374℃and 218atm）. Typical operating conditions are 400 to 650℃ and 24.1 to 34.5MPa. At these temperatures and pressures，materials of construction become critical and solubility of salts can decrease dramatically，causing fouling.

8.4 吸附法工艺及应用

许多工业废物含有难降解的有机物，例如一些杂环有机物，很难或不可能通过常规生物处理工艺（如ABS）去除。这些物质通常可以通过在活性固体表面或吸附剂上吸附去除。最常用的吸附剂是活性炭。

国内习惯将吸附（absorption）和吸收（adsorption）视作两个概念。吸收是一种物质穿过相界面，进入另一种物质的体相内的现象。吸附指一种物质附着在另一种物质的表面，不进入体相内。区别在于吸收伴随传递过程，吸附只是表面的附着，但这种附着可以是单纯的物理作用（物理吸附），也可以是吸附质和吸附剂表面生成了化学键（化学吸附）。由于相界面处的分子受力不均衡，倾向于吸附另外的分子，以减小两相之间的界面，使自身也达到像体相内分子那样的平衡状态，因此，表面吸附是一种热力学上的自发过程。

8.4.1 吸附等温式

在等温吸附过程中，当两相在一定温度下充分接触，最后能使吸附质分子到达吸附剂表面的数量和吸附剂表面释放吸附质的数量相等，即达到吸附平衡。众多学者从不同的模型和学说出发，推导和修正出各种吸附等温式。由于吸附机理比较复杂，这些吸附等温式只能适用于特定的吸附情况。较常用的吸附等温式包括亨利吸附等温式、朗缪尔吸附等温式、弗罗因德吸附等温式及 BET 方程。

1. 亨利吸附等温式

对于低浓度吸附质的水溶液，吸附分子如不缔合或解离，保持分子状态的单分子层吸附于均一表面的吸附剂时，单位吸附剂的吸附量和液体中吸附质质量浓度呈线性关系：

$$q=y/m=Hc \tag{8-1}$$

式中 q——单位吸附剂的吸附量，mg/mg；

y——吸附剂吸附的物质总量，mg；

m——投加的吸附剂量，mg；

H——亨利常数，L/mg；

c——吸附质在液体中的质量浓度，mg/L。

2. 朗缪尔（Langmuir）吸附等温式

朗缪尔假设在吸附剂表面具有均匀的吸附能力，所有的吸附机理相向，被吸附的吸附质分子之间没有相互作用力，也不影响分子的吸附，在吸附剂表面只形成单分子层吸附，因此，吸附速率和解吸速率与吸附剂表面被吸附分子的覆盖率 θ 和裸露率（$1-\theta$）有关，由此导出：

$$\text{吸附速率}=k_a(1-\theta)c \tag{8-2}$$

$$\text{解吸速率}=k_d\theta \tag{8-3}$$

式中 k_a——吸附速率常数；

k_d——解吸速率常数；

c——溶液中溶质的质量浓度，mg/L。

设 q_m 为单位质量或单位体积吸附剂覆盖满一层单分子层时的吸附量。q 为达到任一平衡状态时的吸附量，覆盖率 $\theta=q/q_m$。在平衡状态时，吸附速率等于解吸速率：

$$k_ac(1-\theta)=k_d\theta \tag{8-4}$$

$$\theta=\frac{q}{q_m}=\frac{k_1c}{1+k_1c}\text{或}q=\frac{y}{m}=\frac{q_mk_1c}{1+k_1c}=\frac{k'c}{1+k_1c} \tag{8-5}$$

式中 k_1——朗缪尔常数，$k_1=k_a/k_d$，$k'=q_mk_1$。

在吸附力弱或浓度很低时，$k_1c\ll1$，式（8-5）中分母的 k_1c 可以忽略不计，则式（8-5）改成：

$$q=k_1q_mc=k'c \tag{8-6}$$

与亨利吸附等温式相似，吸附量与吸附质在液相中的浓度成正比，k_1q_m 相当于亨利常数。

如果吸附力比较强，浓度较高时，$k_1c\gg1$，则式（8-5）分母的 1 可以略去，则成为：

$$q=q_m \tag{8-7}$$

吸附量趋于一极限值，吸附等温线趋于一条渐近线。

朗缪尔吸附等温式通过变形，可写成直线式：

$$\frac{1}{q}=\frac{q}{q_{\mathrm{m}}k_{1}c}+\frac{1}{q_{\mathrm{m}}} \tag{8-8}$$

将试验数据以 $1/q$ 为纵坐标，$1/c$ 为横坐标，可求得常数 q_{m} 和 k_1，如图 8-1 所示。

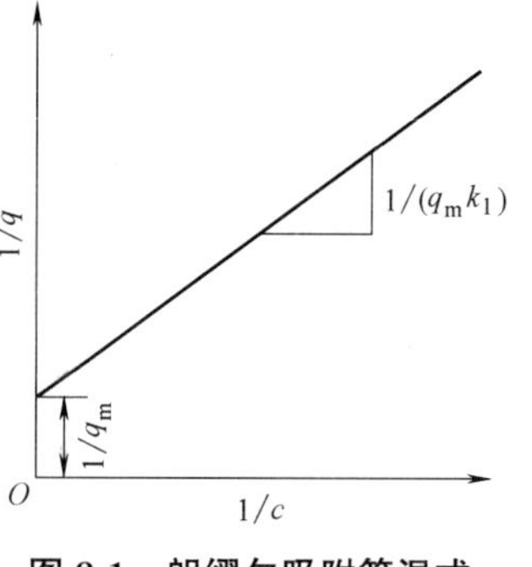

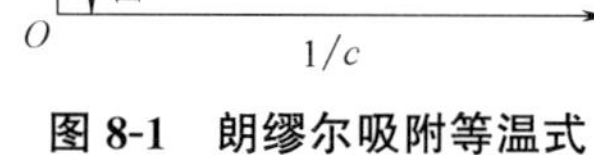

图 8-1　朗缪尔吸附等温式

3. 弗罗因德利希（Freundlich）吸附等温式

弗罗因德利希吸附等温式是一个经验式，该式与不均匀表面吸附理论所得的吸附量和吸附热的关系相符。

$$q=\frac{y}{m}=kc^{\frac{1}{n}} \tag{8-9}$$

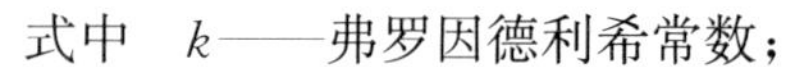

式中　k——弗罗因德利希常数；

n——常数，通常 $n>1$，随温度的升高，吸附指数 $1/n$ 趋于 1，一般认为：$1/n$ 介于 0.1～0.5，则容易吸附；$1/n>2$ 的物质难以吸附。

弗罗因德利希吸附等温式可以表示成对数形式：

$$\lg q=\frac{1}{n}\lg c+\lg k \tag{8-10}$$

以 $\lg q$ 和 $\lg c$ 作图，可得一直线，直线斜率为 $1/n$，截距为 $\lg k$。

4. BET 方程

1938 年 Brunauer，Emmett 和 Teller 三人在朗缪尔模型和假定的基础上提出了多层吸附理论。多层吸附的吸附量等于各层吸附量的总和，由此推出的吸附等温方程式即 BET 方程。

$$\frac{c}{q(c_{\mathrm{s}}-c)}=\frac{1}{kq_{\mathrm{m}}}+\frac{k-1}{kq_{\mathrm{m}}}\frac{c}{c_{\mathrm{s}}} \tag{8-11}$$

$$q=\frac{kq_{\mathrm{m}}c}{(c_{\mathrm{s}}-c)\left[1+(k-1)\frac{c}{c_{\mathrm{s}}}\right]} \tag{8-12}$$

当吸附质的平衡浓度 c 远小于饱和浓度 c_{s} 时，即 $c_{\mathrm{s}}\gg c$ 时，则：

$$\frac{q}{q_{\mathrm{m}}}=\frac{k\frac{c}{c_{\mathrm{s}}}}{1+k\frac{c}{c_{\mathrm{s}}}} \tag{8-13}$$

令 $k_1=k/c_{\mathrm{s}}$，则上式变为朗缪尔方程。

以$\frac{1}{q}\mathrm{g}\frac{c/c_{\mathrm{s}}}{1-c/c_{\mathrm{s}}}$为纵坐标，$\frac{c}{c_{\mathrm{s}}}$为横坐标，则为一直线方程。该直线的斜率为$\frac{k-1}{kq_{\mathrm{m}}}$，截距为$\frac{1}{kq_{\mathrm{m}}}$，由图解法可得 k 和 q_{m} 的值。

8.4.2　吸附速率

上述的吸附平衡值表明了吸附过程中的吸附质在溶液和吸附剂之间的浓度分配的极限，但实际的吸附过程中，可能做不到达到平衡所需要的时间。而现场吸附过程所需要的时间，吸附设备的大小都与吸附速率有关，吸附速率越快，所需要的时间就越短，吸附设备所需要的容积也就越小。

吸附速率取决于吸附剂对吸附质的吸附过程。多孔吸附剂与溶液接触时，在固体吸附剂颗粒表面总存在着一层流体薄层，即液相界膜，吸附剂对吸附质的吸附过程可以理解为吸附质首先要通过液相界膜扩散到吸附剂表面，称为颗粒的外扩散，或称膜扩散。然后吸附质通过细孔向吸附剂内部扩散，称为孔隙扩散，最后吸附质在吸附剂内表面上的吸附，称为吸附反应。吸附速率取决于上述三个过程，通常吸附反应速率非常快，因而吸附速率主要由外扩散和孔隙扩散速率控制。

对于膜扩散而言，界膜内吸附质扩散速率与界膜厚度成反比，与单位体积床层中吸附剂颗粒的外表面积（界膜面积）、流体主体中吸附质的浓度和吸附剂颗粒外表面上流体中吸附质浓度之差成正比。

提高溶液的流速，可使界面膜厚度降低，从而膜传质系数增大，吸附速率提高。吸附剂颗粒变小使界膜面积增加，或提高溶液浓度等都可提高吸附速率。

而孔隙扩散速率与吸附剂孔隙的大小、提高结构、吸附质颗粒的大小等因素有关。吸附反应的吸附速率与吸附质颗粒直径的较高次方成反比，颗粒越小，内扩散阻力越小，扩散速率越快。

8.4.3 影响吸附的因素

吸附是溶剂、溶质和固体吸附剂三者之间的作用，因此，溶质、吸附剂和溶液的性质都对吸附过程产生影响。

1. 溶质（吸附质）的性质

（1）溶质和溶剂之间的作用力：溶质在水中的溶解度越大，溶质对水的亲和力就越强，就越不易转向吸附剂界面而被吸附，反之亦然。有机物在水中的溶解度一般与分子结构和大小有关，并随着链长的增加而降低，如活性炭自水中吸附脂肪族有机酸的量，按甲酸、乙酸、丙酸、丁酸的顺序增加。芳香族化合物较脂肪族化合物容易吸附，苯甲醛在活性炭上的吸附量是丁醛的 2 倍，安息酸是乙酸的 5 倍。

（2）溶质分子的大小：Tranbe 定律认为，大尺寸疏水分子的斥力增加了水－水间的键合，因此，随着吸附质相对分子质量的增加，吸附量增加。但吸附速率受颗粒内扩散速率控制时，吸附速率随相对分子质量的增加而降低，相对分子质量低的有机物反而容易被去除。

（3）电离和极性：简单化合物，非解离的分子较离子化合物的吸附量大，但随着化合物结构的复杂化，电离对吸附的影响减小。有机物的极性是分子内部电荷分布的函数，几乎不对称的化合物都或多或少地带有极性。衡量溶质极性对吸附的影响，服从极性相容的原则，即极性吸附剂能强烈地从非极性溶剂中将极性溶质吸附，但是非极性吸附剂却很难将极性溶剂中的极性溶质吸附。水是很强的极性物质，极性溶质在水中的吸附量随着溶质极性的增强，吸附作用减小。羟基、羧基、硝基、腈基、磺基、胺基等都能增加分子的极性，对吸附是不利的。

2. 吸附剂的性质

吸附量的多少随吸附剂表面积的增大而增加，吸附剂的孔径、颗粒度等都影响比表面积的大小，从而影响吸附性能。不同的吸附剂，或用不同的方法制造的吸附剂，其吸附性能也不相同，吸附剂的极性对不同吸附质的吸附性能也不一样。

3. 溶液的性质

（1）pH：pH 对吸附质在溶液中存在的形态（电离、络合）和溶解度均有影响，因而对吸附性能也产生影响。水中的有机物一般在低 pH 时，电离度较小，吸附去除率高。活性炭吸附剂在低 pH 时，表面上的负电荷将随溶液中氢离子浓度的增加而中和，使活性炭具有更多的活化表面，吸附性能变得更好。

（2）温度：因吸附反应通常是放热的，因此，温度越低对吸附越有利，但是在水处理中，一般温度变化不大，因而温度的影响往往很小，常常可以不加考虑。

（3）共存物质：其影响较复杂，有的可以相互诱发吸附，有的能独立地被吸附，有的则相互起干扰作用。水溶液中有相当于天然水含量的无机离子共存时，对有机物的吸附几乎没有什么影响。当有汞、铬、铁等金属离子在活性炭表面氧化还原发生沉积时，会使活性炭的孔径变窄，其结果妨碍了有机物的扩散。悬浮物会堵塞吸附剂的孔隙，油类物质会在吸附剂表面形成油膜，均对吸附有很大影响。

8.4.4 吸附剂

理论上所有的固体表面都或多或少地具有吸附作用，而作为工业用的吸附剂，必须具有较大的比表面积，较高的吸附容量，良好的吸附选择性、稳定性、耐磨性、耐腐蚀性，较好的机械强度，并且具有价廉易得等特点。常用的工业吸附剂有活性炭，活性白土、漂白土、硅藻土等天然矿物质，硅胶，活性氧化铝，沸石分子筛，吸附树脂，腐殖酸类吸附剂。

（1）活性炭：活性炭是由煤、重油、木材、果壳等含碳类物质加热碳化，再经药剂（如氯化锌、氯化锰、磷酸等）或水蒸气活化，制成的多孔性碳结构的非极性吸附剂。活性炭具有巨大的比表面积和特别发达的微孔，是其吸附能力强，吸附容量大的主要原因；当然，比表面积相同的活性炭，对同一物质的吸附容量也可能不同，这与活性炭的内孔结构和分布以及表面化学性质有关；所以，活性炭的性质会由于原料和制备方式的不同而相差很大。

活性炭因其吸附容量大，化学性质稳定，抗腐蚀，在高温解吸时结构热稳定性好，解吸容易等特点，被广泛用于环境保护和工业领域。按照形态，可将活性炭分为粉末状（PAC）、颗粒状（GAC）、纤维活性炭等。目前工业上大量采用的是粒状活性炭，可以耐强酸、强碱，可吸附解吸多次反复使用，能经受水浸、高温、高压作用，不易破碎。表 8-2、表 8-3 分别介绍了几种活性炭的基本性能。

适合废水处理的 GAC 的性能　　表 8-2

<table>
<tr><th>序号</th><th>项目</th><th>数值</th><th>序号</th><th>项目</th><th>数值</th></tr>
<tr><td>1</td><td>比表面积(m^2/g)</td><td>950～1500</td><td>5</td><td>空隙容积(cm^3/g)</td><td>0.85</td></tr>
<tr><td rowspan="4">2</td><td>密度</td><td></td><td>6</td><td>碘值(最小)(mg/g)</td><td>900</td></tr>
<tr><td>堆积密度(g/cm^3)</td><td>0.44</td><td>7</td><td>磨损值(%)</td><td>70</td></tr>
<tr><td>颗粒密度(g/cm^3)</td><td>1.3～1.4</td><td>8</td><td>灰分(最大)(%)</td><td>8</td></tr>
<tr><td>真密度(g/cm^3)</td><td>2.1</td><td>9</td><td>包装后含水率(最大)(%)</td><td>2</td></tr>
<tr><td rowspan="3">3</td><td>粒径</td><td></td><td rowspan="4">10</td><td>筛径(美国标准)</td><td></td></tr>
<tr><td>有效粒径(mm)</td><td>0.8～0.9</td><td>大于 8 号(最大)(%)</td><td>8</td></tr>
<tr><td>平均粒径(mm)</td><td>1.5～1.7</td><td>小于 30 号(最大)(%)</td><td>5</td></tr>
<tr><td>4</td><td>均匀系数</td><td>≤1.9</td><td></td><td></td></tr>
</table>

活性炭的基本性能 **表 8-3**

活性炭形状	原料	活化法	比表面积 (m^2/g)	平均孔径 (Å)	吸附量 (%)	碘吸附量 (g/g)	焦糖脱色率 (%)
粉末	木材	药品	700～1000	20～50		0.7～1.2	85～98
	木材	气体	800～1500	15～30		0.8～1.2	70～95
	其他	气体	750～1350	15～35		0.8～1.1	60～93
颗粒	煤	气体	850～1250	15～25	30～40	0.7～1.2	30～65
	石油	气体	900～1350	15～25	33～435	0.8～1.2	
纤维状	其他	气体	1000～2000	15～25	33～50	0.8～1.2	

注：1. 碘值：在一定程度上能反映活性炭对废水中小分子有机物的吸附能力。
2. 焦糖值：在一定程度上能反映活性炭对较大分子有机物的吸附能力。

活性炭的吸附以物理吸附为主，但在碳化及活化的过程中，氢和氧与碳的化学键结合，使活性炭的表面上有各种有机官能团，从而促使活性炭与吸附质分子发生化学作用，形成一些化学选择性吸附。如果在活性炭中渗入一些具有催化作用的金属离子（如渗银），可以改善处理效果。

纤维活性炭（ACF）亦称活性炭纤维，是一种新型高效吸附材料。它是由有机碳纤维经活化处理后形成的。其超过 50%的碳原子位于内外表面，构筑成独特的吸附结构，较发达的比表面积和较窄的孔径分布使得它具有更快的吸附脱附速度和更大的吸附容量，并形成更多的官能团，因此，纤维活性炭的吸附性能大幅超过目前普通的活性炭。

（2）活性白土、漂白土、硅藻土等天然矿物质：其主要成分是 SiO_2、Al_2O_3、Fe_2O_3，经适当加工活化处理后即可作为吸附剂使用，虽然吸附容量不大，选择吸附分离能力低，但这些天然材料来源广泛、价廉易得。

（3）硅胶：硅胶是一种坚硬、多孔结构的硅酸聚合物颗粒，其分子式为 $SiO_2 \cdot nH_2O$，是用酸处理硅酸钠水溶液生成的凝胶。控制其生成、洗涤和老化的条件，可调节和控制比表面积、孔体积和孔半径的大小。硅胶是极性吸附剂，对极性的含氮或含氧物质如酚、胺、吡咤、水、醇等易于吸附，对非极性物质吸附较难。

（4）活性氧化铝：一般都不是纯的 Al_2O_3，而是由水合物的无定形凝胶和氢氧化物晶体构成的多孔刚性骨架结构的物质，由铝的水合物加热脱水、活化而制成。水合物的结构和形态以及制备条件都影响产品的性质。活性氧化铝是没有毒性的坚硬颗粒，对多数气体性质稳定，在水或液体中不溶胀、软化或崩碎破裂，抗冲击和耐磨损的能力强，也适合用于吸附剂的载体。

（5）沸石分子筛：沸石分子筛是以通式（M^{2+}、M^{+}）$O \cdot Al_2O_3 \cdot nSiO_2 \cdot mH_2O$ 表示的铝硅酸盐晶体，是一种孔径大小均一的吸附剂，M^{2+}、M^{+} 为二价或一价金属离子，n 为硅铝比，m 为结晶水分子数。沸石分子筛具有许多孔穴和微孔，因此，具有很大的内表面积，吸附容量大。沸石分子筛的孔径大小均匀一致，只能吸附能通过孔道的分子。人工合成沸石是极性吸附剂，对极性分子具有很大的亲和力，能根据溶质极性的不同进行选择性吸附。

分子筛的化学稳定性、耐热稳定性、抗酸碱能力、机械强度、耐磨损性都较差。除了人工合成沸石外，我国天然沸石资源丰富，价格低廉，亦具有沸石分子筛的性能。天然沸石因成因和晶体结构不同，种类繁多，性质相差也较大，天然沸石离子交换容量和选择性较低，工业上常用改性的沸石。

（6）吸附树脂：吸附树脂是具有巨大网状结构的合成大孔径树脂，由苯乙烯、吡啶等单体和乙二烯苯共聚而成。这些大孔径树脂具有非极性到高极性多种类型，除价格较活性炭贵外，它的物理化学性能稳定，品种较多，可按不同的需求选择使用。

树脂吸附剂的结构容易人为控制，因而具有适应性大、应用范围广、吸附选择性特殊、稳定性高等优点，并且再生简单，多数为溶剂再生。在应用上其介于活性炭等吸附剂、离子交换树脂之间，且兼具它们的优点，既具有类似于活性炭的吸附能力，又比离子交换更易再生。树脂吸附剂具有筛选性，能从污水中有选择地吸附有机物，最适宜于吸附处理废水中微溶于水、分子量略大和带有极性的有机物，如脱酚、除油、脱色等。树脂的吸附能力一般随吸附质亲油性的增强而增大。

（7）腐殖酸类吸附剂：主要有天然的富含腐殖酸的风化煤、泥煤、褐煤等，它们可以直接使用或经简单处理后使用；将富含腐殖酸的物质与适当的胶粘剂结合可制备成腐殖酸系树脂。

腐殖酸是一组具有芳香结构的，性质与酸性物质相似的复杂混合物，内含的活性基团有酚羟基、羧基、醇羟基、甲氧基、羰基、醌基、胺基、磺酸基等。这些活性基团有阳离子吸附性能。腐殖酸对阳离子的吸附，包括离子交换、螯合、表面吸附、凝聚等作用。

腐殖酸系吸附剂可用于处理工业废水，尤其是放射性废水以及重金属废水，如汞、铬、锌、镉、铅、铜等。腐殖酸类物质在吸附重金属离子后，可以用 H_2SO_4、HCl、NaCl 等进行解吸。目前，这方面的应用还处于试验、研究阶段，吸附（变换）容量不高，适用的 pH 范围较窄，机械强度低等问题还需要进一步研究和解决。

8.4.5 吸附工艺与设备

吸附的工艺包括吸附和再生两个阶段。吸附剂在达到饱和吸附后，必须进行脱附再生，才能重复使用。脱附是吸附的逆过程，即在吸附剂结构不变化或者变化极小的情况下，用某种方法将吸附质从吸附剂孔隙中除去，恢复它的吸附能力。通过再生使用，可以降低处理成本；减少废渣排放；同时回收吸附质。

目前吸附剂的再生方法有加热再生、药剂再生、化学氧化再生、湿式氧化再生、生物再生等。在选择再生方法时，主要考虑三方面的因素：吸附融化性质；吸附机理；吸附质的回收价值。

在设计吸附工艺和装置时，应首先确定采用何种吸附剂，选择何种吸附和再生操作方式以及废水的预处理和后处理措施。一般需通过静态和动态试验确定处理效果、吸附容量、设计参数和技术经济指标。

吸附操作分间歇和连续两种。前者是将吸附剂投入废水中，不断搅拌，经一定时间达到吸附平衡后，用沉淀或过滤的方法进行固液分离。如果经过一次吸附，出水达不到要求时，则需增加吸附剂剂量和延长停留时间或者对一次吸附出水进行二次或多次吸附。间歇工艺适合于小规模、间歇排放的废水处理。当处理规模大时，需建设较大的混合池和固液分离装置。GAC 吸附床的典型参数见表 8-4。

连续式吸附操作是废水不断地流进吸附床，与吸附剂接触，当污染物浓度降至处理要求时，排出吸附柱。按照吸附剂的充填方式，又分固定床、移动床和流化床三种，目前工程应用的主要是前两者。

（1）固定床吸附装置

固定床是吸附中最常用的方式，其构造与给水处理中使用的快速砂滤池大致相同，如图 8-2 所示。把颗粒状的吸附剂填装在吸附装置（柱、塔、罐）中，使含有吸附质的流体流过吸附装置时进行吸附，从而实现废水水质净化的方法。若吸附剂数量足够，则吸附装置处理出水的吸附质浓度可以降得很低，甚至为零。吸附剂使用一段时间后，出水中的吸附质含量会逐渐增加，当增加到一定程度时，吸附剂需要再生。再生可以和吸附在同一装置内交替进行，也可以将失效的吸附剂卸出进行处理。

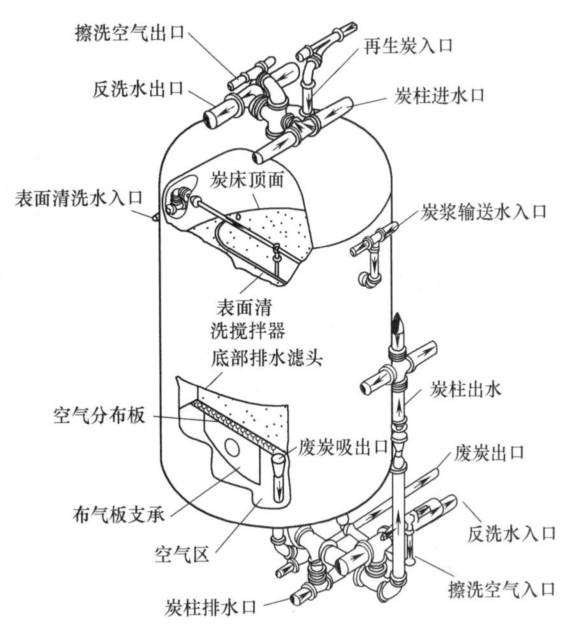

图 8-2　固定床吸附器

固定床按照水流方向又可分为升流式和降流式两种。降流式固定床出水水质较好，但经过吸附层的水头损失较大，特别是含悬浮物较高的废水。为防止吸附剂被悬浮物堵塞，需定期进行反冲洗。对于升流式固定床，当水头损失增大时，可适当提高水流速度，使吸附剂稍有膨胀，降低层内水头损失增长速度，延长运行时间，但流速的增加有可能会造成吸附剂的流失。

固定床根据处理水量、水质和处理要求，可将吸附床分成单床式、多床串联式和多床并联式三种。当处理水量大时，可采用并联方式；而为了提高处理效果，可采用串联方式；当处理水量较少时，可采用单床式。其操作工艺参数建议根据水质水量以及设备装置考虑，推荐参数如下：

塔径（D）：1～3.5m；塔高（H）：3～10m；填充层与塔径比：1∶4～1∶1；吸附时间（t）：10～50min；线速度（V）：2～10m/h。

GAC 吸附床的典型参数　　表 8-4

参数	符号	单位	设计值	参数	符号	单位	设计值
体积流量	V	m^3/h	50～400	有效接触时间	t	min	2～10
炭床体积	V_b	m^3	10～50	空床接触时间	$EBCT$	min	5～30
横断面面积	A_b	m^2	5～30	操作时间	t	d	100～600
长度	D	m	1.8～4	炭床体积	BV	m^3	10～100
空隙比	α	m^3/m^3	0.38～0.42	比通过体积	V_{sp}	m^3/kg	50～200
GAC 密度	ρ	kg/m^3	350～550	通过体积	V_L	m^3/m^3	2000～20000
流速	V_f	m/h	5～15				

（2）移动床吸附装置

图 8-3 为移动床吸附塔构造示意图。原水从下而上流过吸附层，吸附剂由上而下间歇或连续移动。间歇移动床处理规模大时，每天从塔底定时卸炭 1～2 次，每次卸炭量为塔

内总炭量的5%～10%；连续移动床，即饱和吸附剂连续卸出，同时新吸附剂连续从顶部补入。理论上连续移动床层厚度只需一个吸附区的厚度。直径较大的吸附塔的进出水口采用井筒式滤网。移动床比固定床更能有效利用床层吸附容量，出水水质良好，且水头损失较小。由于原水从塔底进入，水中夹带的悬浮物随饱和炭排出，因而不需要反冲洗设备，对原水预处理要求较低，操作管理方便。目前较大规模废水处理时多采用这种操作方式。

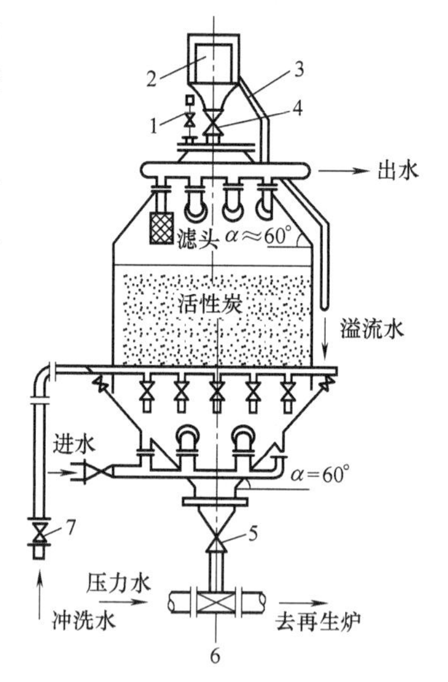

图8-3 移动床吸附塔构造示意图

1—通气阀；2—进料斗；3—溢流管；4、5—直流式衬胶阀；6—水射器；7—截止阀

8.4.6 吸附法在污水处理中的应用

利用吸附作用进行物质分离已有漫长的历史。在水处理领域，吸附法主要用于脱除水中的微量污染物，应用范围包括脱色，除臭味，脱除重金属、各种溶解性有机物、放射性元素等。利用吸附法进行水处理，具有适应范围广、处理效果好、可回收有用物料、吸附剂可重复使用等优点，但对进水的预处理要求较高，运转费用较高，系统庞大，操作较麻烦。

在污水处理中，吸附法处理的主要对象是废水中用生化法难于降解的有机物或用一般氧化法难于氧化的溶解性有机物，包括木质素、氯或硝基取代的芳烃化合物、杂环化合物、洗涤剂、合成染料、除锈剂、DDT等。当用活性炭对这类污水进行处理时，其不但能够吸附这些难分解的有机物，降低COD，还能使废水脱色、脱臭，把污水处理到可重复利用的程度。在处理流程中，吸附法可作为离子交换、膜分离等方法的预处理，以去除有机物、胶体物及余氯等；也可以与其他物理化学法联合，组成所谓物化流程，作为二级处理后的深度处理手段，以保证回用水的质量，比如先用混凝、沉淀、过滤等去除悬浮物和胶体，然后用吸附法去除溶解性有机物。吸附法也可与生化法联合，如向曝气池投加粉状活性炭；利用粒状吸附剂作为微生物的生长载体或作为生物流化床的介质；或在生物处理之后进行吸附深度处理等，这些联合工艺都在工业上得到应用。

8.5 氧化法工艺及应用

对于一些有毒有害的污染物质，当难以用生物法或其他方法处理时，可利用它们在化学反应过程中能被氧化或还原的性质，改变污染物的形态，将其变成无毒或微毒的新物质，或者转化成容易与水分离的形态，从而达到处理的目的，这种方法称为氧化还原法。

废水处理中常用的氧化剂主要包括空气、臭氧、过氧化氢、高锰酸钾、氯气、液氯、次氯酸钠及漂白粉等。除此之外，近年发展起来的高级氧化技术是以羟基自由基（·OH）作为氧化剂实现有机污染物的降解。

以氯气、臭氧为代表的一些强氧化剂也同时充当消毒剂的功能。因为城市污水经二级处理后，水质已经改善，细菌含量也大幅度减少，但细菌的绝对数量仍很可观，并存在有病原菌的可能，必须在去除掉这些微生物以后，废水才可以安全地排入水体或循环再用。

随着居民对生活品质要求的不断提高，污水处理厂的二级处理出水对城市水体造成的影响引起人们对健康和安全问题的更多关注。消毒是灭活这些致病生物体的基本方法之一，因此，污水处理厂的排出水消毒已经成为污水深度处理中必须的一道工序。

8.5.1 化学氧化法

1. 空气氧化法

空气氧化法是利用空气中的氧气作为氧化剂，使一些有机物和还原性物质氧化的一种处理方法。空气氧化法既可用于水溶液体系，也可用于气相及固相体系，因此，可以用空气氧化法处理废水、废气及固体废弃物。因为空气氧化能力较弱，所以其主要用于含还原性较强物质的废水处理，如硫化氢、硫酸、硫的钠盐和铵盐等。

例如，炼油厂含硫废水中的硫化物即可用空气氧化处理。该废水中的硫化物，一般以钠盐（$NaHS$、Na_2S）或铵盐［NH_4SH、$(NH_4)_2S$］形式存在。废水中的硫化物与空气中的氧发生的氧化反应如下：

$$2HS^- + 2O_2 \rightarrow S_2O_3^{2-} + H_2O \tag{8-14}$$

$$2S^{2-} + 2O_2 + H_2O \rightarrow S_2O_3^{2-} + 2OH^- \tag{8-15}$$

$$S_2O_3^{2-} + 2O_2 + 2OH^- \rightarrow 2SO_4^{2-} + H_2O \tag{8-16}$$

废水中有毒的硫化物和硫氢化物被氧化为无毒的硫代硫酸盐和硫酸盐。上述第三步反应进行得比较缓慢。如果向污水中投加少量的氧化铜或氯化钴为作为催化剂，则几乎全部 $S_2O_3^{2-}$ 被氧化成 SO_4^{2-}，但应注意催化剂可能引起的重金属污染问题。

空气氧化脱硫设备多采用氧化塔，处理含硫废水的工艺流程如图 8-4 所示，废水、空气及蒸汽经射流混合后，送至空气氧化脱硫塔。通蒸汽是为了提高温度，加快反应速率。氧化脱硫塔使用拱板分为数段，拱板上安装喷嘴。当废水和空气以较高的速度冲出喷嘴时，空气被粉碎为细小的气泡，增大气液两相的接触面积，使氧化速度加快。在气液并流上升到该段顶的拱板时，气泡会破裂和合并，产生气液分离现象。喷嘴底部缝隙的作用是使气体能够再度均匀地分布在废水中，然后经过喷嘴进一步混合，这样就消除了气阻现象，使塔内压力稳定。有实例表明使用该氧化塔时脱硫效率可在 94.3%～98.3%之间，处理费用为 0.6～0.9 元/m^3（废水）。

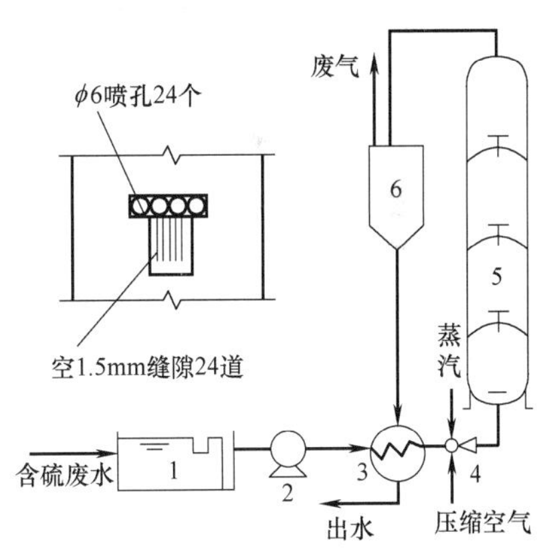

图 8-4 氧化塔处理含硫废水的工艺流程

1—隔油池；2—泵；3—换热器；4—射流器；5—空气氧化塔；6—气液分离器

制革工业中常将石灰、硫化钠等用作脱毛剂，由此产生碱性含硫废水。此类废水 pH=11～13，硫化物含量在 2000～4000mg/L（以 S^{2-} 计），也可采用空气氧化法处理。为提高氧化速度，缩短处理时间，常添加锰盐（如 $MnSO_4$）作催化剂。在国外，催化氧化法是处理含硫废水使用较广泛的方法，在国内所调查的 50 家制革厂中，有 8 家制革厂进行了硫的单项治理，其处理方法也大多是催化氧化法。某制革厂处理流程为：含硫废水经过格栅后，用泵抽入装有充氧器的曝气池氧化，投加 500mg/L 左右的 $MnSO_4$ 作催化剂，曝气时间为 3～6h，气水比约为 15。处理后出水的 S^{2-} 可以降到 5～10mg/L，并且在废水 pH=11～13 的条件下，Mn^{2+} 的残留量在 10^{-7}mol/L 以下，低于排放标准 1mg/L 的要求。

2. 臭氧氧化法

由于臭氧不稳定，因此，通常在现场随制随用。以空气为原料制造臭氧，由于原料来源方便，所以采用比较普遍。典型臭氧处理闭路系统如图 8-5 所示。

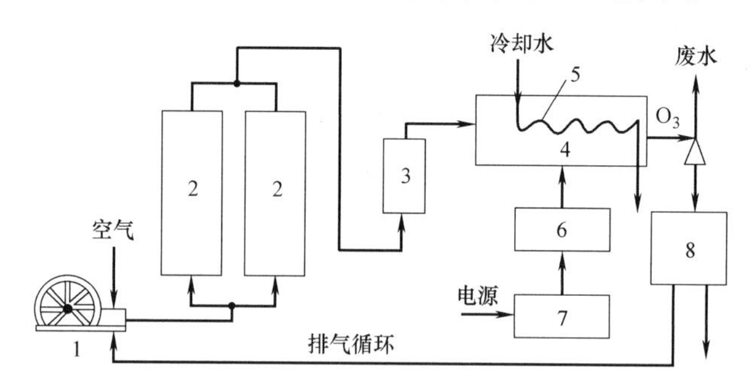

图 8-5 臭氧处理闭路系统

1—空气压缩机；2—净化装置；3—计量装置；4—臭氧发生器；
5—冷却系统；6—变压器；7—配电装置；8—接触器

空气经压缩机加压后，经过冷却及吸附装置除杂，得到的干燥净化空气再经计量装置进入臭氧发生器。要求进气露点在－50℃以下，温度不能高于 20℃，有机物含量小于 15×10^{-6}。

臭氧发生器有板式和管式两种。因板式发生器只能在低压下操作，所以目前多采用管式发生器。管式发生器的外形像列管式换热器，内有几十根甚至上百根相同的放电管。放电管的两端固定在两块管板上，管外通冷却水。每根放电管均由两根同心圆管组成，外壳为金属管（不锈钢管或铝管），内管为玻璃管作介电体。内管一端封闭，管内壁镀有银膜或铝膜作电极。不锈钢管及玻璃管内膜与高压电源相连。内、外管之间留有 1～3mm 的环形放电间隙。管式发生器可承受 0.1MPa（表）的压力，当以空气为原料，采用 50Hz 的电源时，臭氧浓度为 15～20g/m^3，电能比耗为 16～18kWh/kgO_3。

水的臭氧处理在接触反应器内进行，常用鼓泡塔、螺旋混合器、蜗轮注入器、射流器等。选择何种反应器取决于反应类型。当过程受传质速度控制时，如无机物氧化、消毒等，应选择传质效率高的螺旋反应器、蜗轮注入器、喷射器等；当过程受反应速度控制时，如有机物和 NH_4-N 的去除，应选用鼓泡塔，以保持较大的液相容积和反应时间。

水中污染物种类和浓度、臭氧的浓度与投量、投加位置、接触方式和时间、气泡大小、水温与水压等因素对反应器性能和氧化效果都有影响。

（1）O_3 氧化处理废水

水经臭氧处理，可达到降低 COD、杀菌、增加溶解氧、脱色除臭、降低浊度的目的。臭氧之所以表现出强氧化性，是因为臭氧分子中的氧原子具有强烈的亲电子或亲质子性，臭氧分解产生的新生态氧原子也具有很高的氧化活性。某炼油厂利用 O_3 处理重油裂解废水，废水含酚 4～5mg/L，CN^- 6mg/L，S^{2-} 4～5mg/L，油 3～5mg/L，COD 400～500mg/L，pH 为 11，水温为 45℃。投加 O_3 280mg/L，接触 12min。处理出水含酚 0.005mg/L，CN^- 0.1～0.2mg/L，S^{2-} 0.3～0.4mg/L，COD 90～120mg/L，油 2～3mg/L。

将混凝或活性污泥法与臭氧氧化法联合，可以有效地去除色度和难降解的有机物。紫

外线照射以激活 O_3 分子和污染物分子，加快反应速度，增强氧化能力，降低臭氧消耗量。目前臭氧氧化法存在的缺点是电耗大，成本高。

（2）O_3 消毒

臭氧极不稳定，分解时产生初生态氧。初生态氧［O］具有极强的氧化能力，对微生物如病毒、细菌、甚至芽孢等都有强大的杀伤力。还具有很强的渗入细胞壁的能力，从而破坏细菌有机体结构，导致细菌的死亡。图 8-6 为常规臭氧消毒工艺流程。

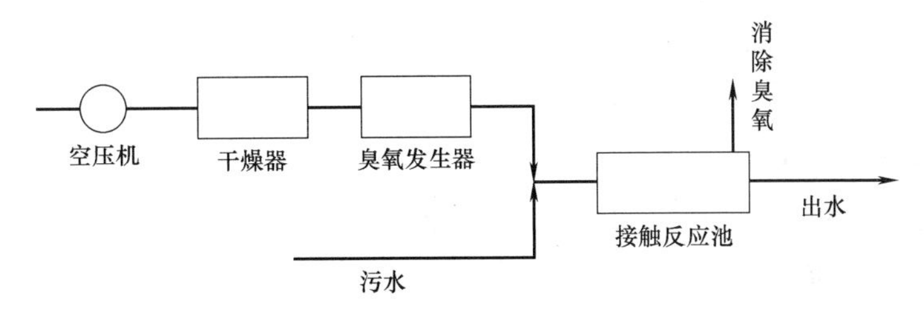

图 8-6　常规臭氧消毒工艺流程

臭氧在水中的溶解度仅为 10mg/L，因此，通入污水中的臭氧往往不可能全部被利用，为了提高臭氧的利用率，接触反应池最好建成水深为 5～6m 的深水池，或建成封闭的几格串联的接触池，设置管式或板式微孔臭氧扩散器。扩散器由陶瓷或聚氯乙烯微孔塑料或不锈钢制成。臭氧消毒迅速，接触时间可采用 15min，可维持剩余臭氧量为 0.4mg/L。接触池排出剩余臭氧，具有腐蚀性，因此，排出的剩余臭氧需做消除处理。臭氧不能贮存，需现场边生产边使用。

臭氧的消毒能力比氯更强。对脊髓灰质炎病毒，用氯消毒，保持 0.5～1mg/L 余氯量需 1.5～2h，而达到同样效果，用臭氧消毒，保持 0.045～0.45mg/L 剩余 O_3 只需 2min。若初始 O_3 超过 1mg/L，经 1min 接触，病毒去除率可达到 99.99%。

3. 氯氧化法

（1）氯氧化剂特性

可在氯氧化法中使用的氯系氧化剂包括液氯，氯的含氧酸及其钠盐、钙盐以及二氧化氯等。氯系氧化剂均为氧化性较强的氧化剂。

各种氯系氧化剂中所含有的氯并不能全部起到氯化作用，因此，采用有效氯的概念表示药剂的氧化能力，具体含义是药剂所含氯中可起氧化作用的氯的比例。由于氯及其大多数衍生物的还原最终都要产生氯化物离子，因此，有效氯指的是化合物中化合价大于氯化物离子（即负一价）的那部分氯。一般以 Cl_2 作为 100%有效氯的基准进行比较。在这一比较基准的反应式中，每个分子的 Cl_2 发生了两个电子的变化。表 8-5 为根据上述定义和基准计算的各种氯系氧化剂的有效氯含量。

各种氯系氧化剂的有效氯含量　　表 8-5

物质名称	分子量	氯当量(mol)	含氯量(%)	有效氯(%)
液氯 Cl_2	71	1	100	100
氧化二氯 Cl_2O	87	2	81.7	163.2
二氧化氯 ClO_2	67.5	2.5	52.5	263
次氯酸钠 NaClO	74.5	1	47.7	95.3

续表

物质名称	分子量	氯当量(mol)	含氯量(%)	有效氯(%)
漂白粉 Ca(ClO)Cl	127	1	56	56
次氯酸钙 $Ca(ClO)_2$	143	2	49.6	99.3
亚氯酸钠 $NaClO_2$	90.5	2	39.2	157
次氯酸 HClO	52.5	1	67.7	135.2
二氯胺 $NHCl_2$	86	2	82.5	165.1
一氯胺 NH_2Cl	51.5	1	69	138

（2）液氯氧化处理废水

在废水处理中，可以利用氯系氧化剂氧化分解废水中酚类、醛类、醇类以及洗涤剂、油类、氰化物等，利用氯氧化法还可进行脱色、除臭、杀菌等处理。

以氯系氧化法除酚为例。酚在氯的氧化作用下开始降解为邻苯二酚及邻苯醌，然后再分解为顺丁烯二酸等。生成的顺丁烯二酸还可被进一步氧化为 CO_2 和 H_2O。同时，还会发生取代反应，生成有强烈异臭和潜在危害的氯酚（主要是 2,6-二氯酚）。为消除氯酚的危害，一方面可投加过量氯（视含酚量情况增大 1.25～2.0 倍），或改用 O_3、ClO_2 等氧化作用更强的氧化剂，以防止氯酚生成；另一方面可采用活性炭对出水进行后处理，除去水中的氯酚及其他氯代有机物。

瓶装液氯在实际中应用较多。由液氯蒸发产生的氯气一般不直接加入水中，而是用加氯器配成氯的水溶液再加入水中。考虑使用安全，常采用真空加氯器（如文丘里水射流加氯器）。

液氯的沸点较低，易气化，在运输和贮藏过程中要防止液氯温升过高，以免引起爆炸，氯气是有毒气体，当液氯瓶大量漏氯不能制止时，可把整个氯瓶投入水池中，或者用大量的水喷淋，让泄漏氯气溶解在水里，也可把氯气通入碱性溶液进行吸收。

4. 氯化消毒

（1）氯系消毒剂的种类及性能

氯系消毒剂包括液氯、漂白粉、漂白精、次氯酸钠、氯胺等。漂白粉主要成分为次氯酸钙，外观呈灰白色颗粒状粉末，有氯气味，其化学成分约为 $3Ca(ClO)Cl \cdot Ca(OH) \cdot nH_2O$，一般以 Ca(ClO)Cl 代表其分子式。漂白粉在空气中易吸收水分和 CO_2，不稳定，有效氯含量约为 28%～32%。漂白精为较纯的次氯酸钙，外观呈白色粉末，其化学分子式为 $Ca(ClO)_2$，性能比较稳定，有效氯含量约为 60%～70%。次氯酸钠的分子式为 NaClO，在水溶液中的含量约为 8%～12%，性质不稳定，宜保存在 pH 大于 12 的碱性溶液中。除商品型水溶液制剂外，次氯酸钠还可由次氯酸钠发生器电解食盐水现场制备，氨胺为有机型氯消毒剂，外观呈白色或淡黄色结晶，氯味及刺激性小，稳定，耐贮存，有效氯含量约为 35%。

氯系消毒剂中“有效氯”这个概念指氯化合物中以正价形式存在的具有消毒作用的氯，通常以重量百分比表示。在消毒过程中，正价态的氯在与细菌酶系统的氧化还原反应中被还原成负一价，从而失去继续消毒的作用。有效氯含量越高，则相应的消毒剂用量就越少。常见的氯系消毒剂除二氧化氯（ClO_2）中的氯为正四价外，其他如次氯酸钠、次氯酸钙、氯胺中的氯都是正一价。氯气（Cl_2）可看作由 Cl^+ 和 Cl^- 组成。

（2）氯化消毒原理

一般认为次氯酸是氯的衍生物中杀菌力最强的成分，氯系消毒剂主要通过水解产物次氯酸（HClO）起作用。其原因在于 HClO 为很小的中性分子，只有它才能比较容易地扩散到带负电的细菌表面，穿过细胞壁渗入菌体内部，进而与细菌的酶系统发生不可逆的氧化反应，使细菌由于酶系统遭到钝化破坏而被灭活。ClO^- 虽亦具有杀菌能力，但因带有负电荷，难以接近呈负电性的细菌表面，杀菌能力比 HClO 差得多。工艺操作的 pH 越低，氯系消毒剂的消毒作用越强，这一结果间接证明 HClO 是消毒的主要因素。

（3）加氯量、需氯量、剩余氯量及折点加氯

氯化消毒操作的加氯量包括需氯量和剩余氯量两个部分。需氯量指用于达到指定的消毒指标（大肠菌数指标）以及氧化水中所含的有机物和还原性物质等所需的有效氯量。除此之外，为抑制水中残存的细菌再度繁殖，在水中还需维持少量残余有效氯量，即为剩余氯量，或简称余氯。剩余氯量用 10min 接触后的游离性有效余氯量或 60min 接触后的综合性有效余氯量（游离氯和氯胺）表示。不同工艺操作条件下，加氯量与剩余氯量之间的关系不尽相同。其具体情况如下：

1）对于洁净水，即水中无微生物、有机物、还原性物质、氨、含氮化合物的理想状况，其需氯量为零，加氯量等于剩余氯量，两者间的关系如图 8-7 及图 8-8 中 45°的倾斜虚线①所示。

2）当水中只含有消毒对象细菌以及有机物、还原性无机物等需氧物质时，加氯量为需氯量与剩余氯量之和，两者间的关系如图 8-8 中的实线②所示。

3）当水中的需氯杂质主要是氨和含氮化合物时，加氯量与剩余氯量之间的关系如图 8-8 中的 OMABP 曲线所示，即曲线②。在该图中，曲线与虚线之间的垂直距离为需氯量，曲线与坐标之间的垂直距离为剩余氯量。

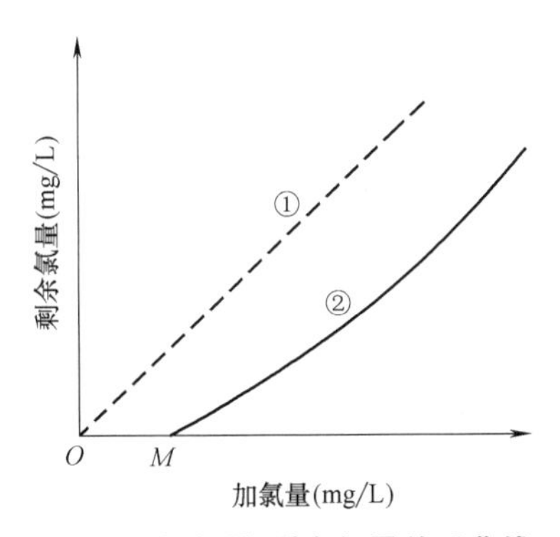

图 8-7　加氯量-剩余氯量关系曲线

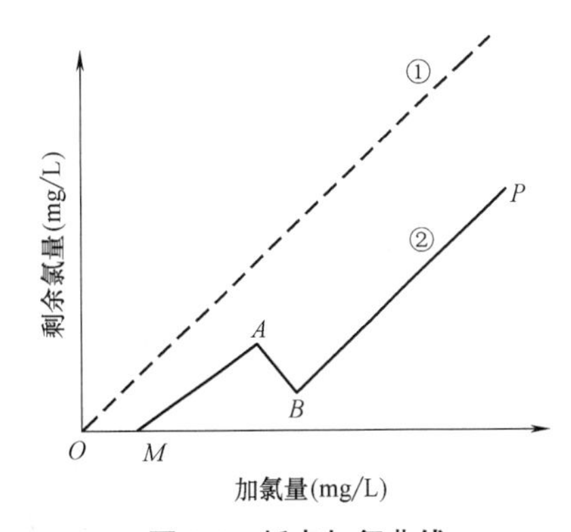

图 8-8　折点加氯曲线

由图 8-8 可以看出，在 OM 段，所投加的有效氯均被水中的细菌及其他杂质消耗，余氯为零，此时的消毒效果不可靠，细菌有再度繁殖的可能；在 MA 段，有效氯与氨生成氯胺（主要成分为一氯胺），余氯以化合性氯的形式存在，所以有一定的消毒效果；在 AB 段，仍然产生化合性余氯，但随加氯量的继续增加，部分氯胺被氧化分解为不起消毒作用的 N_2、NO、N_2O 等，反而导致化合性余氯逐渐减少，其含量由峰点 A 直降至最小

值，即折点 B 的位置；自 B 点之后，水中的需氯杂质已基本消耗殆尽，BP 段所投加的有效氯全部用于增加游离态的余氯量，在此阶段的消毒效果稳定可靠。所谓折点加氯，即加氯量超过折点需要量时的氯化消毒操作。

由于氯化消毒过程具有上述种种情形，在实际加氯操作中，应视原水水质、消毒要求等情况，控制适宜的加氯量。例如，当原水游离氨含量低于 0.3mg/L 时，通常将控制加氯量超过折点 B，以维持一定的游离态余氯量；而当原水游离氨含量在 0.5mg/L 以上时，则将加氯量控制在峰点 A 以前即可，此时的化合性余氯量已满足消毒要求。通常，原水游离氨含量在 0.3～0.5mg/L 范围时，加氯量往往难以把握，如控制在峰点 A 以前，有时不能达到化合性余氯量的要求，而控制在峰点 A 以后，又造成加氯量的浪费。

（4）氯化消毒工艺设备

氯化消毒系统由消毒剂贮存或发生设备、投加设备、混合池、接触池和自动控制设备等组成。消毒处理系统应具备关键设备安全可靠、定比投加，能够保证消毒剂与水的快速混合与充分的接触时间等特点。

氯气（Cl_2）在氯化消毒中的使用最为广泛。为便于贮藏和运输，商品化氯气通常是将 Cl_2 液化后以液氯形式灌入钢瓶，在减压条件下操作使用。由于 Cl_2 的密度约是空气的 2.5 倍，一旦发生泄漏不易散发，易造成操作人员中毒；在环境温度升高、液氯瓶内的压力过度增加时，有可能引起爆炸，因此，运输及使用过程中的安全问题必须予以足够的重视。

8.5.2 高级氧化法

高级氧化工艺（Advanced Oxidation Process，AOP）的概念是 Glaze W. H 等 1987 年提出的。其指利用强氧化剂羟基自由基（·OH）有效地破坏水相中污染物的化学反应，可通过加入氧化剂、催化剂或借助紫外光或可见光、超声波等方法产生羟基自由基。

目前研究的高级氧化技术有 Fenton 试剂氧化法、光催化氧化法、紫外光氧化法、湿式氧化与催化混式氧化法、超声波催化氧化法、超临界氧化法、微波氧化法、臭氧氧化法等。

1. Fenton 试剂氧化法

Fenton 试剂是亚铁离子和过氧化氢的组合，当 pH 低时（一般要求 pH 在 3 左右），在 Fe^{2+} 的催化下过氧化氢会分解产生·OH，从而引发链式反应。作为强氧化剂，Fenton 试剂已有 100 多年的历史，在精细化工、医药化工、医药卫生、环境污染治理等方面得到广泛的应用。

（1）Fenton 试剂氧化法原理

$$Fe^{2+} + H_2O_2 \rightarrow Fe^{3+} + \cdot OH + OH^- \tag{8-17}$$

$$Fe^{2+} + \cdot OH \rightarrow Fe^{3+} + OH^- \tag{8-18}$$

$$Fe^{3+} + H_2O_2 \rightarrow Fe^{2+} + HO_2 + H^+ \tag{8-19}$$

$$HO_2 + H_2O_2 \rightarrow O_2 + H_2O + \cdot OH \tag{8-20}$$

$$RH + \cdot OH \rightarrow \cdots \rightarrow CO_2 + H_2O \tag{8-21}$$

$$4Fe^{2+} + O_2 + 4H^+ \rightarrow 4Fe^{3+} + 2H_2O \tag{8-22}$$

$$Fe^{3+} + 3OH^- \rightarrow Fe(OH)_3\text{(胶体)} \tag{8-23}$$

Fe^{2+} 与 H_2O_2 间反应很快，生成·OH 自由基，由表 8-6 可见，·OH 氧化能力很强，仅次于 Fe^{2+}，有三价铁共存时，由 Fe^{3+} 与 H_2O_2 缓慢生成 Fe^{2+}，Fe^{2+} 再与 H_2O_2 迅

速反应生成·OH，与有机物 RH 反应，使其发生碳链裂变，最终氧化为 CO_2 和 H_2O，从而使废水的 COD 大幅降低。同时 Fe^{2+} 作为催化剂，最终可被 O_2 氧化为 Fe^{3+}。在一定 pH 下，可有 $Fe(OH)_3$ 胶体出现，其有絮凝作用，可大幅降低水中的悬浮物。

基团与普通氧化剂分子的氧化单位 **表 8-6**

氧化剂	Fe^{2+}	·OH	O_3	H_2O_2	HOO	HOCl	Cl_2
氧化电位(V)	3.06	2.80	2.07	1.77	1.70	1.49	1.39

Fenton 法是一种高级化学氧化法，常用于废水高级处理，以去除 COD 色度和味道等。Fenton 试剂氧化一般在 pH<3.5 的条件下进行，在该 pH 时其自由基生成速率最大。

Fenton 试剂及各种改进系统在废水处理中的应用可分为两个方面：一是单独作为一种处理方法氧化有机废水；二是与其他方法联用，如与混凝沉降法、活性炭法、生物法、光催化法等联用。

(2) 类 Fenton 试剂法

在常规 Fenton 试剂法中引入紫外光（UV）、光能、超声、微波、电能等可以提高 H_2O_2 催化分解产生·OH 的效率，增强 Fenton 试剂的氧化能力。例如：$H_2O_2/Fe^{2+}/UV$、$H_2O_2/Fe^{2+}/O_2$、$H_2O_2/Fe^{2+}/UV/O_2$ 等组合工艺，其优点就是可降低 H_2O_2 的用量，紫外光和 Fe^{2+} 对 H_2O_2 的分解具有协同作用。

2. 光催化氧化法

所谓光化学反应，就是在光的作用下进行的化学反应。在自然环境中有一部分近紫外光（190～400nm），他们极易被有机污染物吸收，在有活性物质存在时就会发生强烈的光化学反应使有机物降解。天然水体中存在大量的活性物质，如氧气、自由基以及有机还原物质等，因此，在光照的河水、海水表面发生着复杂的光化学反应。光降解通常指有机物在光作用下，逐步氧化成低分子中间产物，最终生成二氧化碳、水及其他离子（如 NO_3^-、PO_4^{3-}、卤素等）。利用光化学反应治理污染，包括无催化剂和有催化剂参与的光化学氧化。前者多采用臭氧和过氧化氢等作氧化剂，在紫外线的照射下使污染物氧化分解；后者又称为光催化氧化，一般可分为均相和多相（非均相）催化两种类型。

(1) 光催化氧化法的作用机理

光催化反应原理是以半导体能带理论为基础的。半导体粒子一般由填满电子的低能价带（valence band，VB）和空的高能导带（conduction band，CB）构成，价带和导带之间存在禁带（E_g）。当用能量等于或大于禁带宽度（$h_v \geq E_g$）的光照射时，半导体价带上的电子可被激发跃迁到导带，同时在价带上产生相应的空穴，这样就在半导体内部生成电子（e^-）-空穴（h^+）对。空穴具有很强的氧化活性，能够与溶液中的氢氧根离子和水分子反应生成羟基自由基，然后羟基自由基进一步与污染物反应，将污染物降解成二氧化碳和无机物；而光电子则与溶液中的溶解 O_2 发生还原反应，生成过氧化氢，过氧化氢与污染物反应生成水分子和无机物。Fujishima A 和 Honda K 于 1972 年首先发现了 TiO_2 在光照条件下可将水分解为 H_2 和 O_2，使这一技术被迅速应用于废水治理中，已有大量研究证明众多难降解有机物在光催化氧化的作用下可有效去除或降解。以 TiO_2 为例，该过程可用下式描述：

$$TiO_2 + h_v \rightarrow h_{vb}^+ + e^- \quad (8\text{-}24)$$

$$TiO_2(h_{vb}^+) + H_2O \rightarrow TiO_2 + \cdot OH + H^+ \quad (8\text{-}25)$$

$$TiO_2(h_{vb}^+) + OH^- \rightarrow TiO_2 + \cdot OH \quad (8\text{-}26)$$

$$TiO_2(h_{vb}^+) + RH + OH^- \rightarrow TiO_2 + R \cdot + H_2O \quad (8\text{-}27)$$

Carey 等较详细地描述了 TiO_2 光降解水中污染物的历程：光催化剂在光照下产生电子-空穴对；羟基或水在光催化剂表面吸附后形成表面活性中心；表面活性中心氧化水中有机物；氢氧自由基形成，有机物被氧化；氧化产物脱离。其中有机物在光催化剂表面的反应最慢，是光催化氧化过程的控制步骤。

光催化剂就是在光子的激发下能够起到催化作用的化学物质的统称，可加速化学反应，其本身并不参与反应。实验室常用的光催化剂有 TiO_2、ZnO、WO_3、CdS、ZnS、Sr-TiO_3、SnO_2、Ag_3PO_4 等。其中 TiO_2 氧化能力强，化学性质稳定、无毒，是研究最广泛的催化剂。但其带隙较宽（3.2eV），只能在紫外光（仅占太阳辐射总量 5%，而可见光占 43%）照射下产生光催化活性，从而大幅限制其对太阳光的利用率。目前对 TiO_2 的研究主要集中在通过掺杂（C、N 等）、金属沉积、与小于其带隙的半导体构成异质结等方式增强可见光响应范围。Ag_3PO_4 是近年发现的一种可见光响应型催化剂，其量子产率（产氧率）高达 90%，远远大于目前已知的半导体光催化剂（20%）。但由于其结构特征，使其在反应过程中存在光腐蚀现象，目前对其研究主要集中在通过 Ag_3PO_4 与其他材料的复合有效降低光腐蚀，提高该光催化剂的稳定性。

（2）光催化氧化技术影响因素

1）光催化剂类型、粒径与用量

一般选用锐钛矿型 TiO_2 作光催化剂；粒径越小，反应速率越大；催化剂用量，一般认为在 2～4g/L 较合适。

2）光源强度与光照

同等波长下，一般光越强，效率越高；同等光强下，一般波长越短，效率越高。

3）溶液 pH

不同类型、不同结构的污染物降解有各自的最适 pH。

4）污染物初始浓度

光催化剂对污染物的降解都有一个最适宜的初始浓度，浓度过高会存在竞争的关系。

5）氧化剂和还原剂

O_2、H_2O_2、O_3、$S_2O_8^{2-}$ 等均是良好的电子捕获剂，能有效地使电子和空穴分离，提高催化效率；废水中 Cl^-、NO_2^-、SO_4^{2-}、PO_4^{3-} 能与有机物竞争空穴，将会显著降低光催化效率，尤其 PO_4^{3-} 对光催化效率影响很大。

3. 紫外光氧化法

水银灯发出的紫外光能穿透细胞壁并与细胞质反应而达到消毒目的。紫外光波长为 250～360nm 的杀菌能力最强。因为紫外光需照进水层才能起消毒作用，故污水中的悬浮物、浊度、有机物和氨氮都会干扰紫外光的传播，因此，处理水水质越好，光传播系数就越高，紫外线消毒的效果也越好。

紫外线光源是高压石英水银灯，杀菌设备主要有浸水式和水面式两种。浸水式是把石英灯管置于水中，此法的特点是紫外线利用率较高，杀菌效能好，但设备的构造较复杂，水面式的构造简单，但由于反光罩吸收紫外光线以及光线散射，杀菌效果不如前者。紫外线消毒的照射强度为 0.19～0.25W・s/cm^2，污水层深度为 0.65～1.0m。

在表 8-7 中对前述的几种消毒技术进行了综合比较。

常用消毒技术的比较　　表 8-7

消毒方式	优点	缺点	适用条件
紫外线消毒	符合环境保护要求，不会产生三卤甲烷等"三致"物质；杀菌迅速，无化学反应；接触时间短，土建费用少；运行成本低，占地面积小	杀菌效果受出水水质影响较大；没有持续杀菌能力	大、中、小型污水处理厂二级生化处理后的污水
液氯消毒	效果可靠，成本较低；投配设备简单，投量准确	易产生"三致"物质，氧化形成的余氯及某些含氧化合物对水生物有毒害	大、中、小型污水处理厂二级生化处理后的污水、再生水
二氧化氯消毒	具有较好的消毒效果，不会产生"三致"物质	不能储存，只能现制现用	中、小型污水处理厂二级生化处理后的污水、再生水
臭氧消毒	消毒效率高，不产生难处理的生物积累性余物	投资大，成本高，设备管理复杂	常规二级生化处理后的污水、再生水

4. 湿式氧化与催化湿式氧化法

（1）工艺流程及设备

湿式氧化（Wet Oxidation，WO）是在高温（125～320℃）和高压（0.5～20MPa）条件下，以空气或氧气为氧化剂，氧化废水中溶解和悬浮的有机物和还原性无机物的一种方法。在 WO 工艺基础上添加适当的催化剂即成为催化湿式氧化法（CWO）工艺。因氧化过程在液相中进行，故湿式氧化与一般方法相比，具有适用范围广（包括对污染物种类和浓度的适应性）、处理效率高、二次污染低、氧化速度快、装置小、可回收有用物料和能量等优点。

湿式氧化工艺最初由美国的 Zimmermann 研究提出，20 世纪 70 年代以前主要用于城市污水处理的污泥和造纸黑液的处理。20 世纪 70 年代以后，湿式氧化技术发展很快，应用范围扩大，装置数目和规模增大，并开始了催化湿式氧化的研究与应用。20 世纪 80 年代中期以后，湿式氧化技术向三个方向发展：第一，继续开发适于湿式氧化的高效催化剂，使反应能在比较温和的条件下，在更短的时间内完成；第二，将反应温度和压力进一步提高至水的临界点以上，进行超临界湿式氧化；第三，回收系统的能量和物料，目前湿式氧化法在国外已广泛用于各类高浓度废水及污泥处理，尤其是毒性大，难以用生化方法处理的农药废水、染料废水、制药废水、煤气洗涤废水、造纸废水、合成纤维废水及其他有机合成工业废水的处理，还可用于还原性无机物（如 CN^-、SCN^-、S^{2-}）和放射性废物的处理。

废水和空气分别从高压泵和压缩机进入热交换器，与已氧化液体换热，使温度上升到接近反应温度。进入反应器后，废水中有机物与空气中氧气反应，反应热使温度升高，并维持在较高的温度下反应。反应后，液相和气相经分离器分离，液相进入热交换器预热进料，废气排放。在反应器中维持液相是该工艺的特征，因此，需要控制合适的操作压力。在装置初开或需要附加热量的情况下，直接用蒸汽或燃油作热源。由基本流程出发，可得多种改进流程，以回收反应尾气的热能和压力能。用于处理浓废液（浓度≥10%）并回收能量的湿式氧化流程如图 8-9 所示。图 8-10 与图 8-9 的不同在于对反应尾气的能量进行二次回收。首先由废热锅炉回收尾气的热能产生蒸汽或经热交换器预热锅炉进水，尾气冷凝水由第二分离器分离后送回反应器以维持反应器中液相平衡，防止浓废液氧化时释放的大量反应热将水分蒸干。第二分离器后的尾气送入透平机产生机械能和电能，该系统对能量实行逐级利用，减少了有效能损失。

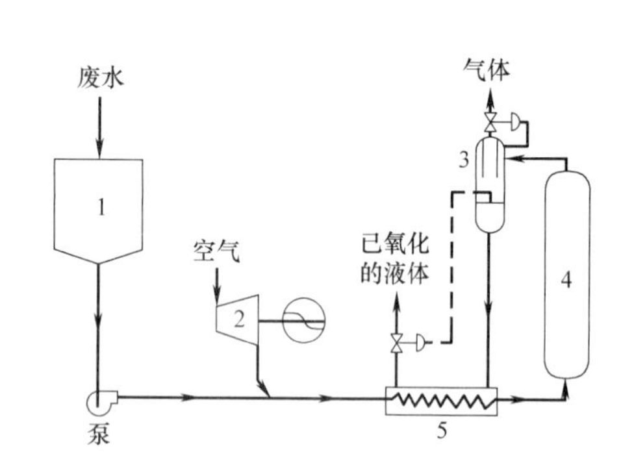

图 8-9　湿式氧化法基本流程

1—贮存罐；2—空压机；3—分离器；
4—反应器；5—热交换器

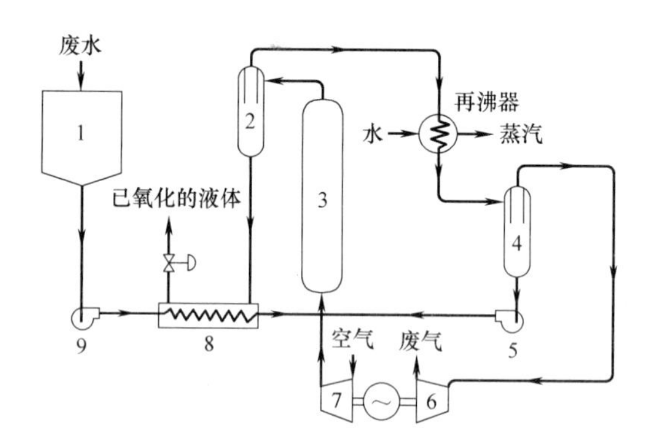

图 8-10　湿式氧化处理废液流程

1—贮存罐；2,4—分离器；3—反应器；5—循环泵；
6—透平机；7—空压机；8—热交换器；9—高压泵

湿式氧化系统的主体设备是反应器，除了要求其耐压、防腐、保温和安全可靠以外，同时要求器内气液接触充分，并有较高的反应速度，通常采用不锈钢鼓泡塔。反应器的尺寸及材质主要取决于废水性质、流量、反应温度、压力及时间。

（2）影响因素

湿式氧化的处理效果取决于废水性质和操作条件（温度、氧分压、时间、催化剂等），其中反应温度是最主要的影响因素。温度对氧化效果的影响如图 8-11 所示。

由图可见：1）温度越高，时间越长，去除率越高；当温度高于 200℃，可达到较高的有机物去除率，当温度低于某个限值，即使延长氧化时间，去除率也不会显著提高，一般认为，湿式氧化温度不宜低于 180℃；2）达到相同的去除率，温度越高，所需时间越短，相应地反应容积便越小；3）湿式氧化过程大致可以分为两个速度段，前 30min，因反应物浓度高，氧化速度快，去除率增加快；此后，因反应物浓度降低或中间产物更难以氧化，致使氧化速度趋缓，去除率增加不多。由此分析，若将湿式氧化作为生物氧化的预处理，则应控制湿式氧化时间为 0.5h 为宜。

气相氧分压对过程有一定影响，因为氧分压决定了液相溶解氧浓度。实验表明，氧化速度与氧分压成 0.3～1.0 次方关系。但总压影响显著，控制一定总压的目的是保证呈液相反应。温度、总压和气相中的水气量三者是偶合因素，如图 8-12 所示。

由图 8-12 可知，在一定温度下，压力越高，气相中水汽量就越小。总压的低限为该温度下水的饱和蒸汽压。如果总压过低，大量的反应热就会消耗在水的汽化上，当进水量低于汽化量时，反应器就会被蒸干。湿式氧化的操作压力一般不低于 5.0～12.0MPa，超临界湿式氧化的操作压力已达 43.8MPa。

不同的污染物，其湿式氧化的难易程度是不同的。对于有机物，其可氧化性与有机物

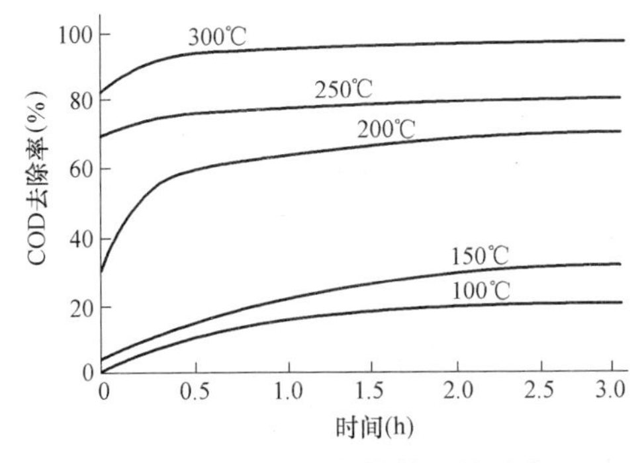

图 8-11 温度对氧化效果的影响

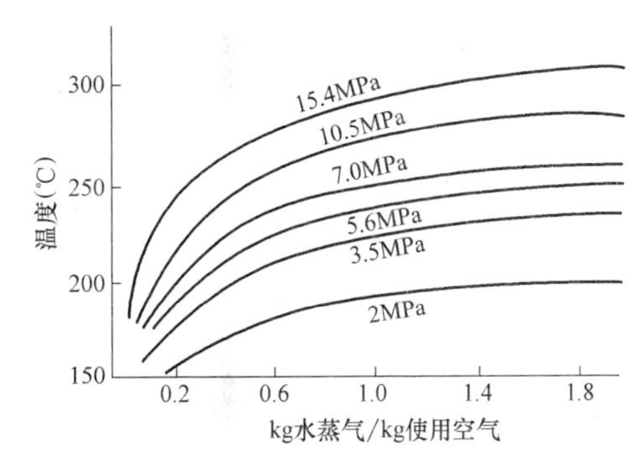

图 8-12 饱和水蒸气量与温度、压力关系

中氧元素含量（O）在分子量（M）中的比例或者碳元素含量（C）在分子量（M）中的比例具有较好的线性关系，即 O/M 越小、C/M 越大、氧化越易。研究指出，低分子量的有机酸（如乙酸）的氧化性较差。

催化剂的运用大幅提高了湿式氧化的速度和程度。有关湿式氧化催化剂的研究，每年都有多项专利注册。对有机物湿式氧化，多种金属具有催化活性，其中贵金属系（如 Pd、Pt、Ru）催化剂的活性高、寿命长、适用广，但价格昂贵，应用受到限制。目前人们多致力于非贵金属催化剂的开发，已获得应用的主要是过渡金属和稀土元素（如 Cu、Mn、Co、Ce）的盐和氧化物。

（3）工程应用

湿式氧化可以作为完整的处理阶段，将污染物浓度一步处理到排放标准值以下，但是为了降低处理成本，也可以作为其他方法的预处理或辅助处理。常见的组合流程是湿式氧化后进行生物氧化。国外多家工厂采用此两步法流程处理丙烯腈生产废水。经湿式氧化处理，COD 由 42000mg/L 降至 1300mg/L、BOD_5 由 14200mg/L 降至 1000mg/L、氰化物由 270mg/L 降至 1mg/L、BOD_5/COD 由 0.2 提高至 0.76 以上。再经活性污泥法处理，总去除率达到：COD 99%；BOD_5 99.9%；氰化物 99.6%。

与活性污泥法相比，处理同一种废水，湿式氧化法的投资高约 1/3，但运转费用却低得多。若利用湿式氧化系统的废热产生低压蒸汽，产蒸汽收益可以抵偿 75%的运转费，则净运转费只占活性污泥法的 15%。若能从湿式氧化系统回收有用物料，其处理成本将更低。

近年来研究最广泛的高级氧化技术除了类 Fenton 试剂法、光催化氧化法、湿式氧化法之外，还包括超声氧化法。超声降解有机物的机理是在超声波（频率一般为 $2\times10^4\sim5\times10^8$ Hz）作用下液体发生声空化，产生空化泡，空化泡崩溃的瞬间，在空化泡内及周围极小空间范围内产生高温（1900～5200K）和高压（5×10^7 Pa），并伴有强烈的冲击波和时速高达 400km/s 的射流，这使泡内水蒸气发生热分解反应，产生具有强氧化能力的自由基，易挥发有机物形成蒸气直接热分解，而难挥发的有机物在空化泡气液界面上或在本体溶液中与空化产生的自由基发生氧化反应得到降解。超声氧化法具有设备简单、易操作、无二次污染等优点，但也存在降解效果差、超声能量转化率及利用率低、处理量小、处理费用高和处理时间长等问题。目前，超声常常作为其他氧化剂或处理技术的辅助和强化技术，形成了 US/O_3、US/H_2O_2、US/Fenton、US/UV/TiO_2、US/WAO（湿式空气氧化）等组合工艺。

8.6 其他相转移分离法

吹脱和汽提都属于气-液相转移分离法，即将气体（载气）通入废水中，使之相互充分接触，使废水中的溶解气体和易挥发的溶质穿过气液界面，向气相转移，从而达到脱除污染物的目的。常用空气或水蒸气作载气，习惯上把前者称为吹脱法，后者称为汽提法。

水和废水中有时会含有溶解气体。例如，用石灰石中和含硫酸废水时产生大量 CO_2；某些工业废水中含有 H_2S、HCN、NH_3、CS_2 及挥发性有机物等。这些物质可能对系统产生侵蚀，或者本身有害，或对后续处理不利，因此，必须分离除去。产生的废气根据其浓度高低，可直接排放、送锅炉燃烧或回收利用。

将空气通入水中，除了吹脱作用以外，还伴随充氧和化学氧化作用，例如：

$$H_2S+\frac{1}{2}O_2 \rightarrow S+H_2O \tag{8-28}$$

8.6.1 吹脱法

吹脱法的基本原理是气液相平衡和传质速度理论。在气液两相系统中，溶质气体在气相中的分压与该气体在液相中的浓度成正比。当该组分的气相分压低于其溶液中该组分浓度对应的气相平衡分压时，就会发生溶质组分从液相向气相的传质。传质速度取决于组分平衡分压和气相分压的差值。气液相平衡关系和传质速度随物系、温度和两相接触状况而异。对给定的物系，通过提高水温，使用新鲜空气或负压操作，增大气液接触面积和时间，减少传质阻力，可以达到降低水中溶质浓度、增大传质速度的目的。

吹脱设备一般包括吹脱池（也称曝气池）和吹脱塔。前者占地面积较大，而且易污染大气，对有毒气体常用塔式设备。

1. 吹脱池

依靠池面液体与空气自然接触而脱除溶解气体的吹脱池称自然吹脱池，其适用于溶解气体极易挥发、水温较高、风速较大、有开阔地段和不产生二次污染的场合。

为强化吹脱过程，通常向池内鼓入空气或在池面以上安装喷水管，构成强化吹脱池。喷水管安装高度离水面 1.2～1.5m。池子小时，还可建在建筑物顶上，高度为 2～3m。为防止风吹损失，四周应加挡木板或百叶窗。喷水强度可采用 $12m^3/(m^2 \cdot h)$。

国内某厂的酸性废水经石灰石滤料中和后，废水中产生大量的游离 CO_2，pH 为 4.2～4.5，不能满足生物处理的要求，因此，中和滤池的出水经预沉淀后，进行吹脱处理。吹脱池为矩形水池，如图 8-13 所示，水深 1.5m，曝气强度为 $25～30m^3/(m^2 \cdot h)$，气水体积比为 5，吹脱时间为 30～40min。空气用塑料穿孔管由池底送入，孔径为 10mm，孔距为 5cm。吹脱后，游离 CO_2 由 700mg/L 降到 120～140mg/L，出水 pH 为 6.0～6.5。存在的问题是布气孔易被中和产物 $CaSO_4$ 堵塞，当废水中含有大量表面活性物质时，易产生泡沫，影响操作和环境。可用高压水喷射或加消泡剂除泡。

2. 吹脱塔

为提高吹脱效率，回收有用气体，防止二次污染，常采用填料塔、板式塔等高效气液分离设备。

填料塔的主要特征是在塔内装设一定高度的填料层，废水从塔顶喷下，沿填料表面呈薄膜状向下流动。空气由塔底鼓入，呈连续相由下而上同废水逆流接触。塔内气相和水相组成沿塔高连续变化，系统如图 8-14 所示。

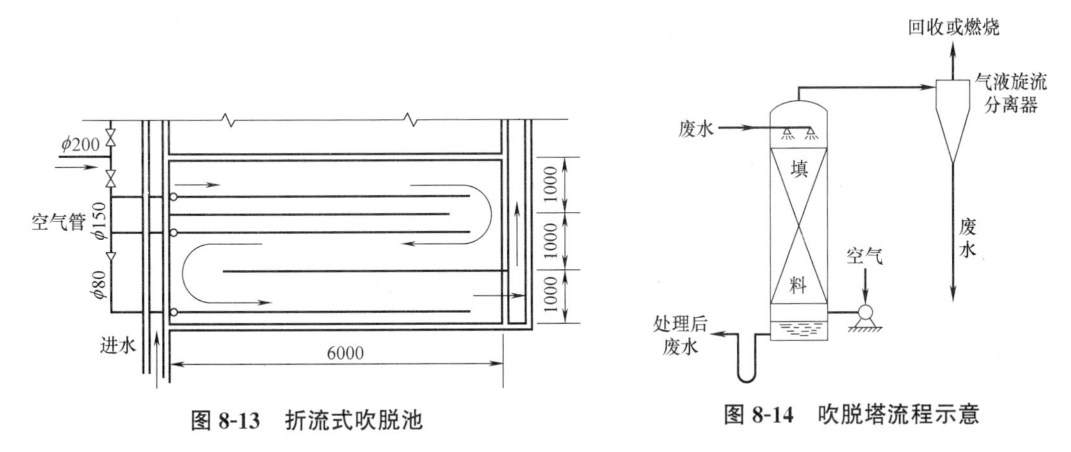

图 8-13　折流式吹脱池

图 8-14　吹脱塔流程示意

板式塔的主要特征是在塔内装设一定数量的塔板，废水水平流过塔板，经降液管流入下一层塔板。空气以鼓泡或喷射方式穿过板上水层，相互接触传质。塔内气相和水相组成沿塔高呈阶梯变化。

吹脱塔的设计计算同吸收塔相仿，单位时间吹脱的气体量，正比于气液两相的浓度差（或分压差）和两相接触面积。在吹脱过程中，影响因素很多，主要有以下几点：

（1）温度。在一定压力下，气体在水中的溶解度随温度升高而降低，因此，升温对吹脱有利。

（2）气水比。空气量过小，气液两相接触不够；空气量过大，不仅不经济，还会发生液泛，使废水被气流带走，破坏操作。为使传质效率较高，工程上常采用液泛时的极限气水比的 80%作为设计气水比。

（3）pH。在不同 pH 条件下，气体的存在状态不同。废水中游离 H_2S 和 HCN 的含量与 pH 的关系见表 8-8。因为只有以游离的气体形式存在才能被吹脱，所以对含 S^{2-} 和 CN^- 的废水应在酸性条件下进行吹脱。

游离 H_2S、HCN 与 pH 的关系　　　　表 8-8

pH	5	6	7	8	9	10
游离 H_2S(%)	100	95	64	15	2	0
游离 HCN(%)		99.7	99.3	93.3	58.1	12.2

8.6.2　汽提法

汽提法用于脱除废水中的挥发性溶解物质，如挥发酚、甲醛、苯胺、硫化氢、氨等。其实质是废水与水蒸气的直接接触，使其中的挥发性物质按一定比例扩散到气相中去，从而达到从废水中分离污染物的目的。常用的汽提设备有填料塔、筛板塔、泡罩塔、浮阀塔等。

1. 汽提法处理含酚废水

汽提法最早用于从含酚废水中回收挥发酚，其典型流程如图 8-15 所示。汽提塔分上下两段，上段叫汽提段，通过逆流接触方式用蒸汽脱除废水中的酚；下段叫再生段，同样通过逆流接触，用碱液从蒸汽中吸收酚。其工作过程如下：废水经换热器预热至100℃后，由汽提塔的顶部淋下，在汽提段内与上升的蒸汽逆流接触，在填料层中或塔板上进行传质。净化的废水通过预热器排走。含酚蒸汽用鼓风机送到再生段，相继与循环碱液和新碱液（含10% NaOH）接触，经化学吸收生成酚钠盐回收其中的酚，净化后的蒸汽进入汽提段循环使用。碱液循环在于提高酚钠盐的浓度，待饱和后排出，用离心法分离酚钠盐晶体，加以回收。

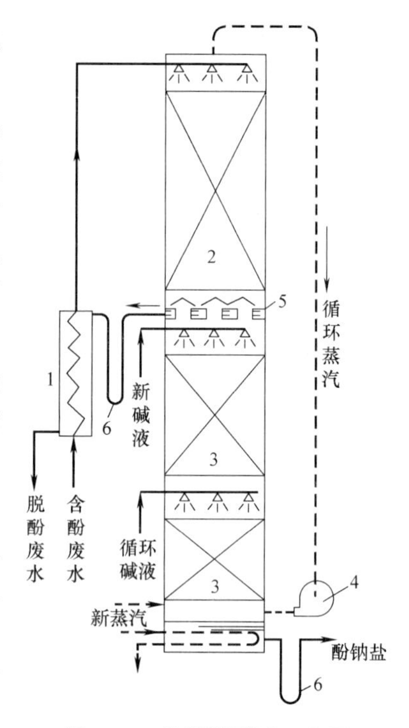

图 8-15　汽提塔脱酚示意图

1—预热器；2—汽提段；3—再生段；4—鼓风机；5—集水槽；6—水封

汽提脱酚工艺简单，对处理高浓度（含酚1g/L以上）废水，可以达到经济上收支平衡，且不会产生二次污染。但是，经汽提后的废水中一般仍含有较高浓度（约400mg/L）的残余酚，必须进一步处理。另外，由于再生段内喷淋碱液的腐蚀性很强，必须采取防腐措施。

2. 汽提法处理含硫废水

石油炼厂的含硫废水（又称酸性水）中含有大量H_2S（高达10g/L）、NH_3（高达5g/L），还含有酚类、氰化物、氯化铵等。一般先用汽提回收处理，然后再用其他方法进行处理。处理流程如图8-16所示。

含硫废水经隔油、预热后从顶部进入汽提塔，蒸汽则从底部进入。在蒸汽上升过程中，不断带走H_2S和NH_3。脱硫后的废水，利用其余热预热进水，然后送出进行后续处理。从塔顶排出的含H_2S及NH_3的蒸汽，经冷凝后回流至汽提塔中，不冷凝的H_2S和NH_3进入回收系统，制取硫黄或硫化钠，并可产生氨水。

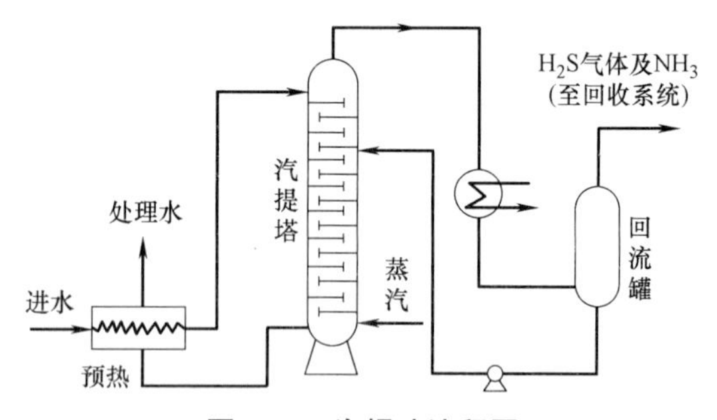

图 8-16　汽提法流程图

国外某公司采用双塔汽提法处理含硫废水，工艺流程如图8-17所示。酸性废水经脱气（除去溶解的氢、甲烷及其他轻质烃）后进行预热，送入H_2S汽提塔，塔内温度约38℃，压力为0.68MPa（表）。H_2S从塔顶汽提出来，水和氨从塔底排出。塔顶气相仅含50mg/L NH_3，可直接作为生产硫或硫酸的原料。水和氨进入氨汽提塔，塔内温度为94℃，压力为0.34MPa（表）。氨从塔顶蒸出，进入氨精制段，除去少量的H_2S和水，在38℃、1.36MPa下压缩，冷凝下来的NH_3含H_2O<1g/L、含H_2S<5mg/L，可作为液氨出售。氨汽提塔底排出的水可重复利用。

据报道，该公司用此流程处理含硫废水，流量为45.6m³/h，每天可回收H_2S 72.6t、NH_3 36.3t，2年多即可回收全部投资。

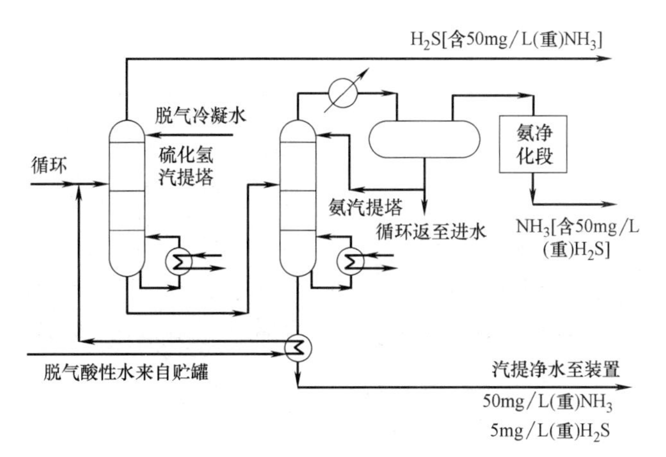

图 8-17　双塔汽提法流程图

国内也有多家炼油厂采用类似的双塔汽提流程处理含硫废水，将含 H_2S 290～2170mg/L、含 NH_3 365～1300mg/L 的原废水净化至含 H_2S 0.95～12mg/L。运转表明，该系统操作方便，能耗低。

8.7　膜生物反应器

膜生物反应器（Membrane Bio-Reactor，MBR）是利用相关设施将生物处理工艺和膜分离工艺组合在一起的新型污水处理装置系统。在装置中，主要利用生化工艺降解和去除污水中的有机污染物，利用膜组件取代传统生物处理技术末端二沉池，既截留水中的悬浮物和未生物降解的大分子有机物，满足有关排水标准的要求，同时又保证生物反应器中维持高活性污泥浓度，显著减少污水处理设施占地面积。此外，膜生物反应器因其有效的截留作用，可保留世代周期较长的微生物，使硝化菌在系统内能充分繁殖，为深度脱氮除磷提供可能。

8.7.1　膜生物反应器分类

膜生物反应器的分类方式很多，比如按照膜材料的不同，可分为微滤膜 MBR 和超滤膜 MBR；按照压力驱动形式的不同，可分为外压式和抽吸式；按照生物反应器的不同，可分为好氧和厌氧。主流上通常按照膜元件结构形式或膜组件的作用进行分类。

1. 按照膜元件结构形式不同

膜元件是构成膜组件的要素，膜组件由多个膜元件组合而成，按膜元件结构形式分类，膜组件形式有中空纤维膜组件、平板型膜组件、管式膜组件及螺旋型组件等。目前污水处理工程应用较多的膜组件为前三者。

（1）中空纤维膜组件

中空纤维膜组件较多应用于浸没式反应器。膜组件使用的中空纤维膜丝一般为不对称（非均向）、自身支撑的滤膜。膜丝可根据工艺和相关的使用要求设计成帘式、束式等形式。中空纤维膜的这些几何设计形式能使膜丝的填充密度最大化，增大处理能力，同时又结构紧凑，有利于长时间的稳定运行。相关的中空纤维膜组件产品如图 8-18 所示。

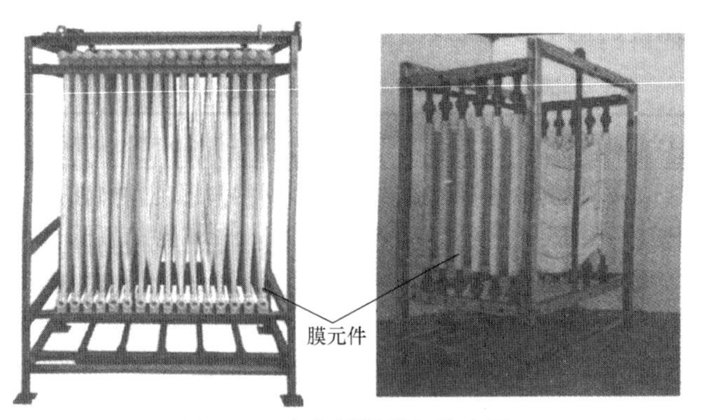

图 8-18　中空纤维膜组件产品

（2）平板型膜组件

平板型膜组件主要由过滤膜片和支撑板构成。一定数量的板框式膜元件通过组合形成平板型膜组件。板框式膜元件以及安装成的组件产品如图 8-19 所示。

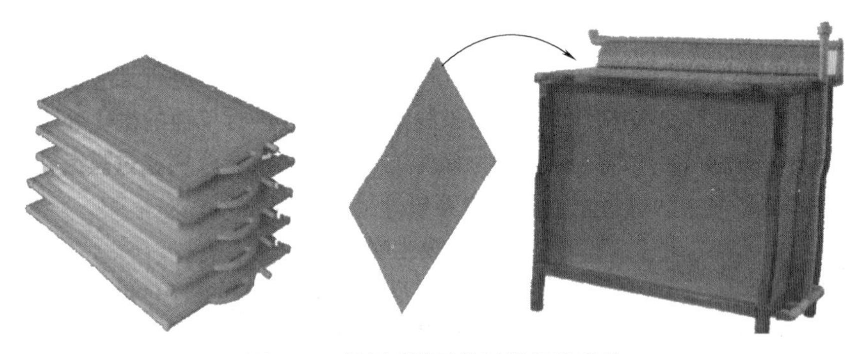

图 8-19　板框式膜元件以及组件产品

平板型膜组件在污水处理工程中也有广泛的应用，相比于中空纤维膜，平板膜的膜通量大，没有断丝问题，具有较强的抗污染性，不易结垢，膜清洗周期长，运行中无需反冲洗，能长期稳定地运行。但平板膜的填充密度一般不大，容积利用率较低，在大型项目的应用中，需要对膜组件的填充方式进行改进，提高膜组件的填充密度。

（3）管式膜组件

管式膜元件是把滤膜和支撑体均制成管状，使二者组合；或将滤膜直接刮制在支撑管的内侧或外侧。将数根膜管元件（直径 10～20mm）组装在一起构成管式膜组件。

管式膜有内压型和外压型两种运行方式，实际中多采用内压型，即进水从膜管中流入，渗透液从管外流出。外置式 Airlift MBR 管式膜元件与膜组件如图 8-20 所示。

图 8-20　外置式 Airlift MBR 管式膜元件与膜组件

2. 按膜组件的作用不同

根据膜组件在膜生物反应器中所起作用不同，可将 MBR 分为膜分离生物反应器、曝气膜生物反应器、萃取膜生物反应器三种。膜分离生物反应器的膜组件相当于传统生物处理系统中的二沉池，用于混合液的固液分离，污泥被截留在膜池中，透过水通过收集系统外排。曝气膜生物反应器用于气体质量传递，为需氧降解工艺供氧，可以实现处理工艺无泡曝气，可实现提高反应器的传氧效率。萃取膜生物反应器，利用膜将有毒工业污水中的优先污染物萃取后进行单独的生化处理。

（1）曝气膜生物反应器

曝气膜生物反应器（Membrane Aerationbiofilm Reactor），简称 MABR，由中空纤维膜组件和供气设备等组成。图 8-21 是曝气膜生物反应器的示意图。

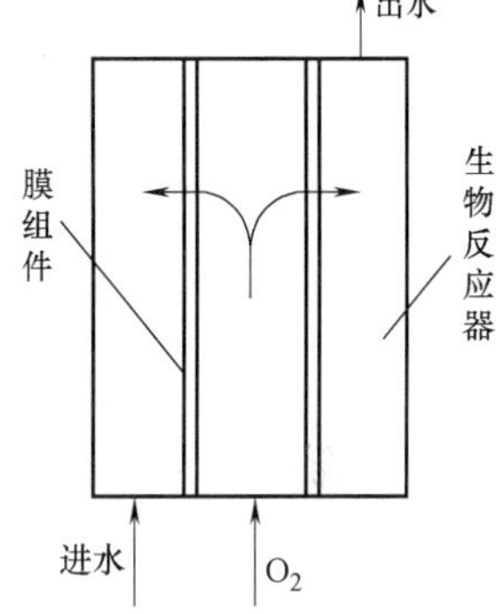

图 8-21　曝气膜生物反应器示意图

在曝气膜生物反应器中，生物膜所需要的氧气通过膜纤维束供给和分配，中空纤维膜不仅起供氧的作用，同时又是生物膜固着的载体。因此，在 MABR 中，空气通过中空纤维膜的微孔为附着在其上的生物膜提供无泡曝气，同时纤维膜外侧具有活性的生物膜与污水充分接触，污水中所含的有机物被生物膜吸附和氧化分解，可实现高效低耗降解污水中污染物的目的。

MABR 所使用的纤维膜材料一般为微孔疏水有机膜或致密硅橡胶膜，微孔直径在 0.1～0.5μm，空气可以以极小的肉眼难以观察到的气泡甚至无泡的形式进入水体中，因此，又称为无泡供氧，优势是可以获得很高的氧利用率。

MABR 主要适用于处理可生化性较高的污水，对高浓度污水处理效果良好，并可用于同时处理污水中的 NH_3-N。

1）原理和特点

在 MABR 中，由于透气膜表面处于富氧条件，微生物极易在膜表面进行积累从而形成生物膜。此时，氧气由膜内腔直接透过膜传递至生物膜，被微生物所利用，传氧效率高。废水中的污染物由液相主体向生物膜传递，与氧气形成反向传递，形成 MABR 的生物膜特有的传质特点（图 8-22a）。对于附着在一般载体上传统的生物膜，溶解氧由表及里浓度逐渐降低，相应生长着好氧微生物、兼性微生物和厌氧微生物（图 8-22b）。而对于 MABR 生物膜，好氧微生物富集在生物膜/透气膜界面，溶解氧从里往外降低。当供氧条件控制得当时，反应器处于缺氧或者厌氧状态，生物膜最外层可以生长厌氧微生物，即相对于常规生物膜，好氧层和厌氧层相对位置发生反转。这样的反转使得 MABR 生物膜的微环境有利于进行硝化反应。由于硝化菌的生长速率远低于好氧异氧菌。对于滴滤池等传统生物膜工艺，只有当 BOD 低于一定值时硝化菌才能占优势。而低 BOD 的条件又会限制反硝化菌的活性，从而影响系统的脱氮效果。对于 MABR，在紧靠透气膜载体表面的生物膜底层溶解氧浓度最大，有机物的浓度经过外侧生物膜的降解后降低，适宜发生硝化反应；外层的微溶解氧和无氧环境、充足的有机碳源可以满足反硝化的需要，这两个过程的结合即可完成脱氮的全过程。因此，在操作条件适宜的情况下，可以实现同步去除有机物和氨氮甚至同步硝化反硝化。

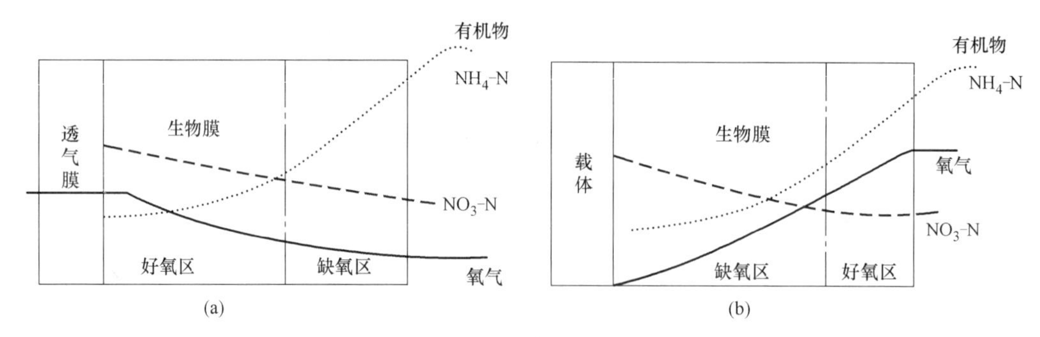

图 8-22　两种生物膜传质特点示意图

(a) MABR 生物膜；(b) 传统生物膜

2）影响 MABR 运行的因素

影响反应器正常运行的主要因素有膜污染、气源压力、液相流速等方面。

① 膜污染

MABR 反应器运行一段时间后，膜组件会被污染物堵塞，反应器处理效果下降，同时，膜使用寿命大幅缩短。膜污染根据发生的位置可以分成两种类型：一种是外部堵塞，即污染物吸附沉积在膜的表面，增加了底物传递阻力；另一种是内部堵塞，即污染物在中空纤维膜壁上的微孔内吸附沉积，减小了膜孔径，从而降低氧的传递速率。这两种膜污染都会严重影响 MABR 的正常运行，为了使 MABR 能够高效运行，经常需要对膜进行反冲洗。

② 气源压力

透过中空纤维膜的气体在表面张力作用下吸附在膜表面，此时如果气液两相压差较大，气体易在膜表面形成气泡，从而降低氧的传质效率，因此，为达到无泡曝气效果，气源压力必须低于起泡点气压。

③ 液相流速

工艺运行过程对液相流速在不同阶段有不同的要求。运行前期即挂膜阶段，液相流速不适宜过高，否则生物膜受到较大的剪切力而很难生长。正常运行期间，为减小液相边界层厚度，加快基质的传递速率，同时控制生物膜的过度积累，应保持较高的液相流速，而维持较高流速所需能量在反应器能耗中占相当大的比例。

3）曝气膜生物反应器特点

① 氧利用效率高；

② 有机物去除率高；

③ 占地小，适合于污泥浓度大、对氧需求大的污水处理；

④ 操作复杂，基建费用高。

到目前为止，曝气膜生物反应器未有实际运行工程案例报道。

（2）萃取膜生物反应器

萃取膜生物反应器又称为 EMBR（Extractive Membrane Bioreactor）。英国学者 Lvingston 研究开发了 EMBR，以解决下面两个污水处理技术难题：

① 高酸碱度污水以及含有对生物有毒物质的污水，都不宜采用与微生物直接接触的方法处理；

② 当污水中含挥发性有毒物质时，若采用传统的好氧生物处理过程，污染物容易随曝气气流挥发，不仅处理效果很不稳定，还会造成大气污染。

萃取膜生物反应器原理为：污水与活性污泥被膜隔开，污水在膜内流动，含某种专性细菌的活性污泥在膜外流动，有机污染物可以选择性透过膜被膜外侧的微生物降解。由于萃取膜两侧的生物反应器单元和污水循环单元各自独立，各单元水流相互影响不大，生物反应器中营养物质和微生物生存条件不受污水水质的影响，使水处理效果稳定。系统的运行条件如 *HRT* 和 *SRT* 可分别控制在最优的范围，维持最大的污染物降解速率。萃取膜生物反应器示意图如图 8-23 所示。

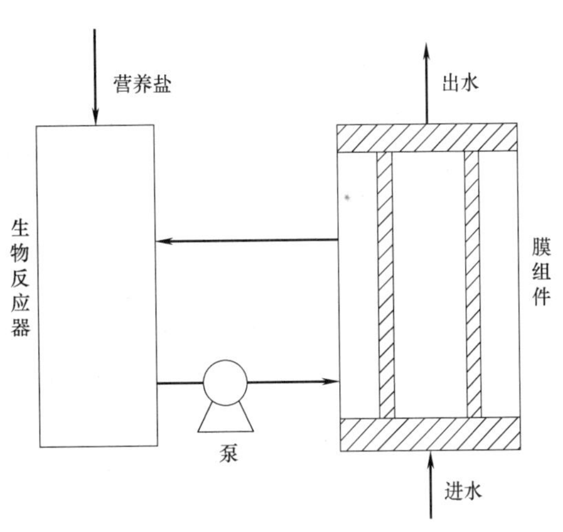

图 8-23 萃取膜生物反应器示意图

萃取膜生物反应器所用膜组件由硅胶或其他疏水性聚合物制成。这种反应器有两种运行方式：一种运行方式是污水流和生物膜被硅橡胶膜隔开，易挥发的有机污染物可很快通过硅橡胶膜，在生物反应器中进行生物降解，而污水中的无机质不能通过硅橡胶膜，因此，污水中的有害离子组分对微生物的降解作用就没有影响。另一种运行方式是由一个传统的生物反应器连接一个具有萃取作用的管式膜组件组成，膜管外侧为生物介质流，管内为污水流，硅橡胶膜按束排列于管内，选择性地将有毒污染物从污水中转移至一个经过曝气的生物介质相，并在其中进行分解。在萃取式膜生物反应器中，污水中的污染物通过膜进入生物反应器，膜外侧流动的营养介质不受膜管内污水的影响，从而使生物降解速率保持在较高水平。另外，萃取膜生物反应器一般存在特征污染物，如果向反应器中投加降解特征污染物的专性细菌，可以提高污染物降解的针对性和效率，还可通过添加无机营养成分促进降解。

(3) 固液分离型膜生物反应器

固液分离型膜生物反应器（Membrane Separation Bioreactor）即膜分离生物反应器（MBR）。MBR 工艺将分离工程中的膜分离技术与传统污水生物处理技术有机结合，提高了固液分离效率，并且由于曝气池中活性污泥浓度的增大，提高了生化反应速率。

与传统的生化水处理技术相比，MBR 工艺具有以下主要特点：

1）高效地进行固液分离，其分离效果远好于传统的沉淀池，出水水质良好，出水悬浮物和浊度接近于零，可直接回用，实现了污水资源化。

2）膜的高效截留作用，使微生物完全截留在生物反应器内，实现反应器水力停留时间（*HRT*）和污泥龄（*SRT*）的完全分离，运行控制灵活稳定。

3）由于 MBR 工艺将传统污水处理的曝气池与二沉池合二为一，并取代了三级处理的全部工艺设施，因此，可大幅减少占地面积，节省土建投资。

4）有利于硝化细菌的截留和繁殖，系统硝化效率高。通过运行方式的改变亦可有脱氮和除磷功能。

5）由于污泥龄可以非常长，从而大幅提高难降解有机物的降解效率。

6）反应器在高容积负荷、低污泥负荷、长污泥龄下运行，剩余污泥产量极低，由于污泥龄可无限长，理论上可实现零污泥排放。

7）系统实现 PLC 控制，操作管理方便、系统，各出水膜组通过自吸泵抽吸出水，并通过 PLC 系统自动控制间断出水。运行期间定期进行在线水反冲洗、在线化学反冲洗以防止和缓解膜污染，保持膜组件的良好出水能力。离线化学清洗彻底清洗污染严重的膜组件，恢复膜通量和产水能力。

8）设备紧凑，占地面积小，由于生物反应器内污泥浓度高，容积负荷可大幅提高，生物反应器体积大幅减小；从形式上看，一体式膜生物反应器可使设备更加紧凑。

9）污泥产率低，同传统活性污泥法相比，膜生物反应器的污泥产率很低。

表 8-9 列出了不同膜生物反应器的主要优点和缺点，可以明显地看出：膜技术与生物污水处理工艺相结合来处理污水与单纯的污水生化处理工艺或者膜过滤工艺相比，有其独到之处，特别是占地面积小、设备集中、模块化，并且具有升级改造的潜力。

不同膜生物反应器的主要优点和缺点 **表 8-9**

反应器	优点	缺点
膜分离生物反应器	1. 占地面积小； 2. 高负荷率； 3. 系统不受污泥膨胀的影响； 4. 模块化、改造与升级容易	1. 膜容易污染； 2. 膜价格高
曝气膜生物反应器	1. 氧利用率高； 2. 效率高； 3. 占地面积小； 4. 氧需要量随时可控； 5. 模块化、改造与升级容易	1. 膜易被污染； 2. 基建投资大； 3. 工艺复杂； 4. 目前无实际工程实例
萃取膜生物反应器	1. 微生物与污水隔离，处理有毒工业污水； 2. 出水流量小； 3. 模块化、改造与升级容易	1. 基建投资大； 2. 工艺复杂

8.7.2 膜生物反应器工艺

1. 膜生物反应器工艺类型

膜生物反应器工艺（简称 MBR 工艺）将生化降解和膜分离过程有机结合，水中的污染物首先经过生化降解得到去除，活性污泥混合液在压力差的作用下，纯水和小于膜孔径的小分子溶质透过膜，成为处理出水，微生物及大分子溶质被膜截留，从而替代沉淀池完成污泥与出水的分离。在生物反应器中保持高活性污泥浓度，提高生物处理有机负荷，从而减少污水处理设施占地面积，并通过保持低污泥负荷减少剩余污泥量。

按照膜组件与生物反应器的组合方式不同，污水处理中常用的膜生物反应器工艺分为两种，浸没式膜生物反应器工艺和外置式膜生物反应器工艺。

（1）浸没式膜生物反应器工艺

浸没式膜生物反应器工艺又称为一体式膜生物反应器，简称 S-MBR，是把膜组件浸没在生化反应池或者单独的膜池中，污染物在系统中先进行生化反应得到降解去除，然后利用膜过滤进行固液分离。在 S-MBR 工艺里，微滤或超滤膜组件直接浸没于生化反应池，并安置在曝气器的上方，利用曝气时气液向上的剪切力来实现膜面的错流效果，也有采用在浸没式膜组件附近进行叶轮搅拌和膜组件自身的旋转来实现膜面错流效应，以减少膜表面物质沉降，从而降低膜的污染。S-MBR 工艺现场证实了曝气流引起的上升气水混合流擦洗膜表面可以去除滤饼层。浸没式膜生物反应器工艺可采用负压产水，也可利用静水压力自流产水。其工艺示意图如图 8-24 所示。

浸没式膜生物反应器工艺近年来在水处理领域应用较多，受到越来越多的关注。其最大特点是运行能耗低，出水水质好，系统耐冲击负荷，运行较稳定；但膜通量一般相对较低，容易发生膜污染，不容易清洗和更换；膜丝也容易发生断丝现象。浸没式工艺的能耗主要来自曝气，占运行总能耗的 90%以上。

（2）外置式膜生物反应器工艺

外置式膜生物反应器工艺把膜组器和生物反应系统分开布置，即膜分离装置置于生物反应器之外，并与生物反应器组成一个回路，污染物在系统中先进行生化处理得到降解去除，生物反应系统中的活性污泥混合液由泵增压后进入膜组件，在压力作用下膜过滤液成为系统处理出水，活性污泥、大分子物质等则被膜截留，随浓缩液回流到生物反应器内，简称 R-MBR。外置式 MBR 工艺示意图如图 8-25 所示。

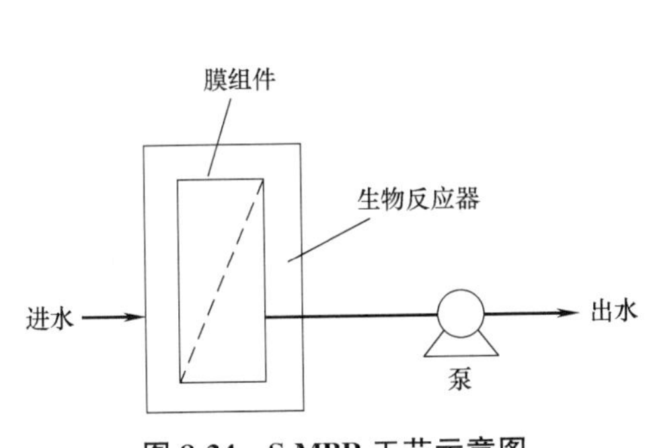

图 8-24 S-MBR 工艺示意图

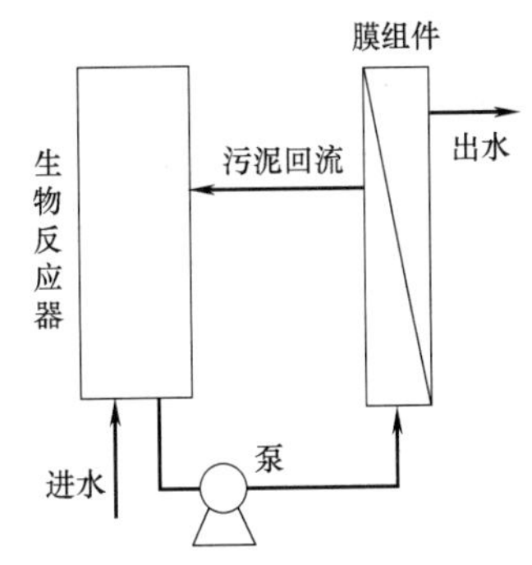

图 8-25 R-MBR 工艺示意图

外置式膜生物反应器的特点是运行稳定可靠，操作管理容易，易于进行膜的清洗、更换及增设，而且膜通量普遍较大。一般条件下，为减少污染物在膜表面的沉积，延长膜的清洗周期，需要用循环泵提供较高的膜面错流流速，水流循环量大、动力费用高，有人还认为泵的高速旋转产生的剪切力可能会导致某些微生物菌体产生失活现象。

外置式工艺需要较高的能耗，为 2～10kWh/m^3 污水，其能耗主要产生在两方面：一方面是因为污染物容易在膜表面沉积，运行中需要较高的错流速度，因此，在循环上需要消耗的能量增大；另一方面，操作压力大，较高的操作压力也带来较大的能耗。外置式工艺中曝气仅占总能耗的 20%左右。

（3）浸没式与外置式膜系统的比较

浸没式与外置式膜系统的有关指标比较见表 8-10。

除了按照膜组件与生物反应器的组合方式划分之外，膜生物反应器工艺还可以分为需氧膜生物反应器和厌氧膜生物反应器两大类。

浸没式与外置式膜系统有关指标比较 **表 8-10**

指标	浸没式膜	外置式膜
系统	开放式系统设计	密闭式系统设计
过滤方式	外压式过滤设计，直流式过滤	内压式或外压式设计，直流和错流式过滤
常用膜材料	PVNF、PVC	PES、PS、PVDF、PVC
预处理要求	单根膜组件装填密度中等，过流通道宽，只要求简单预处理，抗污堵能力强	单根内压式膜组件装填密度高，过流通道小，要求复杂预处理，抗污堵能力差
操作压力	能耗较低，采用虹吸或低压真空抽吸（0.02～0.03MPa）	能耗高，用较高压力过滤（0.2～0.4MPa）
膜寿命	相对较长	相对较短
占地面积	小	较大
适用处理规模	根据膜组件类型不同，适合各种规模的处理设施	适合中、小型处理规模的水处理厂

需氧膜生物反应器主要应用在城市和工业污水处理上，可实现城市污水处理水资源回收利用，在工业中可去除特定污染物，如处理工业含油脂类的废水。

厌氧膜生物反应器是一种低能耗、易操作、更高效的膜生物反应器。其保持了高污泥浓度和长污泥停留时间，缩短了水力停留时间，改善了出水水质。例如把膜单元和 UASB 结合，使固、液分离，不再需要设计三相分离器，膜分离过滤作用可使两相厌氧膜生物反应器产酸菌浓度增加，可实现产酸发酵反应能力速度加快，实现高酸化率。由于厌氧膜生物反应器没有曝气过程，可采用分体式来实现厌氧污泥的悬浮状态，实现高浓度有机污水的厌氧处理。

2. 浸没式膜生物反应器工艺运行控制

浸没式膜生物反应器工艺运行控制参数包括过滤速度、过滤的压力与流速、过滤运行周期等。

（1）过滤速度

过滤速度指单位时间内、单位过滤面积所获得的滤液体积，过滤速度表明反应器的生产强度，代表设备性能的优劣。过滤速度与过滤推动力成正比，而与过滤阻力成反比。在压差过滤中，推动力就是压差，阻力则与滤饼的结构、厚度以及滤液的性质等诸多因素有关，比较复杂。总体上，随着过滤过程的进行，滤饼会逐渐加厚，如果维持过滤压力不变，即采取恒压过滤，则过滤速度将逐步减小。若要维持过滤速度不变，即维持恒速过滤，则必须逐渐增加过滤压力。

（2）恒压过滤与恒速过滤

1）恒压过滤

在恒定压差下进行的过滤称为恒压过滤。采用恒压过滤方式，随过滤时间增长，滤饼厚度会逐渐增加，膜过滤阻力随之上升，过滤速度则不断下降。因此，在恒压操作中，膜的产水量随过滤膜的阻力增大而减少，恒压过滤在实际工程中具有操作不方便的缺点。

浸没式 MBR 工艺有利用水位高差方式进行过滤的。在膜池中，微滤膜组件利用恒定的水压差过滤的形式属于恒压过滤方式，在久保田的膜处理工程中有设计和运行的报道。

2）恒速过滤

在恒定出水流速下进行的过滤称为恒速过滤。随着过滤的进行，过滤阻力不断增大，要维持过滤速度不变，需要增大过滤的推动力。

浸没式 MBR 工艺通常以恒速方式运行，跨膜压差（TMP）将随运行时间逐渐增加，此时通过定期的反冲洗或者气擦洗可以清除污染层，在一定程度上流速能得以恢复。与恒压控制相比，恒速过滤不仅比较容易控制膜污染，在延长膜的清洗周期的同时，还能保持较高膜通量，使系统得以长时间稳定运行。

工程上的恒速过滤设计一般采用压力计控制，在产水抽吸泵吸水管上安装压力变送器，监测膜的跨膜压差，当跨膜压差变化超过设定值时，压力计将信号传到 PLC，PLC 发出调整管道上的抽吸泵、阀门运行状态的指令，停止产水程序，执行清洗程序。

（3）过滤运行周期

浸没式 MBR 工艺一般采取连续进水，周期间歇出水的操作方式。采用间隙抽吸的操作模式旨在通过定期地停止膜过滤，使池中混合液到膜面的净流速为零，以使沉积在膜面上的污泥在曝气鼓泡作用下松弛从而从膜面上脱落下来，使膜的过滤性能得以部分恢复。一般来说，抽吸过程越长，污染物在膜表面的积累越多；停止抽吸时间越长，膜表面可逆污染物脱落越多，膜过滤性能的恢复也就越好。一般过滤出水的运行时间占总运行时间的 80%～90%。

（4）曝气

采用间歇出水操作时可一并进行曝气，曝气的目的除了为膜池中微生物供氧之外，还能让上升的气泡及其产生的紊动水流阻止污泥聚集和清洗膜表面，保持膜通量稳定。曝气强度可根据膜反应器设备提供商提供的数据进行设计。

8.7.3 膜生物反应器的特点及应用

1. 膜生物反应器的工艺特点

1）采用膜组件代替生物处理中的二沉池。省去了二沉池，可以减少占地面积；利用膜对生化反应池内的含泥污水进行过滤，实现泥水分离，同时系统传氧效率由于膜而得到提高，污泥产率低。

2）对污染物的去除率高，出水中悬浮物少；由于膜的截流作用使 SRT 延长，营造有利于增殖缓慢的微生物（如硝化细菌）生长的环境，可以提高系统的硝化能力，同时有利于提高难降解大分子有机物的处理效率，出水水质稳定。净水效果可达到国家污水排放一级 A 标准。

3）实现了污泥龄（SRT）和水力停留时间（HRT）的分离。

4）膜的机械过滤作用避免了微生物的流失，生物反应器内可保持高的污泥浓度，从而能提高有机物体积负荷，降低污泥负荷；在运行过程中，活性污泥会因污水中有机物浓度的变化而变化，并达到一种动态平衡，这使系统具有出水稳定并有耐冲击负荷的优点。

5）工程实践表明，针对病毒的去除，MBR 工艺具有比传统消毒工艺更加明显的优势。

6）易于实现自动控制，操作管理方便。

7）易于从传统工艺进行改造。可以作为传统污水处理工艺的深度处理单元，在城市二级污水处理厂出水深度处理（从而实现城市污水的大量回用）等领域有着较广阔的应用前景。

MBR 工艺与活性污泥法参数对比见表 8-11，MBR 工艺与活性污泥法效果比较见表 8-12。

MBR 工艺与活性污泥法参数对比 **表 8-11**

项目	活性污泥法	MBR 工艺
COD 负荷[kgCOD/(m³·d)]	0.5～0.7	小于 2
MLSS(mg/L)	3000～4000	8000～15000
污泥有机负荷[kgCOD/(kgMLSS·d)]	小于 0.2	小于 0.1
污泥龄(h)	5～15	10 ～30

MBR 工艺与活性污泥法效果比较 **表 8-12**

污染指标名称	进水平均浓度(mg/L)	传统活性污泥法		MBR 工艺	
		出水(mg/L)	去除率(%)	出水(mg/L)	去除率(%)
COD	520	75	85.6	15	97.1
SS	110	40	63.6	接近 0	100
TKN	48.3	48.3	37.5	3.4	93
TP	15	15	47.3	2.25	85
浊度(NTU)	38	38	60.5	0.44	94.5

膜生物反应器也存在一些不足，主要表现在以下几个方面：

① 膜造价高，使膜生物反应器的基建投资高于传统污水处理工艺；

② 膜容易出现污染，给操作管理带来不便；

③ 能耗高。首先 MBR 池中 MLSS 较普通活性污泥法高，要保持足够的传氧速率，必须加大曝气强度，曝气系统能耗相对要大一些，造成 MBR 的能耗要比传统的生物处理工艺高。其次是泥水分离过程必须保持一定的膜驱动压力，也需要一定的能耗。

2. 膜生物反应器的应用与发展方向

(1) 应用

在欧洲、美国、日本等，有关 MBR 对生活污水的处理特性一直是研究的重点，其研究的目的一方面在于改造污水处理厂，使其达到深度处理的要求；另一方面，用于废水处理使其达到回用的目的。目前在北美，MBR 处理生活污水的应用主要是流量在 10～200m³/d 的小型处理装置。同时好氧 MBR 工艺也成功应用于下列行业的工业污水处理：包括医药、纺织、化妆品、食品、造纸与纸浆、饮料、炼油工业与化工厂，在欧洲，垃圾填埋场渗滤液的好氧 MBR 处理厂也正在兴建。

随着水资源回用需求的提升、污水处理排放标准的提高、膜技术应用的成熟、膜性能的提高和膜价格的下降，MBR 工程的应用发展也将会得以巨大推动。

(2) 工艺的研发方向

针对不同的水质，可以研发的 MBR 工艺方向有：

1) 复合 MBR 工艺；

2) 厌氧 MBR 工艺；

3) 好氧颗粒污泥型 MBR 工艺；

4）专用菌种 MBR 工艺。

3. 技术提升方向

（1）提升膜材料与膜组件的性能，包括增大膜通量，以扩大处理能力和规模；

（2）进一步研究膜污染机理和控制膜污染的技术；

（3）降低 MBR 工艺的能耗，拓展应用领域；

（4）膜元件与膜组件的标准化。

本章关键词（Keywords）

吸附	Adsorption
活性炭	Activated carbon
反渗透	Reverse osmosis
电渗析	Electrodialysis
单分子层	Monomolecular layer
界面浓度	Interfacial concentration
过氧化氢	Hydrogen peroxide
自由基	Free radicals
硫酸亚铁	Ferrous sulfate
可生物降解性	Biodegradability
高锰酸钾	Potassium Permanganate
氧化剂	Oxidizing agent
氧化还原电位	Oxidation reduction potential
高级氧化	Advanced oxidation
膜生物反应器	Membrane Bioreactor

思考题

1. 由吸附等温式阐述什么是吸附等温曲线。研究吸附等温线有何实际意义？
2. 如何绘制动态吸附的穿透曲线？它能为吸附设计提供什么依据？
3. 举例说明哪些污染物难以被活性炭吸附。针对它们，可以换用什么吸附剂？
4. 分析活性炭吸附法用于水处理的优点和适用条件，目前存在什么问题？
5. 简述氯消毒的基本原理。什么是需氯量？
6. 臭氧氧化工艺有什么类型，各适合什么情况？
7. 什么是高级氧化技术？常用高级氧化技术有哪些？
8. 比较吹脱法与汽提法的异同。
9. 含有 Na_2S、NaCN 的废水能否用吹脱法去除？若要用此方法，需要采取什么措施？
10. 膜生物反应器的主要类型和特点是什么？

Chapter 9 Sludge Handling and Disposal

9.1 Introduction of Sludge Handling (Treatment)

Most of the treatment processes normally employed in domestic sewage and industrial water pollution control yield a sludge from a solids-liquid separation process (sedimentation, flotation, etc.) or produce a sludge as a result of a chemical coagulation or a biological reaction. These solids usually undergo a series of treatment steps involving thickening, dewatering, stabilization and final disposal. Organic sludges may also undergo treatment for reduction of the organic or volatile content prior to final disposal. Sludges contain the free water removed by thickening, the capillary water removed through dewatering, and the bound water removed only by chemical or thermal means.

In general, gelatinous-type sludges such as alum or activated sludge yield lower concentrations, whereas primary and inorganic sludges yield higher concentrations in each process sequence. The sludge cake must be disposed. This can be accomplished by handling the cake to a land disposal site or by incineration. The processes selected depend primarily on the nature and characteristics of the sludge as well as the final disposal method employed. For example, activated sludge is more effectively concentrated by flotation than by gravity thickening. Final disposal by incineration desires a solids content that supports its own combustion. In some cases, the process sequence is apparent from experience with similar sludges or by geographical or economic constraints. In other cases, an experimental program must be developed to determine the most economical solution to a particular problem.

Typical sludge handling (treatment) flow diagrams with biological digestion and three different sludge dewatering processes: (1) belt-filter press, (2) centrifuge, (3) drying beds, in some plants, flows that are to be returned to the headwork are stored in equalization tanks or basins for return to the treatment process during the early morning hours when the plant load is reduced (Figure 9-1).

Most commonly, sewage sludge is subjected to anaerobic digestion in a digester designed to allow bacterial action to occur in the absence of air. This reduces the mass and volume of sludge and ideally results in the formation of a stabilized humus. Disease agents are also destroyed in the process.

Following digestion, sludge is generally conditioned and thickened to concentrate and stabilize, thus rendering the sludge more dewaterable. The relatively inexpensive proces-

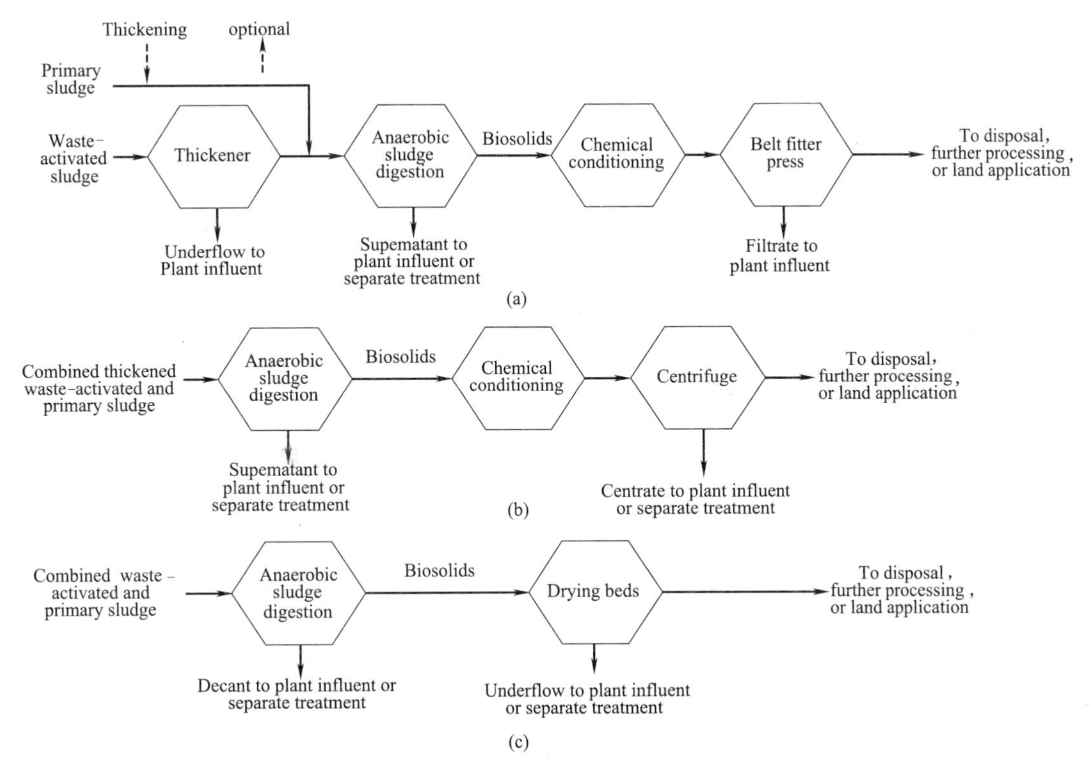

Figure 9-1 Generalized sludge-processing flow diagram

ses, such as gravity thickening, may be employed to lower the moisture content to about 95%. Sludge may be further conditioned chemically by the addition of iron or aluminum salts, lime, or polymers.

9.2 The Basic Processes for Sludge Handling (Treatment)

9.2.1 Thickening

Separate as much water as possible by gravity or flotation. Gravity thickening is usually applied to primary and chemical sludges which thicken well by gravity settling. It is accomplished in a tank equipped with a slowly rotating rake mechanism that breaks the bridge between sludge particles, thereby increasing settling and compaction. Flotation thickening through dissolved air flotation is particularly applicable to gelatinous sludges such as activated sludge. In flotation thickening, small air bubbles released from solution attach themselves to and become enmeshed in the sludge flocs. The air-solid mixture rises to the surface of the basin, where it concentrates and is removed. Experience has shown that in some cases dilution of the feed sludge to a lower concentration increases the concentration of the floated solids. The use of polyelectrolytes will usually increase the solids capture.

9.2.2 Stabilization

Converte the organic solids to more refractory (inert) forms so that they can be handled or used as soil conditioners without causing a nuisance or health hazard through processes referred to as "digestion". (Two basic stabilization processes are use. One is carried out in closed tanks devoid of oxygen and is called anaerobic digestion. The other approach injects air into the sludge to accomplish aerobic digestion. These are biochemical oxidation processes.)

Two types of anaerobic digestion processes used nowadays are the standard-rate process and the high-rate process.

The former process does not employ sludge mixing, but rather the digester contents are allowed to stratify into zones. Sludge feeding and withdrawal are intermittent rather than continuous. The digester is generally heated to increase the rate of fermentation and therefore decrease the required retention time. Retention time ranges between 30 and 60d for heated digesters. The organic loading rate for a standard-rate digester is between 0.48 and 1.6kg total volatile solids per cubic meter of digester volume per day.

The major disadvantage of this process is the large tank volume required because of long retention times, low loading rates, and thick scum-layer formation. Only about a third of the tank volume is used in the digestion process. The remaining two-thirds of the tank volume contains the scum layer, stabilized solids, and the supernatant.

The high-rate process evolves two digesters operating in series, separating the functions of fermentation and solids-liquid separation. The contents of the first-stage equipped with fixed or floating covers, high-rate unit are thoroughly mixed, and the sludge is heated to increase the rate of fermentation. Because the contents are thoroughly mixed, temperature distribution is more uniform throughout the tank volume. Sludge feeding and withdrawal are continuous or nearly so. The retention time required for the first-stage unit is normally between 10 and 15d. Organic loading rates vary between 1.6 and 8.0kg total volatile solids per cubic meter of digester per day. The first-stage digester of this process approximates a completely mixed reactor without solids recycle. Hence, the biological *SRT* and the hydraulic retention time are equal.

The primary functions of the second-stage digester are solids-liquid separation and residual gas extraction. Their covers are often of the floating type, and their units are generally unheated.

The BOD remaining at the end of digestion is still quite high. Likewise, the suspended solids may be as high as 12000mg/L, whereas the TKN may be on the order of 1000mg/L. Thus, the settled sludge needs to be conditioned and dewatered for disposal.

Aerobic digestion when applied to excess biological sludges involves the oxidation of cellular organic matter through endogenous metabolism. The oxidation of cellular organics has been found to follow first order kinetics when applied to the degradable volatile suspended solids. The aerobic digestion of biological sludges is nothing more than a continuation of the activated sludge system. When a culture of aerobic heterotrophs is placed in an environment containing a source of organic material, the microorganisms remove and use most of this material. This process is accomplished by aerating the organic sludges in an open tank resembling an activated sludge aeration tank. Like the activated sludge aeration tank, the aerobic digester must be followed by a settling tank unless the sludge is to be disposed of on land in liquid form. Unlike the activated sludge process, the effluent (supernatant) from the clarifier (settling tank) is recycled back to the head end of the plant. This is because the supernatant is high in suspended solids (100～300 mg/L), BOD_5 (to 500mg/L), TKN (to 200mg/L), and total P (to 100mg/L).

Because the fraction of volatile matter is reduced, the specific gravity of the digested sludge solids will be higher than it was before digestion. Thus, the sludge settles to a more compact mass, and the clarifier underflow concentration can be expected to reach 3%.

9.2.3 Conditioning

Treating the sludge with chemicals or heat so that the water can be readily separated. Several methods are available for conditioning sludge to facilitate the separation of the liquid and solids. One of the most commonly used is the addition of coagulants such as ferric chloride, lime, or organic polymers. Ash from incinerated sludge has also found use as a conditioning agent. When coagulants are added to turbid water, chemical coagulants act to agglomerate the solids together so that they are more easily separated from the water. In recent years, organic polymers, particularly polyacrylamides, have become increasingly popular for sludge conditioning. Polymers are easy to handle, require little storage space, and are very effective even at a much lower dosage than those inorganic ones. The conditioning chemicals are injected into the sludge just before the dewatering process and are mixed with the sludge.

9.2.4 Dewatering

Sand drying beds are the oldest, most commonly used type of drying bed, many design variations are possible, including the layout of drainage piping, thickness and type of gravel and sand layers, and construction materials. Sand drying beds for wastewater sludge are constructed in the same manner as water treatment plant sludge-drying beds. Sand drying beds can be built with or without provision for mechanical sludge removal and with or without a roof. When the cost of labor is high, newly constructed beds are designed for mechanical sludge removal. Separating water by subjecting the sludge to vacuum, pressure, or drying. Mechanical devices may be employed, including vacuum filtration, centrifugation, and filter presses (Figures 9-2, 9-3). Heat may be used to aid drying process.

9.2.5 Reduction

Convening the solids to a stable form by wet oxidation or incineration. (These are chemical oxidation processes; they decrease the volume of sludge, hence the term reduction.) To minimize the amount of fuel used, the sludge must be dewatered as completely as possible before incineration. The exhaust gas from an incinerator must be treated carefully to avoid air pollution.

Figure 9-2 Belt filter press

Figure 9-3 Plate filter press

Although a large number of alternative combinations of equipment and processes are used for treating sludge, the basic alternatives are fairly limited. The ultimate depository of the materials contained in the sludge must either be land, air, or water.

9.3 污泥的性质和组成

生活污水和工业废水在处理过程中分离或截流下来的固体物质统称为污泥。这些污泥的总量很大，约占处理水量的0.3%~0.5%（以含水率97%计），污泥中含有具有利用价值的有机质、氮、磷、钾和各种微量元素，同时污泥里含有大量的有毒、有害和对环境产生负面影响的物质，如合成有机物、重金属、病原菌、寄生虫卵等。如果不对污泥进行无害化处理处置，会对环境造成二次污染。污泥的处理与处置是两个不同的阶段，处理必须满足处置的要求。因此，污泥的处理技术措施，要以达到在最终处置后不对环境产生有害影响为目标。不同的处置方式须对应相应的处理方法。

污泥处理的工艺路线选择需要强调污泥的减量化、稳定化、无害化以及污泥的资源化综合利用。其中污泥的减量化指通过一定的技术措施削减污泥的量和体积；稳定化指将污泥中的有机物（包括有毒有害有机物）降解成为无机物的过程。污泥在环境中的最终消纳方式包括土地利用、做建材的原料或进行无害化填埋等。

污泥的性质和组成主要取决于污水的来源，同时还和污水处理工艺有密切关系。按来源不同，污泥可分为初沉污泥、剩余活性污泥（来自活性污泥法二沉池）、腐殖污泥（来自生物膜法二沉池）、消化污泥（经过厌氧消化或好氧消化处理后的污泥）和化学污泥。其中，初沉污泥、剩余活性污泥、腐殖污泥统称为生污泥或新鲜污泥。初沉污泥、生物膜法及活性污泥法污泥的组成及特点见表9-1、表9-2。

初沉污泥、生物膜法及活性污泥法污泥的组成　　表9-1

污泥对比	有机分(%)	腐殖质(%)	总氮(%)	磷(P_2O_5)(%)	钾(%)
初沉污泥	55~70	33	2.0~3.4	1.0~3.0	0.1~0.3
生物膜法污泥（腐殖污泥）	60	47	2.8~3.1	1.0~3.0	0.11~0.8
活性污泥法污泥（剩余活性污泥）	70~85	41	3.5~7.2	3.3~5.0	0.2~0.4

初沉污泥、生物膜法及活性污泥法污泥的特点 **表 9-2**

污泥对比	来源	颜色	气味	pH	含固量
初沉污泥	初沉池排出	1. 正常情况时：棕褐色略带灰色；2. 发生腐败时：灰色或黑色	有难闻气味	5.5～7.5（典型值：6.5）	2%～4%（典型值：3%）
生物膜法污泥（腐殖污泥）	生物膜法后的二沉池排出				1%～4%
活性污泥法污泥（剩余活性污泥）	活性污泥法后的二沉池排出	黄褐色絮状	土腥味	6.5～7.5	0.5%～0.8%

污泥的性质主要包括含水率、相对密度、可消化程度、污泥肥分、脱水性能等方面。

9.3.1 含水率和含固率

污泥中水分的存在形式有以下几种：颗粒间的间隙水（又称游离水，约占总水分的70%）、毛细水（颗粒间毛细管内的水，约占20%）、污泥颗粒吸附水和颗粒内部水（约占10%）。

污泥中所含水分的质量与污泥总质量之比的百分数称为含水率。与此对应，污泥中固体的质量分数叫含固率。显然，含固率与含水率之间存在如下关系：含固率＋含水率＝100%。污泥的体积、质量及所含固体浓度之间的关系为：

$$\frac{V_1}{V_2}=\frac{W_1}{W_2}=\frac{100-P_2}{100-P_1}=\frac{C_2}{C_1} \tag{9-1}$$

式中 P_1、P_2——污泥处理前后的含水率，%；

V_1、W_1、C_1——污泥含水率为 P_1 时的污泥体积、质量与固体物浓度；

V_2、W_2、C_2——污泥含水率为 P_2 时的污泥体积、质量与固体物浓度。

该公式适用于含水率大于65%的污泥。原因是当 $P\leqslant65\%$ 时，污泥的体积由于颗粒具有弹性，不再收缩。通常认为含水率大于85%的污泥具有流动性，含水率在70%～80%的是塑性污泥，含水率小于60%的可视为干泥。

由于多数污泥都由亲水性固体组成，因此，含水率一般都很高，而不同污泥之间含水率的差异对污泥特性有着重要影响。代表性污泥的含水率见表 9-3。

代表性污泥的含水率 **表 9-3**

名称		含水率(%)	名称		含水率(%)
初沉污泥		95	慢速滤池		93
混凝污泥		93	快速滤池		97
活性污泥	空气曝气	98～99	厌氧消化污泥	初沉污泥	85～90
活性污泥	纯氧曝气	96～98	厌氧消化污泥	活性污泥	90～94

【例 9-1】 某污水处理厂剩余污泥量为 $900m^3/d$，污泥含水率为99%，为减少污泥产生量，对污泥进行浓缩处理，浓缩后污泥含水率为97%，求浓缩后的污泥量、污泥体积变化、污泥的含固率。

【解】 (1) 浓缩后的污泥量

由式 (9-1)，含水率由 99%变为 97%，污泥量变为：

$$V_2=\frac{100-P_1}{100-P_2}V_1=\frac{100-99}{100-97}\times900=300\text{m}^3/\text{d}$$

(2) 浓缩后污泥体积变化

浓缩后污泥体积为 $300\text{m}^3/\text{d}$，浓缩前污泥体积为 $900\text{m}^3/\text{d}$，即浓缩后体积变为原来的 1/3。

(3) 浓缩后污泥的含固率

浓缩后污泥含水率为 97%，由含水率＋含固率＝100%，因此，浓缩后污泥的含固率为 3%。

9.3.2 污泥的相对密度

污泥相对密度指污泥的质量与同体积水质量的比值。污泥相对密度主要取决于含水率和污泥中固体组分的比例。固体组分的比例越大，含水率越低，则污泥相对密度也就越大。城镇污水及其类似污水处理系统排除的污泥相对密度一般略大于 1。工业废水处理系统排出的污泥相对密度往往较大。污泥相对密度与其组分之间存在如下关系：

$$\gamma=\frac{1}{\sum_{i=1}^{n}\left(\frac{w_i}{\gamma_i}\right)} \tag{9-2}$$

式中 w_i——污泥中第 i 项组分的质量分数，%；

γ_i——污泥中第 i 项组分的相对密度。

若污泥仅含有一种固体成分（或者近似为一种成分），且含水率为 P(%)，则上式可简化如下：

$$\gamma=\frac{100\gamma_1\gamma_2}{P\gamma_1+(100-P)\gamma_2} \tag{9-3}$$

式中 γ_1——固体相对密度；

γ_2——水的相对密度。

一般城市污泥中固体的相对密度为 2.5，若含水率为 99%，则由式 (9-3) 可知该污泥相对密度约为 1.006。

9.3.3 污泥的可消化程度

污泥中的有机物是消化处理的对象，通过消化可使污泥达到稳定，可消化程度表示污泥中可被消化降解的有机物数量，污泥可消化程度用式 (9-4) 表示。

$$R_\text{d}=\left(1-\frac{P_\text{v2}P_\text{s1}}{P_\text{v1}P_\text{s2}}\right)\times100\% \tag{9-4}$$

式中 R_d——可消化程度，%；

P_s1、P_s2——生污泥及消化污泥的无机物含量，%；

P_v1、P_v2——生污泥及消化污泥的有机物含量，%。

污泥经厌氧或好氧消化后，污泥中有机物被氧化分解，因此，污泥体积会减小，消化后的污泥体积为：

$$V=\frac{(100-P_1)V_0}{100-P_2}\times\left(1-\frac{P_\text{v1}R_\text{d}}{10000}\right) \tag{9-5}$$

式中　V——消化后的污泥体积，m^3；

P_1——生污泥的含水率，%；

P_2——消化污泥的含水率，%；

V_0——生污泥体积，m^3；

R_d——污泥的可消化程度，%；

P_{v1}——生污泥中有机物含量，%。

【例 9-2】 某污水处理厂初沉池污泥和二沉池污泥混合浓缩后污泥量为 $800m^3/d$，混合污泥含水率为 96%，污泥中有机物含量为 66%，混合污泥经厌氧消化后体积减小，消化后熟污泥中有机物含量为 50%，熟污泥含水率为 97%，消化池无上清液排除设备，求污泥的可消化程度和消化后的熟污泥量。

【解】 （1）污泥的可消化程度

生污泥中有机物含量为 66%，无机物含量为 34%；熟污泥中有机物含量为 50%，无机物含量为 50%，由公式（9-4）得：

$$R_d=\left(1-\frac{P_{v2}P_{s1}}{P_{v1}P_{s2}}\right)\times100\%=\left(1-\frac{50\times34}{66\times50}\right)\times100\%=48.5\%$$

（2）消化后的熟污泥量

污泥可消化程度为 48.5%，消化前污泥量为 $800m^3/d$，由公式（9-5），消化后的污泥量为：

$$\begin{aligned}V&=\frac{(100-P_1)V_0}{100-P_2}\times\left(1-\frac{P_{v1}R_d}{10000}\right)\\&=\frac{(100-96)\times800}{100-97}\times\left(1-\frac{66\times48.5}{10000}\right)\\&=725.2m^3/d\end{aligned}$$

9.3.4 污泥中的重金属和病原微生物

由于一些企业的工业废水排入城市排水管网，城市污水中含有部分工业废水，在污水处理过程中 70%～90%的重金属元素通过吸附或沉淀转移到污泥中，导致城市污水处理厂污泥中含有重金属离子，且一般含量都较高。重金属是制约污泥填埋、土地利用、建材利用的关键因素，我国城市污水处理厂污泥中重金属含量见表 9-4。污水中的病原体和寄生虫经过处理还会进入污泥，新鲜污泥中检测得到的病原体多达千种，其中危害较大的是寄生虫。污水处理厂污泥的微生物污染指标主要包括细菌总数、大肠菌群数和寄生虫卵，表 9-5 给出了城镇污水处理厂污泥中的病原微生物量。

城市污水处理厂污泥中重金属统计分析（单位：mg/kg）　　**表 9-4**

项目	总镉	总铅	总铬	总镍	总锌	总铜	总汞	总砷
最大值	63.7	279.5	6107	1080	10070	4564.4	13.5	134
最小值	0	1.8	2.9	12.1	32	56.9	0	0.2
平均值	3.3	69.8	438.7	97.1	1639.3	614.8	2.4	19.2

城镇污水处理厂污泥中的病原微生物　　表 9-5

污泥名称	细菌总数 (10^6 个/g 干)	总大肠菌群数 (10^6 个/g 干)	粪大肠菌群数 (10^6 个/g 干)	寄生虫卵数 (个/g 干)
初沉池污泥	47.2	20	15.8	233(活卵率 78.3%)
二沉池污泥	73.8	1.8	1.2	170(活卵率 67.8%)
消化污泥	3.8	0.16	0.12	139(活卵率 60%)

9.3.5 污泥的脱水性能和可压缩性能

污泥的脱水性能与污泥性质、调理方法及条件等有关，还与脱水机械种类有关。污泥脱水的难易程度或脱水性能常用污泥比阻和毛细管吸水时间衡量。

污泥比阻可在实验室通过布氏（Buchner）漏斗试验确定。比阻（α_{av}）为单位过滤面积上，滤饼单位干固体质量所受到的阻力，其单位为 m/kg，可以用式（9-6）表示。

$$\alpha_{av}=\frac{2\Delta PA^2K_b}{\mu\omega} \tag{9-6}$$

式中　ΔP——过滤压力（为滤饼上下表面间的压力差），N/m^2；

A——过滤面积，m^2；

K_b——过滤时间/滤液体积的斜率，s/m^6；

μ——滤液动力黏度，$N\cdot s/m^2$；

ω——滤液所产生的干固体质量，kg/m^3。

污泥比阻用来衡量污泥脱水的难易程度，反映污泥中水分通过污泥颗粒所形成的泥饼层时所受阻力的大小。不同的污泥种类，其比阻值差别较大。通常，初沉池污泥比阻为 $20\times10^{12}\sim60\times10^{12}$ m/kg，活性污泥比阻为 $100\times10^{12}\sim300\times10^{12}$ m/kg，厌氧消化污泥比阻为 $40\times10^{12}\sim80\times10^{12}$ m/kg。一般来说，比阻小于 1×10^{11} m/kg 的污泥易于脱水，大于 1×10^{13} m/kg 的污泥难以脱水。

污泥的毛细管吸水时间（capillary suction time，CST）指污泥与滤纸接触时，污泥中水分在滤纸上渗透 1cm 长度所需要的时间。毛细管吸水时间可以用毛细管吸水时间测定仪测量，毛细管吸水时间（CST）越小，表明污泥脱水性能越好。一般 CST 小于 20s 时污泥比较容易脱水。

污泥具有一定的可压缩性，在实际中用压缩系数衡量，一般污泥的压缩系数为 0.6～0.9。压缩系数用来反映污泥的渗滤性质，压缩系数大的污泥说明当压力增加时，污泥的比阻会迅速增加，这种污泥宜采用真空过滤或离心脱水的方法脱水，与此相反，压缩系数小的污泥宜采用板框和带式压滤机脱水。经调理的污泥往往比阻减少，而压缩系数增加，所以在脱水时须选择合适的压力，否则压力过大会使污泥絮体破碎，反而不利于过滤脱水。各种污泥的污泥比阻和压缩系数见表 9-6。

各种污泥的污泥比阻和压缩系数　　表 9-6

污泥类型	比阻(m/kg)	压力(0.1MPa)	压缩系数
初沉污泥	4.61×10^{13}	0.5	0.54
活性污泥	2.83×10^{14}	0.5	0.81
消化污泥	1.39×10^{14}	0.5	0.74
$Al(OH)_3$ 混凝污泥	2.16×10^{13}	3.5	
$Fe(OH)_3$ 混凝污泥	1.47×10^{13}	3.5	

9.3.6 污泥中的挥发性固体与灰分

挥发性固体（VSS）代表污泥中有机物的含量，又称为灼烧减量，是将污泥中的固体物质在550～600℃高温下焚烧时以气体形式逸出的那部分固体量。灼烧残渣（FS）称为固定性固体，又称为灰分，灰分代表污泥中无机物的含量。挥发性固体是污泥最重要的化学性质，决定污泥的热值与可消化性。一般情况下，初沉池污泥 VSS 的比例为 50%～70%，活性污泥 VSS 比例为 60%～85%，经厌氧消化后的污泥 VSS 占比为 30%～50%。挥发性固体与灰分可以通过污泥烘干、高温焚烧称重测得。

9.4 污泥处理的目的和作用

污泥含水率高，体积庞大，污（废）水生物处理法处理产生的污泥含有高浓度有机物，很不稳定，易在微生物作用下腐败发臭，并常常含有病原微生物、寄生虫卵及重金属离子等有害物质，必须进行相应的处理。

污泥处理是污（废）水处理的重要组成部分。对于以活性污泥法为主的城镇污水处理厂，污泥处理系统的建设投资约占污水处理厂总投资的 20%～40%，污泥处理运营费用约占污水处理厂总运营费用的 20%～30%，而污泥处理的投资和运营费用与选择的处理工艺密切相关。因此，对污泥处理工艺的选择应当给予足够的重视。

例如，目前在城镇污水处理中普遍采用生物除磷的工艺，此时所产生的剩余污泥由于富含无机磷，进行重力浓缩时，由于浓缩池内的厌氧状态，会促使磷的释放，使浓缩池上清液中含磷较高，如不通过化学除磷予以去除，则上清液中的磷会回流至提升泵房而返回到污水处理系统。此时常用的方法是经调理后直接进行机械浓缩和脱水，采用浓缩离心脱水一体化装置可避免污泥浓缩厌氧释磷，从而提高脱水污泥的磷含量，利于污泥农用。

污泥处理处置的目的是实现污泥的“稳定化、减量化、无害化”，在处理处置过程中利用污泥中的物质和能量，实现“资源化”。污泥处理的“稳定化”指通过物理、化学或物化处理，使污泥达到不易腐败发臭，控制病原体的目的。污泥处理的“减量化”指采用适当的处理方法，减少污泥的重量和体积。“无害化”指污泥经过处理后，不对环境造成二次污染，不对人体健康产生危害。“资源化”指回收污泥中的有用物质和能源。污泥的调理则是减小污泥的比阻和毛细管吸水时间以提高污泥的脱水性能。表 9-7 列出了常用的各种污泥处理方法的目的和作用。

各种污泥处理方法的目的和作用 **表 9-7**

处理方法		目的和作用
污泥浓缩	重力浓缩	缩小体积
	气浮浓缩	缩小体积
	机械浓缩	缩小体积
污泥稳定	加氯稳定	稳定
	石灰稳定	稳定
	厌氧消化	稳定、减少质量
	好氧消化	稳定、减少质量

续表

处理方法		目的和作用
污泥调理	化学调理	改善污泥脱水性能
	加热调理	改善污泥脱水性能及稳定、消毒
	冷冻调理	改善污泥脱水性能
污泥脱水	自然脱水	缩小体积
	机械脱水	缩小体积
机械加热干燥		减重、缩小体积
污泥堆肥		提高污泥对农业的适用性
污泥焚烧		缩小体积、灭菌
污泥最终处置	卫生填埋	解决污泥最终出路
	农业利用	利用污泥肥分改良土壤

污泥处理的方法与工艺流程的选择取决于当地条件、环境保护要求、投资情况、运行费用及维护管理等多种因素。典型的污泥处理工艺包括“浓缩＋脱水/干化”的污泥减量处理工艺（图 9-4），“消化＋制肥”的污泥土地利用处理工艺（图 9-5），“好氧发酵＋稳定”的污泥填埋处理工艺（图 9-6），“脱水＋焚烧”的污泥建筑建材处理工艺（图 9-7）。

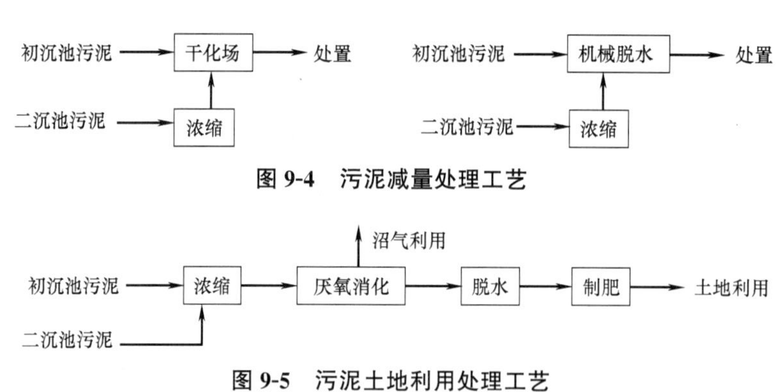

图 9-4　污泥减量处理工艺

图 9-5　污泥土地利用处理工艺

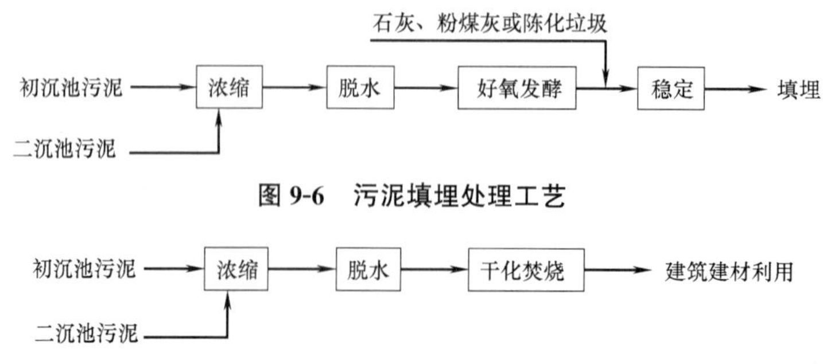

图 9-6　污泥填埋处理工艺

图 9-7　污泥建筑建材处理工艺

以上是典型的污泥处理工艺，各个企业可根据污（废）水的处理工艺、污泥的性质和处理后污泥的用途经过技术经济比较后进行合理组合，使污泥得到妥善的处理处置。表 9-8 给出了各种污泥处理工艺的优缺点和适用范围。

各种污泥处理工艺优缺点和适用范围　　**表 9-8**

项目	干化焚烧	厌氧消化	好氧发酵
优点	处理效率高，占地面积小，污泥减量化、无害化程度高	运行成本低，产生消化气，可以回收能源，沼渣富含有机质及微量元素，可以改良土壤，沼液可以作为肥料	处理时间较短，发酵产物可以作为肥料
缺点	投资及运行成本高，产生飞灰、二噁英等污染物，烟气需净化处理	厌氧微生物生长速率低，对环境要求苛刻，设备体积大，投资较高	发酵过程需要氧气，能耗高，发酵过程中有臭气、气溶胶散发，环境条件差
适用范围	浓缩后的有机污泥，土地紧张，经济发达地区	大、中型污水处理厂，经济相对发达，资源紧缺的地区	中、小型污水处理厂，土地资源丰富，经济欠发达地区

9.5 污泥的单元处理技术

9.5.1 污泥的浓缩和脱水

污泥处理的方法常取决于污泥的含水率和最终的处置方式。例如，含水率大于 98% 的污泥，一般要考虑浓缩，使含水率降至 96%左右，以减少污泥体积，有利于后续处理。为了便于污泥处置时的运输，污泥要脱水，使含水率降至 80%以下，失去流态。某些国家规定，若污泥进行填埋，其含水率要在 60%以下。

二维码 8
污泥脱水

由于污泥厌氧消化前需浓缩，污泥堆肥前需脱水，因此，可以将污泥浓缩、脱水统称为污泥的预处理技术。污泥浓缩主要用于降低污泥中的空隙水，是减小污泥容积的主要方法；污泥浓缩方法包括重力浓缩、离心浓缩和气浮浓缩，目前重力浓缩应用最多，表 9-9 列出了各种污泥浓缩方法比较。

重力浓缩适用于初沉池污泥、二沉池污泥和消化后的污泥，不适用于脱氮除磷工艺产生的剩余污泥。重力浓缩耗能少、缓冲能力强，但占地面积较大，易造成磷的释放，臭味大，需要增加除臭设施。初沉池污泥采用重力浓缩时，污泥固体负荷为 80～120kg/(m^2·d)，停留时间宜为 6～8h，含水率一般可从 97%～98%降至 95%以下。混合污泥采用重力浓缩时，污泥固体负荷为 50～75kg/(m^2·d)，停留时间宜为 10～12h，含水率可由 96%～98.5%降至 95%以下。污泥浓缩池一般宜设置去除浮渣的装置。

气浮浓缩与重力浓缩相反，通过溶于水中的气体突然减压释放出大量微小气泡并捕捉污泥颗粒，使固体颗粒的密度小于水而产生上浮，从而达到污泥浓缩的目的。因此，气浮浓缩对于密度接近于 1g/cm^3 的活性污泥尤其适用。气浮浓缩法操作简便，固体物质的回收率在 99%以上，分离液中的悬浮物（SS）可以降到 100mg/L 以下，浓缩后污泥中的含固率为 5%～7%；气浮浓缩速度快，水力停留时间仅为重力浓缩法的 1/3 左右，构筑物占地面积小；浓缩过程中需要通入空气，能保持污泥中的溶解氧含量，污泥不易腐败发臭，但动力费用高，适用于剩余污泥产量不大的活性污泥法处理系统，尤其是生物除磷系统产生的剩余污泥。

离心浓缩是根据污泥中固体和液体的相对密度不同，在离心浓缩机中所受离心力不同实现污泥的浓缩。离心浓缩时污泥中需要投加高分子混凝剂，与重力浓缩相比离心浓缩电耗较高，具有占地面积小、可以避免厌氧状态磷释放等特点。离心浓缩一般可将剩余污泥的含水率从 99.2%～99.5%降至 94%～96%。

污泥浓缩方法比较　　表 9-9

浓缩方法	优点	缺点	适用范围
重力浓缩	贮泥能力强；动力消耗小；运行费用低；操作简便	占地面积大；浓缩效果差；浓缩后污泥含固率低；易厌氧发酵产生臭气	主要用于浓缩初沉池污泥、初沉池污泥和剩余污泥的混合污泥
气浮浓缩	占地面积小，处理时间短，浓缩后污泥含水率低，能同时去除油脂，产生臭气较少	占地面积、运行费用小于重力浓缩；污泥贮存能力小于重力浓缩；动力消耗、操作要求高于重力浓缩	主要用于初沉污泥和剩余活性污泥的混合污泥；特别是易发生膨胀的剩余污泥、腐殖污泥和生物除磷污泥
离心浓缩	占地面积很小，处理能力强，污泥含水率低；全封闭，无臭气发生	离心机价格高，浓缩的污泥需要加入药剂，电耗是气浮法的 10 倍，操作管理要求高	主要用于难以浓缩的剩余污泥、腐殖污泥、消化污泥，生物除磷系统污泥

浓缩池适用于选矿厂的精矿和尾矿脱水处理，把 20%～30%的矿浆提高到 40%～70%左右，广泛应用于冶金、化工、煤炭、非金属选矿、环保等行业。主要性能特点如下：

（1）添加絮凝剂增大沉降固体颗粒的粒径，从而加快沉降速度。

（2）装设倾斜板缩短矿粒沉降距离，增加沉降面积。

（3）发挥泥浆沉积浓相层的絮凝、过滤、压缩和提高处理量的作用。

（4）配备完整的自控设施。

二维码 9
脱水后的污泥

污泥脱水包括自然干化脱水、热干化脱水和机械脱水。污泥脱水主要是脱除污泥毛细水。污泥自然干化简单易行，但受自然条件如风速、温度、湿度、降水等影响大，占地面积大，自然干化过程中产生臭气，卫生条件差，工人劳动强度大。污泥机械脱水主要有带式压滤机、板框压滤机和离心脱水机。表 9-10 列出了几种污泥机械脱水方法的比较。

几种污泥机械脱水方法的比较　　表 9-10

脱水方法	优点	缺点	泥饼含水率
带式压滤机	连续操作，自动控制，噪声小，能耗少	占地面积和冲洗水量较大，车间环境较差，附属设备多	可在 82%以下
板框压滤机	脱水泥饼含水率低	间歇操作，操作管理复杂，占地和冲洗水量较大，车间环境较差	一般 65%～75%
离心脱水机	占地面积小、不需冲洗水、车间环境好	连续操作，电耗高，药剂消耗量高，噪声大	一般 75%～80%

其他新型处理技术包括污泥碳化技术与叠螺式脱水机。

1. 污泥碳化技术

低温碳化技术利用较广泛：污水工艺优化可降低剩余污泥产量，污泥破壁及强力干化技术能提高污泥的脱水性能；最终通过污泥碳化技术实现污泥的资源化，从源头上解决污泥的产量，最终达到污泥零排放的目的。

所谓污泥碳化，就是通过一定的手段，使污泥中的水分释放出来，同时又最大限度地保留污泥中的碳值，使最终产物中的碳含量大幅提高的过程（Sludge Carbonization）。污泥碳化主要分为3种：（1）高温碳化。碳化时不加压，温度为649～982℃。先将污泥干化至含水率约30%，然后进入碳化炉高温碳化造粒。碳化颗粒可以作为低级燃料使用，其热值约为8360～12540kJ/kg（日本或美国）。该技术可以实现污泥的减量化和资源化，但由于其技术复杂，运行成本高，产品中的热值含量低，当前尚未有大规模的应用。（2）中温碳化。碳化时不加压，温度为426～537℃。先将污泥干化至含水率约90%，然后进入碳化炉分解。工艺中产生油、反应水（蒸汽冷凝水）、沼气（未冷凝的空气）和固体碳化物。另外，该技术是在干化后对污泥实行碳化，其经济效益不明显，除澳洲一家处理厂外，尚无其他潜在的用户。（3）低温碳化。碳化前无需干化，碳化时加压至6～8MPa，碳化温度为315℃，碳化后的污泥成液态，脱水后的含水率在50%以下，经干化造粒后可作为低级燃料使用，其热值约为15048～20482kJ/kg（美国）。该技术通过加温加压使得污泥中的生物质全部裂解，仅通过机械方法即可将污泥中75%的水分脱除，极大地节省了运行中的能源消耗。污泥全部裂解保证了污泥的彻底稳定。污泥碳化过程中保留了绝大部分污泥中热值，为裂解后的能源再利用创造了条件。

2. 叠螺式脱水机

叠螺式脱水机的叠螺主体是由固定环和游动环相互层叠，螺旋轴贯穿其中形成的过滤装置。前段为浓缩部，后段为脱水部（图9-8）。固定环和游动环之间形成的滤缝以及螺旋轴的螺距从浓缩部到脱水部逐渐变小。螺旋轴的旋转在推动污泥从浓缩部输送到脱水部的同时，也不断带动游动环清扫滤缝，防止堵塞。

其工作过程如下：

（1）当驱动电机带动螺旋推动轴转动时，设在推动螺旋轴上的多片动环、定环叠片相对移动，在重力作用下，水从相对移动的叠片间隙中不断滤出，实现污泥快速浓缩脱水工序；

（2）经过浓缩的污泥随着螺旋轴的转动不断往前移动，沿泥饼出口方向，同时螺旋轴的螺距逐渐变小，环与环之间的间隙也逐渐变小，螺旋腔的体积不断收缩；在出口处背压板的作用下，内压逐渐增强，在螺旋推动轴依次连续运转推动下，污泥中的水分受挤压排出，滤饼含固量不断升高，最终实现污泥的连续脱水；

（3）螺旋轴的旋转，推动游动环不断转动，设备依靠固定环和游动环之间的移动实现连续的自清洗过程，从而巧妙地避免了传统脱水机普遍存在的堵塞问题。

它的优点有以下3方面：

（1）配置专有的旋盘预浓缩装置，更善于处理低浓度污泥，改善现有重力式浓缩缺点，实现低浓度污泥的高效浓缩絮凝与浓缩一体化完成，减轻后续脱水压力，结合调节伸缩阀，可将进泥浓度调理到脱水较好状态（图9-9）。

（2）叠螺式脱水机在螺旋轴的旋转作用下，活动板相对于固定板不断错动，从而实现连续的自清洗过程，避免传统脱水机普遍存在的堵塞问题。因此，抗油污能力强，易分离、不堵塞，并配有高压水枪。

（3）叠螺式脱水机依靠容积压脱水、配桶等大型设备，实现2～4r/min的低运行速度。因此，本机能实现低能耗、低耗水量、低噪声。平均能耗仅为带式机的 1/8，离心机的 1/20，单位能耗仅为 0.01～0.1kWh/kg-DS，能够降低污水处理系统的运行成本。

图 9-8　新式叠螺式脱水机

图 9-9　叠螺式脱水机处理效果图

9.5.2　污泥的稳定

污泥的稳定处理技术包括污泥的厌氧消化技术、好氧消化技术、石灰稳定技术等。

1. 污泥厌氧消化

厌氧消化是在没有分子氧的情况下分解有机物和无机物（主要是硫酸盐），类似于第 3 章，主要应用于稳定城市和工业废水处理中产生的浓缩污泥。由于对节能和回收以及对废水生物固体的有益利用的重视，厌氧消化仍然是污泥稳定化的主要工艺手段。此外，在许多情况下，城市污水污泥的厌氧消化可以产生足够的消化气体，以满足工厂运行所需的大部分能源。

厌氧消化后污泥特点（图 9-10）：

（1）污泥含水率下降，含固率上升，含固率能达到 5%。

（2）有机物含量下降。厌氧消化后，污泥中有机物含量减少 30%～50%。

（3）污泥固体总量减少，污泥体积减小。通常厌氧消化使 25%～50%的污泥固体被分解，减少后续污泥处理的费用。

（4）污泥中致病菌减少。消化过程尤其是高温消化过程（在 50℃～60℃条件下），能杀死致病菌。

（5）污泥可作为土壤调节剂。消化后污泥含有一定量的灰分和有机物，能提高土壤的肥力和改善土壤的结构。

（6）消化污泥不易沉淀。污泥颗粒周围有甲烷及其他气体的气泡，使其不易沉淀。

图 9-10　消化污泥图

污泥厌氧消化是利用兼性菌和厌氧菌分解污泥的有机物，使污泥达到稳定的一种处理工艺。污泥厌氧消化后有机物去除率应大于40%，粪大肠菌群值应小于0.5×10^{-6}。根据消化温度的不同，分为污泥高温消化和污泥中温消化。

经过浓缩、均质后的污泥进入厌氧消化池进行消化处理。中温厌氧消化温度一般为29～38℃之间，常采用35℃，污泥龄应大于20d，消化池容积负荷一般为2.0～4.0kgCOD/(m^3·d)，污泥投配率为5%～8%，产气率一般不小于0.4～0.5Nm^3/kgVSS（去除）。污泥中温消化速率较慢，产气率低，需要的能耗较少，沼气产率能够维持在较高水平。对处理规模大于$5\times10^4m^3/d$的城镇二级污水处理厂，其产生的污泥宜通过中温厌氧消化处理，同时对产生的沼气进行综合利用。

污泥高温厌氧消化温度一般为50～60℃，常采用55℃，此温度适合嗜热产甲烷菌生长。高温厌氧消化有机物分解速度快，消化时间一般为10～15d，污泥投配率一般为7%～10%，高温消化可以杀灭高达99%的各种致病菌和寄生虫卵。污泥高温消化有机物分解率和消化气产生量略高于中温消化，但能量消耗较大，运行费用较高，系统稳定性差，总体比较，经济性不高，因而采用较少。当卫生指标有特殊要求或需减少消化时间时，可以采用高温消化。

污泥厌氧消化系统主要包括污泥进出料系统、污泥加热系统、消化池搅拌系统及沼气收集、净化利用系统。

消化池通常有蛋（卵）形（图9-11）和圆柱形等池形，可根据工艺条件、搅拌系统、投资及景观要求进行选择，宜选用蛋形消化池或具有较高高径比的圆柱形消化池。消化池池体可采用混凝土结构或钢结构。由于厌氧消化池内消化液pH在较大的范围内变化，且消化气中含有大量的水分和硫化氢，对池体有腐蚀，各种类型的消化池内部均应采取防腐蚀措施。

图9-11 蛋形消化池

搅拌可使消化物料分布均匀，增加活性微生物与物料的接触，并使消化产物及时分离，从而提高消化效率、增加产气量。污泥厌氧消化可根据系统的要求选择沼气搅拌、机械搅拌、水力循环搅拌，蛋形消化池宜采用机械搅拌器，机械搅拌器应能正反方向转动以使消化池内物料分布均匀。每日全池污泥搅拌或循环的次数不宜少于3次，间歇搅拌时，每次搅拌的时间不宜大于循环周期的一半。

为了维持厌氧微生物所需要的温度，保证污泥消化以较高的速率进行，需对消化池安装加热装置并进行保温处理，保温层外侧应设置保护层。在全年气温高的南方地区，消化池可以考虑不设置保温措施，节省投资。

污泥厌氧消化产生的消化气是一种混合气体，一般由60%～70%的CH_4、30%～40%的CO_2和少量硫化氢、水蒸气、氨、氢、氮等组成，消化气热值为21～25MJ/Nm^3。消化气中的H_2S气体不仅对人的身体健康有很大的危害，而且对管道、仪表及设备还具有很强的腐蚀性。因此，消化气利用前需进行过滤、脱水、净化，去除消化气中水分、二氧化碳、氨、硫化氢等杂质，提高消化气的热值。消化气脱硫方法包括干法脱硫、湿法脱硫和生物脱硫，可采用两种脱硫方法联合使用。表9-11给出了各种污泥厌氧消化气脱硫方法的比较。

污泥厌氧消化气脱硫方法比较　　表9-11

脱硫方法	优点	缺点	适用范围
湿法脱硫	工艺流程简单、操作连续、脱硫效率高	系统一次性投资多、运行管理复杂、脱硫成本高	粗脱硫，适用于高浓度硫化氢消化气
干法脱硫	结构简单、技术成熟，应用广泛	装置占地面积大、操作不连续、脱硫剂再生困难、更换脱硫剂频繁、脱硫效率低	精细脱硫，进气H_2S浓度不宜大于1000mg/m^3，适用于低浓度硫化氢消化气
生物脱硫	设备简单、投资小、能耗低、二次污染少	脱硫效果受温度、pH、微生物种类、浓度影响大	适用于各种浓度硫化氢消化气

2. 污泥的好氧消化

污泥的好氧消化又称为好氧生物稳定，是指在有氧状态下对污泥进行长时间的曝气，使污泥中有机物氧化为CO_2、NH_3和H_2O的过程。污泥好氧消化时间短，反应速度快，消化程度高，污泥产量小，基建费用比厌氧消化低；但污泥好氧消化需长时间曝气，电耗大，运行成本高，且不能回收沼气，消化后污泥脱水性能差。污泥好氧消化适用于小型污水处理厂产生的污泥。目前，城镇污水处理厂普遍采用脱氮除磷工艺，污泥好氧消化可使排放污泥中的磷酸盐重新溶出，因此，生物除磷污泥采用好氧消化时上清液应采用化学除磷措施。污泥好氧消化设计参数见表9-12。

污泥好氧消化设计参数　　表9-12

名称		设计参数
停留时间(d)	活性污泥	10～15
	初沉污泥、初沉污泥与活性污泥混合	15～25
有机(容积)负荷[kgVSS/(m^3·d)]	经重力浓缩后的原污泥	0.7～2.8
	机械浓缩后的原污泥	≤4.2
鼓风曝气所需空气量[m^3/(m^3·min)]	活性污泥	0.02～0.04
	初沉污泥、初沉污泥与活性污泥混合	0.04～0.06
机械曝气所需功率(kW/m^3池容)	所有污泥	0.02～0.04
最低溶解氧(mg/L)	所有污泥	>2
温度(℃)	所有污泥	≥15
污泥含水率(%)	所有污泥	<98
挥发性固体去除率(VSS,%)	所有污泥	≥40

3. 污泥的石灰稳定

污泥的石灰稳定指向污泥中投加一定比例的生石灰并均匀掺混，生石灰与脱水污泥中的水分发生反应，生成氢氧化钙和碳酸钙并释放热量，达到污泥稳定、杀菌、脱水的目的。污泥石灰稳定反应生成氢氧化钙呈碱性，可以结合污泥中的部分金属离子，钝化重金属；石灰稳定可对污泥起到固化作用改善储存和运输条件，避免渗滤液泄漏。污泥加入石灰后 pH 在 12 以上的持续时间应大于 2h 或 pH 在 11.5 以上的持续时间应大于 24h。在投加石灰过程中，应监测 pH 的变化，防止石灰投加量不足造成 pH 的降低影响稳定效果。当污泥的含固率大于 30%时，应延长停留时间和提高反应温度。污泥石灰稳定后体积增加量控制在 5%～12%之间，表 9-13 给出了对于不同污泥石灰的投加量。

石灰稳定法石灰投加量 **表 9-13**

污泥类型	污泥固体浓度(%)		$Ca(OH)_2$ 投加量(g/kgSS)	
	变化范围	平均值	变化范围	平均值
初沉污泥	3～6	4.3	60～170	120
活性污泥	1～1.5	1.3	210～430	300
消化污泥	6～7	6.5	140～250	190
生物膜法污泥	1～4.5	2.7	90～510	200

石灰稳定的污泥可以采用建材利用、土地利用、卫生填埋等污泥处理处置措施。采用石灰稳定技术应考虑当地石灰来源的稳定性、经济性和质量方面的可靠性。

9.5.3 污泥的调理

污泥的调理主要是对污泥颗粒表面的有机物进行改性，或对污泥的细胞和胶体结构进行破坏，降低污泥的水分结合容量。经调理后污泥的比阻和毛细管吸水时间减小，脱水性能大幅改善。污泥调理方法主要有化学调理、物理调理和热工调理 3 种类型。

污泥化学调理指加入调理剂使小颗粒絮凝并释放吸附水，降低污泥碱度。无机调理剂最有效、最便宜的是铁盐，如氯化铁、氯化亚铁、硫酸铁、聚合硫酸铁（PFS）。三氯化铁和石灰同时使用，不但能调节 pH，而且由于石灰和污水中的重碳酸盐生成碳酸钙能形成颗粒结构增加污泥的孔隙率。常用的铝盐如明矾、氯化铝、碱式氯化铝，聚合氯化铝（PAC）等。

无机调理剂价格低廉，适用于板框压滤机，用量通常为污泥中干固体重量的 5%～20%。污泥经无机调理剂处理后污泥量增加，污泥中无机成分的比例提高，污泥的燃烧价值降低。有机调理剂适用于带式压滤机和离心脱水机，有机调理剂如阳离子型聚丙烯酰胺（PAM）和阴离子型聚丙烯酰胺（PAM），用量通常为污泥干固体重量的 0.1%～0.5%。有机调理剂受污泥 pH 的影响较小，综合运用 2～3 种混凝剂混合投配或依次投配，能提高效果。污泥经有机调理剂处理后，有机成分提高，污泥的燃烧价值提高。

污泥调理前后的变化见表 9-14。

污泥调理前后的变化 **表 9-14**

		污泥调理前	污泥调理后
组成	破解胞外聚合物	胞外聚合物(EPS)，微生物在一定环境下代谢分泌在细胞壁外的聚合高分子有机物，包含了羟基、氨基和磷酸盐等功能团。按照与细胞相的结合程度可以分为 S-EPS(上清液层胞外聚合物)、LB-EPS(松散结合的胞外聚合物)和 TB-EPS(紧密结合的胞外聚合物)	EPS 破解

续表

		污泥调理前	污泥调理后
组成	混凝形成絮体	污泥的固体颗粒一般很细小，多为胶体颗粒	破坏污泥胶体颗粒的稳定状态，使颗粒脱稳，相互聚集，形成较大的絮体，实现固液分离
	加强絮体的强度	强度不一定能达到深度脱水要求的絮体	石灰是最常用作骨架的助滤剂，钙离子可与污泥中的腐殖酸反应，形成多网格状的骨架

污泥物理调理指向被调理的污泥中投加不发生化学反应的物质，降低污泥比阻和黏度。物理调理方法主要有添加惰性助滤剂和淘洗等方法。向污泥中投加无机助滤剂，可在滤饼中形成较大的絮体，减少污泥过滤比阻，常用的无机助滤剂有污泥焚烧时产生的烟道灰、飞灰、硅藻土、锯末等，以烟道灰效果最好。

在各种物理调理方法中，淘洗是较常用的方法。污泥淘洗是利用清水或污水处理厂出水与污泥混合，然后再澄清分离，降低消化污泥碱度，带走细小固体。此外，污泥中的细小固体不仅消耗化学药剂，而且增加过滤阻力，通过淘洗可大幅提高污泥的过滤性能。淘洗工艺通常采用多级逆流方式运行，淘洗液中的有机物浓度较高，需要回流重新进入污水处理系统。

热工调理包括冷冻、中温加热调理和高温加热调理等方式，常用高温加热调理。高温加热调理指污泥在反应器内经180～220℃和1～2.5MPa的高温高压水蒸气30～90min处理后，水蒸气使污泥中生物体的细胞壁破碎，释放结合水，降低污泥黏滞性，热工调理后脱水污泥含固率可以达到40%，而且高温能杀灭细菌，解决卫生问题。但污泥热工调理气味大，设备易腐蚀，需高温高压蒸汽，能耗高。

污泥经反复冷冻—融化能破坏污泥中固体与结合水的联系，提高过滤能力，使污泥结合水含量大幅降低。但人工冷冻调理成本较高，自然冷冻法受气候条件的限制，这两种技术难以推广使用。

9.6 污泥的处置

污泥最终处置的一些替代方案包括土地填埋、海洋倾倒和焚烧。这些方法中的每一个都有其缺点，例如，在土壤中填埋的污泥中存在有毒物质，或者焚烧所带来的高燃料成本。

经浓缩、消化及脱水等处理之后的污泥，不仅体积大幅减小，而且在一定程度上得到了稳定，对其最终处置的主要方法有土地利用、堆肥、焚烧、填埋与投海等。

9.6.1 土地利用

城市污水处理厂污泥含有丰富的氮、磷、钾和有机物，可作为农业肥料使用，其中的有机物还可起到土壤改良剂的作用。我国城市污水处理厂污泥营养成分见表9-15。污泥土地利用维持了有机物→土壤→农作物→城市→污水→污泥→土壤的良性大循环，无疑是污泥处置最合理的方式。过去，国外污泥大量用于填埋，但近年来呈显著下降趋势，污泥土地利用则呈急剧上升趋势。目前，在美国污泥土地利用已经代替填埋成为最主要的污泥处置方式。加拿大土地利用的污泥数量，占了将近一半，显著高于其他技术。

我国城市污水处理厂污泥营养成分（以干污泥计，%）　　表 9-15

污泥类型	总氮(TN)	磷(P_2O_5)	钾(K)
初沉池污泥	2.0～3.4	1.0～3.0	0.1～0.3
生物膜法污泥	2.8～3.1	1.0～2.0	0.1～0.8
剩余活性污泥	3.5～7.2	3.3～5.0	0.2～0.4

污泥必须经过厌氧消化、好氧发酵等稳定化及无害化处理后，才能进行土地利用。污泥土地利用包括园林绿化、林地利用、农地利用 3 种类型。污泥作为农业肥料使用时，需严格控制污泥中的重金属浓度，氮、磷、钾营养物质的平衡和污泥的施用量；同时，应采用工业排水预处理和使之与城市排水分流，以及对污泥进行有效预处理的措施，积极控制污泥中的有毒有机物、病原菌和盐分含量，避免对周围环境和人类食物链安全造成负面影响。当污泥用作农田肥料时，需要符合《城镇污水处理厂污染物排放标准》GB 18918—2002 和《农用污泥污染物控制标准》GB 4284—2018 中的无害化指标、稳定化指标和污染物浓度限值（表 9-16 和表 9-17）。

农用污泥的理化指标和卫生学指标　　表 9-16

项目		相关参数要求
无害化指标	含水率(%)	≤60
	pH	5.5～8.5
	粒径(mm)	≤10
	有机质(以干基计)(%)	≥20
卫生学指标	蛔虫卵死亡率(%)	≥95
	粪大肠菌群值	≥0.01

农用污泥污染物指标及限值　　表 9-17

项目	污染物限值	
	A 级污泥产物	B 级污泥产物
总镉（以干基计）(mg/kg)	<3	<15
总汞(以干基计)(mg/kg)	<3	<15
总铅(以干基计)(mg/kg)	<300	<1000
总铬(以干基计)(mg/kg)	<500	<1000
总砷(以干基计)(mg/kg)	<30	<75
总铜(以干基计)(mg/kg)	<500	<1500
总锌(以干基计)(mg/kg)	<1200	<3000

续表

项目	污染物限值	
	A级污泥产物	B级污泥产物
总镍(以干基计)(mg/kg)	<100	<200
矿物油(以干基计)(mg/kg)	<500	<3000
苯并(a)芘(以干基计)(mg/kg)	<2	<3
多环芳烃(PAHs)(以干基计)(mg/kg)	<5	<6

注：1. A级污泥产物：用于耕地、园地、牧草地；

2. B级污泥产物：用于园地、牧草地、不种植食用农作物的耕地。

9.6.2 污泥堆肥

污泥作为污水处理过程中的副产物，最后被堆肥处理是农业利用的有效途径。堆肥是一种将有机物质经过生物降解生成性质稳定产物的处理工艺，经过堆肥处理的产物是一种无害的、类似腐殖质的物质。污泥堆肥一般分为厌氧堆肥和好氧堆肥。厌氧堆肥由于其堆肥时间长、卫生条件差，不能用作污泥大规模处理处置；好氧堆肥利用嗜温菌、嗜热菌的作用，分解污泥中有机物并杀死传染病菌、寄生虫卵与病毒，提高污泥肥分，是生产高质量的堆肥产品的手段。

好氧堆肥的工艺过程主要包括前处理、一次发酵、二次发酵和后处理4个过程。

前处理包括含水率、pH和粒度的调整以及接种。好氧堆肥应调节含水率至60%左右，调整方法有：添加辅料、成品回流、干燥和二次脱水等。添加的辅料以木屑、米糠和稻草为主。成品回流法是在1体积泥饼中加入3～5体积的成品，使堆料的含水率在50%左右，适用于含水率较低的泥饼，具有不需要供给辅料、难分解物质少、可不进行二次发酵、发酵时间短等优点。干燥和二次脱水一般被认为是成品回流方式的辅助手段，由于干燥费用高和二次脱水机性能没有解决，目前应用很少。调整后的pH在8.0以下，粒度在2～30mm可满足发酵的要求。在使用机械装置快速发酵时，为加速反应必须进行接种。接种通常用成品回流实现，一般成品回流比为20%～30%（体积比）。

一次发酵阶段分为升温、高温消毒及腐熟3个过程。升温过程：在强制通风条件下，堆肥中有机物开始分解，嗜温菌迅速繁殖，堆温上升到45～55℃。高温消毒过程：有机物分解所释放的能量，一部分合成新细胞，一部分使堆温继续上升到55～70℃，此时嗜温菌受到抑制，嗜热菌大量繁殖，病原菌、寄生虫卵与病毒被杀灭，由于大部分有机物被分解，需氧量逐渐减少，温度开始回落。腐熟过程：温度降低到40℃左右，一次发酵基本完成。一次发酵堆肥阶段耗时约7～9d，在堆仓（或槽）内完成，主要设备有鼓风机、抽风机或者采用人工翻堆好氧。

二次发酵阶段是在一次发酵完成后停止强制通风，采用自然堆放方式使其进一步熟化、干燥、成粒，主要设备有卸料机、输送机、装料机等。二次发酵期长，其时间在30d左右。

后处理是将堆肥物进一步加工，使之成为粒径为3～5mm左右的粒状产品。

堆肥成熟的标志是物料呈黑褐色，无臭味，手感松散，颗粒均匀，蚊蝇不繁殖，病原菌、寄生虫卵与病毒及植物种子均被杀灭，氮、磷、钾等肥效增加且易被作物吸收，符合卫生部颁布的《粪便无害化卫生要求》GB 7959—2012的要求以及农业有机肥料的要求，如含水率小于20%，有机质含量大于30%，氮磷钾总含量大于4%，pH为6~8，重金属含量应控制在相应含量范围之内。

9.6.3 污泥焚烧

焚烧是污泥最终处置的最有效和彻底的方法。污泥焚烧指污泥所含水分被完全蒸发、有机物质被完全焚烧，焚烧的最终产物是 CO_2、H_2O、N_2 等气体及焚烧灰渣。焚烧时借助辅助燃料，使焚烧炉内温度升至污泥中有机物的燃点以上，令其自燃，如果污泥中的有机物的热值不足，则须不断添加辅助燃料，以维持炉内的温度。

污泥具有一定的热值，但仅为标准煤的30%~60%，低于木材，与泥煤、煤矸石接近，故污泥焚烧工艺可以在一定程度上借鉴煤矸石焚烧工艺。各种污泥的热值见表9-18。

各种污泥的热值 **表9-18**

污泥种类		热值(kJ/kg)	污泥种类		热值(kJ/kg)
初沉污泥	新鲜污泥	15876~18249	初沉污泥和腐殖污泥混合	新鲜污泥	14952
	消化污泥	7224		消化污泥	6762~8148
初沉污泥和活性污泥混合	新鲜污泥	17010	新鲜活性污泥		14952~15263
	消化污泥	7476			

污泥的燃烧热值也可用式(9-7)计算得出。

$$Q=2.3a\left(\frac{100P_{\mathrm{v}}}{100-G}-b\right)\left(\frac{100-G}{100}\right) \tag{9-7}$$

式中 Q——污泥的燃烧热值，kJ/kg(干)；

P_{v}——有机物质(即挥发性固体)含量，%；

G——机械脱水时，所加的无机混凝剂量(以占污泥干固体质量%计)，当用有机高分子混凝剂或未投加混凝剂时，$G=0$；

a，b——经验系数，与污泥性质有关(新鲜初沉污泥与消化污泥：$a=131$，$b=10$；新鲜活性污泥：$a=107$，$b=5$)。

污泥焚烧前需通过脱水或热干化等工艺减少污泥中水分，以提高污泥热值，降低运输和贮存成本，减少燃料和其他物料的消耗。

污泥焚烧包括污泥单独焚烧、混合焚烧技术。污泥单独焚烧指单独建设焚烧设施对污泥进行的焚烧。当污泥含固率在35%~45%之间，有机物含量大于50%，热值为4.8~6.5MJ/kg时，可采用单独焚烧工艺。

污泥单独焚烧设备有回转窑式燃烧炉、立式多段焚烧炉及流化床焚烧炉等。立式多段焚烧炉的焚烧能力低、污染物排放较难控制；回转窑式焚烧炉的炉温控制困难、对污泥发热量要求较高；流化床焚烧炉结构简单、操作方便、运行可靠、燃烧彻底、有机物破坏去除率高。目前，多采用循环流化床锅炉，并要求进泥含水率小于或等于10%，预热温度为136℃，污泥焚烧温度为850~950℃，炉内烟气有效停留时间大于2s，焚烧炉出口烟气中氧含量为6%~10%(干烟气)，灰渣热灼减率应小于5%；由于污泥焚烧烟气中湿度较大，为防止积灰和腐蚀，排烟温度应大于180℃。污泥焚烧系统通常包括储运系统、干化系统、焚烧系统、余热利用系统、烟气净化系统、电气自控仪表系统及其辅助系统等。为有效控制二次污染，焚烧泥质须满足《城镇污水处理厂污泥处置 单独焚烧用泥质》

GB/T 24602—2009 的规定。

由于污泥的热值一般偏低，单独焚烧具有一定难度，故宜考虑与热值较高的垃圾或燃料煤同时焚烧。污泥混合焚烧技术包括污泥与生活垃圾混烧、污泥的水泥窑协同处置、污泥的燃煤电厂协调处置。

污泥与生活垃圾混烧指将污泥与生活垃圾混合在生活垃圾焚烧设备中焚烧。混合焚烧前应采用干化技术将污泥含水率降至与生活垃圾相似的水平，不宜将脱水污泥与生活垃圾直接掺混焚烧。污泥和生活垃圾混合焚烧，应选择流化床焚烧炉进行焚烧，污泥与生活垃圾的质量之比不应超过 1：4，混烧的焚烧温度不得低于 850℃。污泥与生活垃圾混合焚烧时会增加飞灰产生量，降低灰渣热灼减量率，增加烟气净化系统的投资和运行成本，降低生活垃圾发电厂的发电效率和垃圾处理能力。污泥与生活垃圾混烧烟气经处理后必须符合《生活垃圾焚烧污染控制标准》GB 18485—2014 的相关规定。

污泥的水泥窑协同处置指利用水泥窑产生的高温烟气干化污泥、干化后的污泥和水泥原料混合送入水泥窑炉焚烧，使焚烧产物固化在水泥熟料的晶格中，成为水泥熟料的一部分，从而达到污泥安全处置的目的。利用水泥窑炉混烧的污泥中汞含量应小于 3mg/kg，最大进料比例不超过混合物料总量的 5%。利用水泥窑协同处置固体废物时，水泥窑及窑尾余热利用系统排出的大气污染物中颗粒物、二氧化硫、氮氧化物和氨的排放限值按《水泥工业大气污染物排放标准》GB 4915—2013 中的要求执行，其他污染物如氯化氢、氟化氢、重金属、二噁英等必须符合《水泥窑协同处置固体废物污染控制标准》GB 30485—2013 的相关规定。

污泥的燃煤电厂协同处置指利用电厂余热对污泥干化后将污泥与煤混合送入燃煤电厂的循环流化床锅炉、煤粉锅炉和链条炉等焚烧炉焚烧。燃煤电厂协同处置污泥时，污泥需干化至半干化（含水率 40%以下），干化后污泥形态应疏松，入炉污泥的掺入量不宜超过燃煤量的 8%；燃煤电厂直接掺烧污泥会降低锅炉内温度，增加飞灰产生量，增加除尘装置和烟气净化系统负荷，降低系统热效率 3%～4%，并引起低温腐蚀等问题。

污泥热干化的尾气、污泥焚烧的烟气、污泥焚烧灰含有较多的污染物质，如重金属、放射性物质、二噁英等，处置不当会造成二次污染，需进行必要处理与处置。

污泥焚烧炉渣与飞灰应分别收集、贮存、运输，并妥善处置，符合要求的炉渣首先考虑综合利用，经鉴别属于危险废物的底渣和飞灰，应纳入危险固体废物进行管理，不属于危险废物，按一般固体废物进行管理。

在大、中型城市且经济发达的地区、大型城镇污水处理厂或部分污泥中有毒有害物质含量较高的城镇污水处理厂，可采用污泥焚烧技术处置污泥。

9.6.4 污泥填埋

当污泥中含有的重金属或其他有毒有害物质浓度超过土地利用标准时，可以采用填埋作为污泥的最终处置方法。污泥可单独填埋，也可与城市生活垃圾混合填埋。目前我国主要将污泥与垃圾混合填埋。另外，污泥经处理后还可作为垃圾填埋场覆盖土。

污泥脱水后含水率在 80%左右，污泥填埋前，应进行稳定化处理，稳定化后的污泥根据经验添加适量硬化剂（如粉煤灰、石灰、烟道灰等）使其含水率低于 60%，pH 在 5～10 之间。污泥与生活垃圾混合填埋参照城市生活垃圾卫生填埋的有关技术规范，并达到《城镇污水处理厂污泥处置　混合填埋用泥质》GB/T 23485—2009 和《生活垃圾填埋场污染控制标准》GB 16889—2008 的相关要求。

污泥与生活垃圾混合填埋是先将污泥撒布在城市生活垃圾上面，混合均匀后铺放于填埋场内，然后压实覆土。含固率大于 3%的污泥均可混合填埋，但在实际操作中，填埋的污泥的含固率通常在 20%以上。两者的混合比见表 9-19。

污泥与城市生活垃圾、泥土的混合比例 **表 9-19**

填埋方法	混合物料	污泥含固率(%)	混合物料:湿污泥(质量比)
与垃圾混合	垃圾	3～10	7∶1
		10～17	6∶1
		17～20	5∶1
		≥20	4∶1
与泥土混合	泥土	≥20	1∶1

污泥用于垃圾填埋场覆盖土时，必须经过稳定化和卫生化处理，同时采用石灰、水泥基材料、工业固体废弃物等对污泥进行改性提高污泥强度。也可通过在污泥中掺入一定比例的泥土或矿化垃圾，混合均匀并堆置 4d 以上，以提高污泥的承载能力并消除其膨润持水性。用作覆盖土的污泥泥质含水率应小于 45%，臭气浓度应小于 6 级臭度，横向剪切强度大于 25kN/m^2，粪大肠菌群值大于 0.01，蛔虫卵死亡率大于 95%。有毒有害物质满足《城镇污水处理厂污泥处置　混合填埋用泥质》GB/T 23485—2009 中相关要求。

污泥填埋操作相对简单，投资费用较低，适应性强，但侵占土地严重，对环境的潜在危害较大。若防渗技术没有达到相关技术标准要求，可能会导致较为严重的土壤污染和地下水污染。因为成本低，污泥填埋曾经是长期以来的最主要污泥处置技术，但近年来随着对污泥填埋管理要求的提高，填埋成本已经有显著的上升，超过了污泥土地利用（包括利用前的消化稳定等预处理）成本。

9.7　污泥处理与处置的设计计算

9.7.1　污泥重力浓缩设计计算

1. 设计规定

（1）重力浓缩池面积应按污泥沉降曲线确定的固体通量计算。无污泥沉淀试验资料时，可参考表 9-20 选取。浓缩池的上清液应重新回流到污水处理系统进行处理。

（2）污泥浓缩时间一般不宜小于 12h，同时也不应该超过 24h。

（3）浓缩池有效水深一般为 4m，最低不小于 3m。

（4）重力浓缩池根据运行方式的不同分为连续式和间歇式。连续式重力浓缩池一般为竖流式和辐流式。辐流式浓缩池可以采用刮泥机和吸泥机集泥。采用刮泥机时，刮泥机的回转速度为 0.75～4r/h，其外缘线速度一般宜为 1～2m/min，池底向泥斗的坡度不宜小于 0.05。

（5）当采用生物除磷工艺进行污水处理时，不应采用重力浓缩。

（6）重力浓缩池刮泥机上应设置浓缩栅条。污泥浓缩池一般应有去除浮渣的装置。

重力浓缩池固体通量经验值　　表 9-20

污泥类型	污泥含水率(%)	固体通量[kg/(m² · d)]	浓缩污泥含水率(%)
初沉污泥	95～97	80～120	90～92
活性污泥	99.2～99.6	20～30	97～98
腐殖污泥	98～99	40～50	96～97
混合污泥	99～99.4	30～50	97～98

2. 重力浓缩池设计计算

（1）污泥浓缩池表面积

$$A=\frac{QC_0}{G} \tag{9-8}$$

式中　A——污泥浓缩池表面积，m^2；

Q——进入浓缩池的污泥量，m^3/d；

C_0——污泥固体浓度，kg/m^3；

G——污泥池固体通量，$kg/(m^2 \cdot d)$。

（2）浓缩池直径

$$D=\sqrt{\frac{4A}{n\pi}} \tag{9-9}$$

式中　D——浓缩池直径，m；

n——浓缩池数量，个。

（3）浓缩池工作部分高度

$$h_2=\frac{QT}{24A} \tag{9-10}$$

式中　h_2——浓缩池工作部分高度，m；

T——为浓缩时间，h。

（4）浓缩池总高度

$$H=h_1+h_2+h_3+h_4+h_5 \tag{9-11}$$

式中　H——浓缩池总高度，m；

h_1——浓缩池超高，m，一般为 0.3m；

h_3——浓缩池缓冲层高度，m，一般为 0.3～0.5m；

h_4——浓缩池池底坡度造成的高度，m；

h_5——浓缩池污泥斗高度，m。

3. 例题

【例 9-3】 某城市污水处理厂二沉池排放剩余污泥量 $Q_1=1500m^3/d$，含水率为 99.5%；初沉池排放的污泥量 $Q_2=200m^3/d$，含水率为 96.5%，要求浓缩后的污泥含水率为 97%，设计计算重力浓缩池。

【解】 辐流式重力浓缩池计算草图如图 9-12 所示。

（1）计算污泥量和污泥固体浓度

二沉池排放的剩余污泥量 $Q_1=1500m^3/d$，含水率为 99.5%，固体浓度 $C_1=5g/L$；

初沉池排放的污泥量 $Q_2=200\mathrm{m}^3/\mathrm{d}$，含水率为 96.5%，固体浓度 $C_2=35\mathrm{g/L}$；

总污泥量为：$Q=Q_1+Q_2=1500+200=1700\mathrm{m}^3/\mathrm{d}$。

混合后的污泥浓度为：$C=\dfrac{C_1Q_1+C_2Q_2}{Q_1+Q_2}=\dfrac{5\times1500+200\times35}{1500+200}=8.53\mathrm{g/L}$

（2）污泥浓缩池表面积

浓缩污泥为活性污泥和剩余污泥混合污泥，污泥固体通量选用 $40\mathrm{kg/(m^2\cdot d)}$，则浓缩池表面积为：

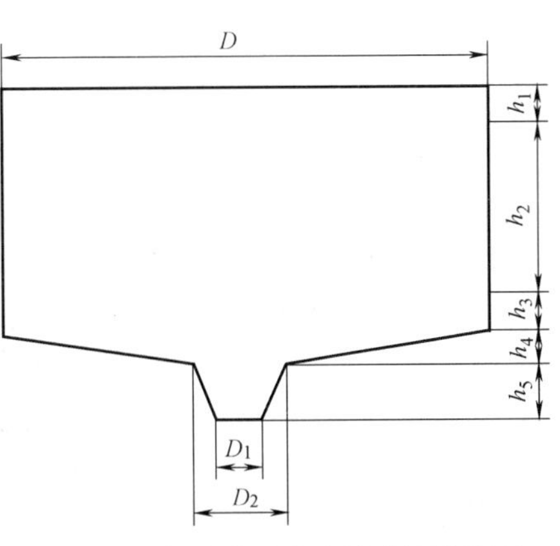

图 9-12 辐流式重力浓缩池计算草图

$$A=\frac{QC_0}{G}=\frac{1700\times8.53}{40}=362.53\mathrm{m}^2$$

（3）浓缩池直径

设计采用 2 个圆形辐流式污泥浓缩池，单池面积为：$A_1=\dfrac{362.53}{2}=181.27\mathrm{m}^2$。

浓缩池直径为：$D=\sqrt{\dfrac{4A_1}{\pi}}=\sqrt{\dfrac{4\times181.27}{3.14}}=15.2\mathrm{m}$

（4）浓缩池工作部分高度

取浓缩时间 $T=16\mathrm{h}$，则浓缩池工作部分高度为：

$$h_2=\frac{QT}{24A}=\frac{1700\times16}{24\times362.53}=3.13\mathrm{m}$$

（5）浓缩池总高度

取浓缩池超高 $h_1=0.3\mathrm{m}$，缓冲层高 $h_3=0.3\mathrm{m}$，污泥浓缩池中机械刮泥机，池底坡度 $i=0.05$，污泥斗下底直径 $D_1=1.0\mathrm{m}$，上底直径 $D_2=2.0\mathrm{m}$，污泥斗与水平面的夹角为 55°。

池底坡度造成的高度为：$h_4=\left(\dfrac{D}{2}-\dfrac{D_2}{2}\right)i=\left(\dfrac{15.2}{2}-\dfrac{2.0}{2}\right)\times0.05=0.33\mathrm{m}$

污泥斗高度为：$h_5=\left(\dfrac{D_2}{2}-\dfrac{D_1}{2}\right)\times\tan55°=\left(\dfrac{2.0}{2}-\dfrac{1.0}{2}\right)\times\tan55°=0.71\mathrm{m}$

浓缩池总高度为：$H=h_1+h_2+h_3+h_4+h_5=0.3+3.13+0.3+0.33+0.71=4.77\mathrm{m}$

9.7.2 污泥气浮浓缩设计计算

1. 设计规定

（1）气浮浓缩池分为圆形和矩形两类，处理能力小于 $100\mathrm{m}^3/\mathrm{h}$ 多采用矩形气浮浓缩池，处理能力大于 $100\mathrm{m}^3/\mathrm{h}$ 多采用辐流式气浮浓缩池。

（2）气固比应通过气浮试验确定。对于活性污泥，气固比一般在 0.01～0.04 之间；无试验资料时，一般采用 0.005～0.04，入流污泥浓度较低时，取上限；反之，入流污泥浓度较高时，取下限。

（3）溶气罐的容积，一般按加压水停留时间 1～3min 计算，罐内溶气压力一般为 0.2～

0.4MPa，溶气效率一般为50%～80%。溶气罐的直径根据过水断面负荷确定，溶气罐高度一般为2.5～3m，溶气罐直径与高度比值一般为1∶(2～4)。

(4) 矩形气浮浓缩池的长宽比通常为(3～4)∶1，有效水深一般为3～4m，池宽度与深度比值不小于3，大型气浮浓缩池一般单格池宽不超过10m、池长不超过15m。辐流式气浮浓缩池深度应大于3m。

(5) 气浮浓缩池的表面水力负荷和表面固体负荷可参考表9-21。

气浮浓缩池表面水力负荷和表面固体负荷 **表9-21**

污泥种类	表面水力负荷[$m^3/(m^2 \cdot h)$]		表面固体负荷[$kg/(m^2 \cdot h)$]	气浮污泥固体浓度(%)
	有回流	无回流		
活性污泥的混合液	1.0～3.6	0.5～1.8	1.04～3.12	3～6
空气曝气的活性污泥			2.08～4.17	
纯氧曝气的活性污泥			2.5～6.25	
初沉污泥与活性污泥			4.17～8.34	
初沉污泥			<10.8	

2. 设计计算

(1) 气固比

$$\alpha=\frac{A_a}{S} \tag{9-12}$$

式中 α——气固比，即气浮单位质量固体物质所需空气质量的比值；

A_a——所需空气量，g/h；

S——进入气浮池的固体总量，g/h。

(2) 回流比

$$R=\frac{\alpha\rho_0}{S_a\left(f\frac{P}{P_0}-1\right)} \tag{9-13}$$

式中 R——回流比，%；

ρ_0——进入气浮池的污泥浓度，g/m^3；

S_a——一定温度和压力下空气在水中的溶解度，g/m^3，S_a=空气在水中的溶解度×空气密度，空气在水中的溶解度与温度的关系见表9-22。

空气在水中的溶解度与温度的关系 **表9-22**

温度(℃)	空气密度(g/m^3)	溶解度(m^3/m^3)	温度(℃)	空气密度(g/m^3)	溶解度(m^3/m^3)
0	1252	0.0288	30	1127	0.0161
10	1206	0.0226	40	1092	0.0142
20	1164	0.0187			

f——溶气效率，即回流加压水中已达到的空气饱和系数，%，一般为50%～80%；

P——溶气罐的表压力，MPa；

P_0——当地绝对压力，MPa。

（3）气浮浓缩池表面积

$$A=\frac{Q(R+1)}{q} \tag{9-14}$$

式中　A——气浮浓缩池表面积，m^2；

q——表面水力负荷，$m^3/(m^2 \cdot h)$。

（4）校核表面固体负荷

$$q_g=\frac{Q\rho_0}{A} \tag{9-15}$$

式中　q_g——表面固体负荷，$kg/(m^2 \cdot h)$。

3. 例题

【例 9-4】 某城市二级污水处理厂采用活性污泥法工艺，二沉池排放剩余污泥量为 1000 m^3/d，污泥含水率为 99.5%，水温 20℃，当地大气压为 0.1MPa。拟采用回流加压气浮浓缩二沉池污泥，压力溶气罐的表压为 0.4MPa，要求浓缩后的污泥含水率为 96%，设计计算气浮浓缩池表面积 A 和回流比 R。

【解】 设计一座矩形的平流式气浮浓缩池

污泥小时流量为：$Q=\frac{1000}{24}=14.7m^3/h$

（1）气固比

对于活性污泥，A_a/S 一般在 0.01～0.04 之间，本设计取 A_a/S 为 0.03。

（2）回流比

当水温为 20℃时，查表 9-22 得，计算空气在水中溶解度 $S_a=1164\times0.0187=21.77g/m^3$；污泥含水率为 99.5%，污泥浓度 ρ_0 取 5000g/m^3，溶气效率 f 取 0.7；溶气罐表压力 P 取 0.4MPa，代入以上数值，则回流比为：

$$R=\frac{\alpha\rho_0}{S_a\left(f\frac{P}{P_0}-1\right)}=\frac{0.03\times5000}{21.77\times\left(0.7\times\frac{0.4}{0.1}-1\right)}=383\%$$

（3）气浮池设计流量

$$Q'=Q(1+R)=14.7\times(1+3.83)=71.0m^3/h$$

（4）气浮浓缩池表面积

表面水力负荷 q 取 2.0$m^3/(m^2 \cdot h)$，则气浮浓缩池表面积为：

$$A=\frac{Q(R+1)}{q}=\frac{14.7(3.83+1)}{2}=35.0m^2$$

（5）校核表面固体负荷

$$q_g=\frac{Q\rho_0}{A}=\frac{14.7\times5000}{35.0\times1000}=2.1kg/(m^2 \cdot h)$$

表面固体负荷符合设计规定。

9.7.3 厌氧消化池设计计算

1. 厌氧消化池池型

厌氧消化池按池体形状分为圆柱形和蛋（卵）形，分别如图 9-13 和图 9-14 所示。

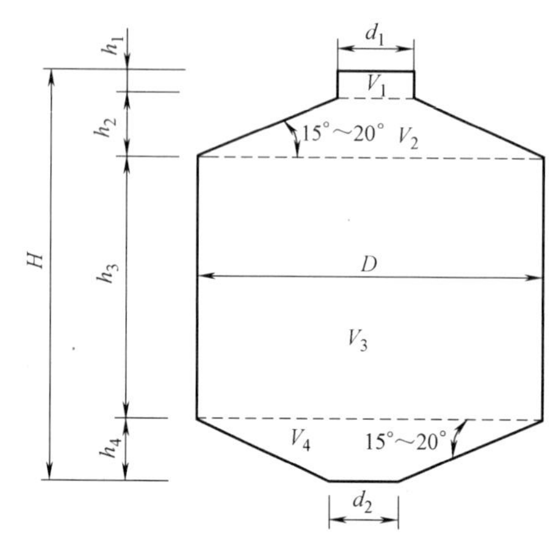

图 9-13　圆柱形厌氧消化池

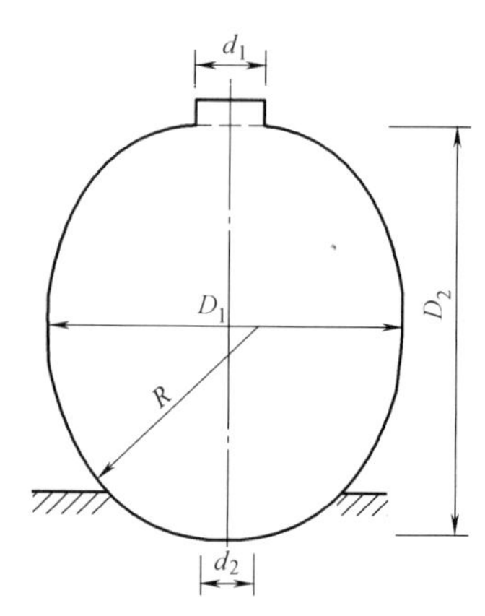

图 9-14　蛋（卵）形厌氧消化池

圆柱形消化池结构较蛋形消化池简单，易选择搅拌系统，建造施工容易。但底部面积大，易造成粗砂及淤泥的堆积，因此，需定期停池进行清理，重新启动系统时间较长，一般需要几个月的时间。

蛋形消化池有效地消除了粗砂和浮渣的堆积，池内不产生死角，可保证生产的稳定性和连续性，有利于底部沉淀物的排除和浮渣的消除，容积利用率高。污泥混合性能好，用于搅拌的能量消耗最少。单位体积的池表面积小，散热量较低，系统温度容易维持。壳体形状使池体结构受力分布均匀，但蛋形消化池结构和施工复杂，造价昂贵。

2. 厌氧消化设计一般规定

污泥厌氧消化工艺设计包括消化池池形的选择及容积计算，搅拌方式选择，热量平衡计算及加热方式选择，储气柜及沼气净化系统选择等。厌氧消化设计一般规定如下：

（1）为检修和维护方便，消化池的数量至少为 2 座。

（2）圆柱形消化池直径一般为 6～35m；集气罩直径一般为 2～5m，高 1～3m，池底及池盖与水平面倾角一般取 15°～20°；池底直径一般取 1～2m，圆柱体高度与直径之比应大于 0.5，消化池总高与直径之比为（1.25～2）：1；考虑检修和观察方便，池顶设直径大于 0.7m 的人孔 1～2 个，观察窗 1 个。

（3）蛋形厌氧消化池短轴直径可达 22m，长轴直径可在 45m 以上。短轴直径与长轴直径比一般为 1：(1.4～2.0)。

（4）当污泥采用两级消化时，一级消化池容积与二级消化池容积比可采用 1：1、2：1 或 3：2，一级消化池加热并搅拌；二级消化池可不加热、不搅拌，但应有上清液排出设施。

（5）污泥厌氧消化主要设计参数见表 9-23。

污泥厌氧消化主要设计参数　　　表 9-23

项目	中温消化	高温消化
消化温度(℃)	33～35	50～55
污泥龄(d)	20～30	10～15
污泥投配率(%)	3.33～5	6.67～10
容积负荷[kgVSS/(m^3·d)]	0.6～1.5	2.0～2.8
产气量[m^3/(m^3·d)]	1.0～1.3	3.0～4.0

3. 厌氧消化池容积计算

污泥厌氧消化池容积可采用污泥龄、污泥投配率或容积负荷法计算。

（1）按污泥龄或污泥投配率计算

$$V=QT \tag{9-16}$$

式中 V——消化池容积，m^3；

Q——每日进入消化池的污泥量，m^3/d；

T——污泥龄，d。

$$\text{或} V=\frac{Q}{\eta}\times 100 \tag{9-17}$$

式中 η——污泥投配率，%。

（2）按容积负荷计算

$$V=\frac{QS_0}{L_v} \tag{9-18}$$

式中 S_0——污泥中挥发性悬浮固体（VSS）浓度，kg/m^3；

L_v——消化池容积负荷，$kgVSS/(m^3 \cdot d)$。

4. 消化池热量计算

污泥消化池加热主要采用消化池外间接加热法，加热时所需热量优先考虑污泥消化过程所产生的沼气进行加热，当所产生的沼气热量不能满足加热污泥所需的热量时，则需要考虑其他能源进行补充。

污泥厌氧消化池总耗热量包括：把生污泥加热到消化温度所需热量；消化池壳体散热热量损失；输泥管道与热交换器的热量损失。

（1）加热生污泥所需热量

$$Q_1=\frac{Q}{24}(T_1-T_2)\times 4168.8 \tag{9-19}$$

式中 Q_1——把生污泥温度加热到消化温度所需热量，kJ/h；

T_1——消化污泥温度，℃；

T_2——生污泥温度，计算最大所需热量采用日平均最低污泥温度，℃。

（2）消化池壳体耗热量

$$Q_2=\sum FK(T_1-T_3)\times 1.2 \tag{9-20}$$

式中 Q_2——消化池壳体耗热量，kJ/h；

F——消化池壳体散热面积，m^2；

K——消化池池体的传热系数，与消化池壳体材料和保温状况有关。为简化计算，可取 $K=7.98kJ/(m^2 \cdot h \cdot ℃)$；

T_3——消化池池外介质温度，池外介质为大气时，采用冬季室外最低温度计算，池外介质为土壤时，采用全年平均温度计算，℃。

（3）输泥管道、热交换器等热量损失

输泥管道和热交换器的耗热量取生污泥加热所需热量（Q_1）与消化池壳体耗热量（Q_2）之和的5%～15%，即：

$$Q_3=(0.05\sim 0.15)(Q_1+nQ_2) \tag{9-21}$$

式中 n——消化池的数量，个。

5. 例题

【例 9-5】 已知某城市污水处理厂采用二级生物处理，污泥产生量为 $500m^3/d$，初沉池污泥和二沉池剩余污泥经浓缩后含水率为 96%，挥发性固体（VSS）含量为 65%；冬季污泥平均温度为 11.5～14.5℃，夏季污泥平均温度为 14.5～20℃；当地冬季室外日平均最低温度为−12℃，夏季最高温度为 37℃。混合污泥拟采用中温消化处理，设计消化池各部分尺寸和消化池所需热量。

【解】（1）消化池容积计算

1）按污泥龄计算

污泥龄取 20d，则消化池容积为：

$$V=QT=500\times24=12000m^3$$

2）消化池容积的校核

浓缩后的混合污泥含水率为 96%，则污泥中固体含量为 4%，其中挥发性固体（VSS）占 65%，则污泥中挥发性悬浮固体浓度为：

$$S_0=4\%\times65\%\times1000=26kg/m^3$$

消化池容积负荷为：$L_v=\dfrac{QS_0}{V}=\dfrac{500\times26}{12000}=1.08kgVSS/(m^3\cdot d)$

消化池容积负荷满足所规定的容积负荷要求，容积确定基本合理。

污泥消化采用中温两级消化，一级消化池容积与二级消化池容积之比为 2∶1，则一级消化池总容积为 $8000m^3$，设计 2 座一级消化池，单池容积为 $4000m^3$；设计 1 座二级消化池，二级消化池容积为 $4000m^3$。

（2）圆柱形消化池池体设计

一级消化池、二级消化池均采用圆柱形消化池，消化池容积相同，计算简图如图 9-13 所示。

1）消化池直径及高度计算

消化池直径 D 取 18m，集气罩直径 $d_1=2m$，集气罩高 $h_1=2.0m$，池底锥底直径 $d_2=2m$，锥角采用 15°。故：

$$h_2=h_4=\frac{18-2}{2}\times\tan15°=2.1m$$

消化池柱体高度 h_3 与消化池直径相同，取 18m。

消化池总高度为：

$$H=h_1+h_2+h_3+h_4=2.0+18.0+2.1+2.1=24.2m$$

2）消化池各部分容积计算

集气罩容积为：

$$V_1=\frac{\pi d_1^2}{4}h_1=\frac{\pi\times2^2}{4}\times2.0=6.28m^3$$

上锥体容积为：

$$V_2=\frac{1}{3}\times\pi h_2\left(\frac{D^2}{4}+\frac{Dd_2}{4}+\frac{d_2^2}{4}\right)$$
$$=\frac{1}{3}\times\pi\times2.1\times\left(\frac{18^2}{4}+\frac{18\times2}{4}+\frac{2^2}{4}\right)=200.0m^3$$

下锥体尺寸与上锥体尺寸相同，因此，下锥体体积与上锥体体积相等，即 $V_2=V_4=200.0\text{m}^3$

圆柱体容积为：

$$V_3=\frac{\pi D^2 h_3}{4}=\frac{\pi\times 18^2\times 18}{4}=4578.1\text{m}^3$$

消化池总有效容积为：

$$V=V_2+V_3+V_4=4578.1+200+200=4978.1\text{m}^3>4000\text{m}^3$$

满足设计要求。

3）消化池各部分表面积计算

集气罩表面积为：

$$A_1=\frac{\pi}{4}d_1^2+\pi d_1 h_1=\frac{\pi}{4}\times 2^2+\pi\times 2\times 2=15.7\text{m}^2$$

上锥体表面积为：

$$A_2=\frac{\pi}{2}(D+d_1)\times\frac{h_2}{\sin\alpha}=\frac{\pi}{2}\times(18+2)\times\frac{2.1}{\sin 15^\circ}=254.8\text{m}^2$$

下锥体表面积为：

$$\begin{aligned}A_4&=\frac{\pi d_2^2}{4}+\frac{\pi}{2}(D+d_2)\times\frac{h_4}{\sin\alpha}\\&=\frac{\pi\times 2^2}{4}+\frac{\pi}{2}(18+2)\times\frac{2.1}{\sin 15^\circ}=258\text{m}^2\end{aligned}$$

消化池圆柱体表面积为：

$$A_3=\pi D h_3=\pi\times 18\times 18=1017.4\text{m}^2$$

消化池总表面积为：

$$A=A_1+A_2+A_3+A_4=15.7+254.8+258+1017.4=1545.9\text{m}^2$$

（3）蛋（卵）形消化池池体设计

一级消化池、二级消化池均采用蛋（卵）形消化池，消化池容积相同，计算简图如图 9-14 所示。

消化池短轴直径 D_1 取 18m，短轴半径 a 为 9m。长轴直径 D_2 与短轴直径 D_1 之比一般为（1.4～2.0）∶1，取 $D_2=1.5D_1=1.5\times 18=27\text{m}$，则长轴半径 b 为 13.5m。

壳体弧线半径 $R=(0.74\sim 0.84)D_2$，取 $R=0.75\times 27=20.25\text{m}$。

蛋形消化池池体体积为：

$$V=\frac{4\pi}{3}a^2 b=\frac{4\pi}{3}\times 9^2\times 13.5=4578.1\text{m}^3$$

消化池壳体表面积为：

$$\begin{aligned}F_1&=2\pi\frac{\sqrt{b^2-a^2}}{a}\times\frac{a^2 b^2}{b^2-a^2}\left(\frac{\pi}{2}-2\arcsin\frac{a}{b}+\frac{a}{b^2}\sqrt{b^2-a^2}\right)\\&=2\pi\frac{\sqrt{13.5^2-9^2}}{9}\times\frac{9^2\times 13.5^2}{13.5^2-9^2}\left(\frac{\pi}{2}-2\arcsin\frac{9}{13.5}+\frac{9}{13.5^2}\sqrt{13.5^2-9^2}\right)\\&=1372.1\text{m}^2\end{aligned}$$

注：arcsin $\frac{a}{b}$ 用弧度 π 表示。

（4）消化池热量计算

消化池体积相同时，蛋形消化池表面积较小，向外界散发热量较少，冬季所需热量较多，所以本例题以蛋形消化池冬季所需热量进行计算。

1）加热生污泥所需热量

消化污泥温度 T_1 取 35℃，冬季新鲜污泥最低温度 T_2 取 11.5℃，加热生污泥所需热量为：

$$
\begin{aligned}
Q_1 &= \frac{Q}{24}(T_1 - T_2) \times 4168.8 \\
&= \frac{500}{24}(35 - 11.5) \times 4168.8 \\
&= 2040975 \text{kJ/h}
\end{aligned}
$$

2）消化池壳体耗热量

蛋形消化池总外表面积为 1372.1m²，当地冬季室外最低温度为－12℃，消化池池体的传热系数，K 取 7.98kJ/(m²·h·℃)，消化池壳体耗热量为：

$$
\begin{aligned}
Q_2 &= \sum FK(T_1 - T_3) \times 1.2 \\
&= 1372.1 \times 7.98 \times [35 - (-12)] \times 1.2 \\
&= 617543.8 \text{kJ/h}
\end{aligned}
$$

3）输泥管道、热交换器等热量损失

输泥管道和热交换器的耗热量取 Q_1 与 Q_2 之和的 5%～15%，设计取 10%，共设计 3 座消化池，输泥管道、热交换器等热量损失为：

$$
\begin{aligned}
Q_3 &= (0.1)(Q_1 + nQ_2) \\
&= 0.1 \times (2040975 + 3 \times 617543.8) \\
&= 389360.6 \text{kJ/h}
\end{aligned}
$$

则冬季消化池所需总热量为三部分热量之和：

$Q = Q_1 + Q_2 + Q_3 = 2040975 + 617543.8 + 389360.6 = 2492089.4 \text{kJ/h}$

本章关键词（Keywords）

最终处置	Final disposal
自由水	Free water
毛细管水	Capillary water
结合水	Bound water
重力浓缩	Gravity thickening
稳定化	Stabilization
浮渣层	Scum layer
发酵	Fermentation
内源代谢	Endogenous metabolism
挥发性物质	Volatile matter
消化污泥	Digested sludge

调理、调节	Conditioning
调节剂	Conditioning agent
聚丙烯酰胺	Polyacrylamides
干化床	Drying bed
真空过滤	Vacuum filtration
离心	Centrifugation
压滤机	Filter presses
焚烧	Incineration
腐殖质	Humus

思考题

1. 污泥含水率从 97%降至 94.5%，求污泥体积的变化。
2. 简述污泥比阻的含义。
3. 常用的污泥脱水方法有哪些?
4. 简述污泥浓缩与污泥脱水的区别。
5. 讨论能否先对污泥进行调理然后再消化处理。
6. 污泥厌氧消化池由哪六部分构成？并简述各部分的功能。
7. 污泥最终处置有哪些方法？各有什么优缺点?

Chapter 10 Wastewater Treatment Processes Cases Analysis

10.1 生活污水处理工艺案例分析

10.1.1 污水处理厂案例分析

案例一：某大学采用膜生物反应器处理宿舍区、教学区生活污水，其工艺流程如图10-1所示。

图 10-1 膜生物反应器处理生活污水的工艺流程

二维码 10 生物膜的清洗过程

该工艺日处理水量为1100t，采用一体化的 A^2O 工艺配合膜生物反应器技术，可有效降解生活废水中的氨氮和有机物；缺氧段和活性污泥回流系统，可提高氨氮硝化效果。通过控制溶解氧，使氨氮在好氧段发生亚硝化反应，通过污泥回流，使亚硝酸盐在缺氧段发生反硝化反应，产生氮气并消耗部分有机污染物，实现脱氮功能。在好氧剩余氨氮和亚硝酸盐氮进行硝化反应，膜过滤系统放在好氧段后，确保出水氨氮和亚硝酸盐氮达标，同时可使剩余有机物降解成二氧化碳和水。膜的高效截留作用使污泥浓度达到较高水平，增加去除氨氮和有机物的效果，提高产水水质。产生的中水用于冲洗厕所、绿化和景观。

由于采用了中空纤维膜组件来实现泥水分离，膜生物反应器中的水力停留时间和污泥停留时间不再相互影响，生物反应器中的污泥浓度可以大幅高于传统活性污泥处理工艺，因此，可以得到更好的COD、BOD去除效率，并且还有占地面积小；彻底去除出水中的固体物质；出水无需消毒；COD、固体和营养物可以在一个单元内被去除；高负荷率；低污泥产率（或零污泥产率）；系统不受污泥膨胀的影响等优点。

为更明晰设备内部结构和相互关系，某校以Visual Basic 6.0为主要设计工具开发仿真界面，包括中水站俯视流程图（图10-2）；中水站五月份原理图（图10-3），并且各个界面能够互相转换观察，实现软件的启停操作，以及参数的实时显示功能。

仿真界面包括曝气、停气、显示数据、隐藏数据、退出5个按钮。单击曝气按钮，界面内会有气泡上下浮动以实现曝气的仿真；单击停气按钮，气泡消失可实现停气功能；还实现了出水水质在线监测的模拟，使用户能够对出水水质有一个比较直观的了解。单击显示数据按钮，界面内会出现相应的水质数据；单击隐藏数据则数据消失。最终的仿真软件以可安装的应用软件的形式出现，可以完全脱离VB的开发环境，在Windows操作系统下独立运行，并且达到了界面友好，方便使用，既能用于仿真实验，又能达到仿真运行培训的目的，极大地提高了教学效果。

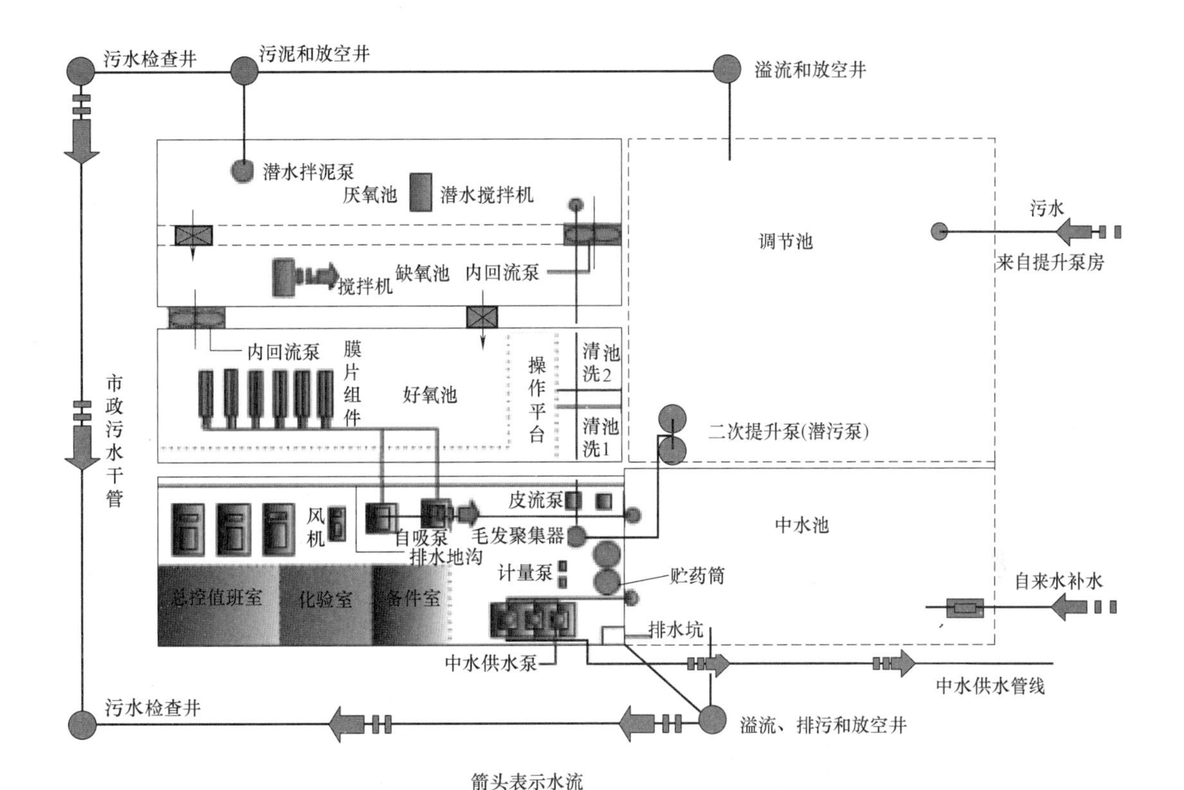

图 10-2　某校中水站俯视流程图

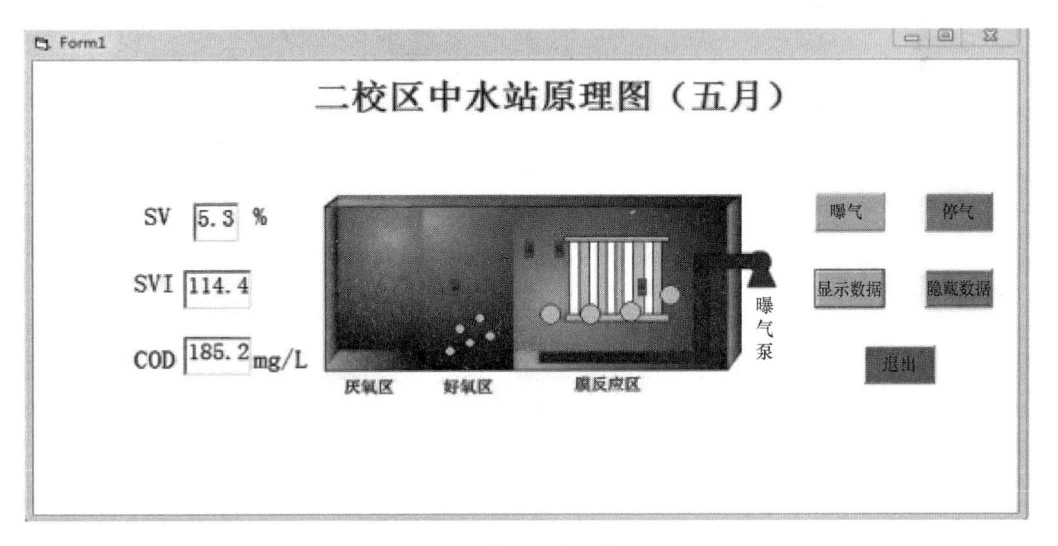

图 10-3　某校中水站原理图

案例二：某再生水处理厂设计处理能力达到 10 万 t/d，采用膜生物反应器（MBR）工艺+反渗透（RO）工艺，通过管网收集的污水进入厂区，经粗格栅间进水泵房。生产的再生水一部分输入再生水管网，供给市政杂用与园林绿化等用户，剩余部分作为河道的补充水源。

生物反应池为 UCT 工艺，由厌氧段、缺氧段、好氧段和变化区组成，共分 4 个系列，每座池均可独立运行。每座池分 3 个廊道，每个廊道宽 6.5m，池长 75m，池中水深 5.5m，厌氧段和缺氧段均加盖以减少异味散发，并设置气体收集和输送管路，利用生物除臭池对气体进行处理。变化区内同时安装潜水推进器和曝气头，可以根据进水水质、处理效果、季节变化以及出水水质需求，转换为缺氧区或好氧区。为了实现脱氮除磷功能，生物处理单元为 UCT 工艺，设计污泥龄为 16.6d，缺氧段至厌氧段的回流比为 100%～120%，好氧段至缺氧段内回流比为 450%～700%（变频调节），设内回流泵 4 台，外回流泵 4 台，剩余污泥泵 5 台。MBR 膜生物反应池标准供气量为 37600m^3/h，采用单级高速离心鼓风机 3 台，2 用 1 备。

膜池位于 MBR 生物池北侧，中间是膜池的配水渠道和混合液回流渠道，如图 10-4 所示。生物池内的混合液用泵提升到膜池配水渠道内，并通过配水管进入膜池内，膜池共 8 座，膜池长 16m，宽 8m，水深 3.5m，每个膜池设计安装 38 个膜组件，预留 4 个膜组件位置，膜出水泵共设 9 台，泵的功率为 8.6kW，每台泵最大出水流量为 442m^3/h，扬程为 10m。膜系统配套的擦洗鼓风机 4 用 1 备，单台风量为 $Q=228m^3/min$，风压为 35kPa。

图 10-4　膜池

案例三：某再生水处理厂处理规模 7 万 m^3/d，核心工艺采用“超滤＋部分反渗透＋臭氧＋液氯消毒”，反渗透装置运行前后的超滤膜池如图 10-5～图 10-7 所示，出水水质满足国家污水再生利用城市杂用、景观、工业用水相关标准。

10.1.2　燃煤电厂生活污水处理厂案例分析

1. 工程概况

大唐某电厂生活污水处理系统 2012 年开始建设，2013 年投入使用，该污水处理站主要处理电厂各车间、浴室及办公楼排放的生活污水。生活污水排放总量为 240m^3/d，全天 24h 连续排放。该生活污水可生化性较好，适于生物法处理，进水水质指标见表 10-1。处理后的水质要求达到《城市污水再生利用　景观环境用水水质》GB/T 18921—2019 水质要求，主要指标见表 10-2。

图 10-5 反渗透装置

图 10-6 超滤膜池

图 10-7 运行时的超滤膜池

污水处理站进水水质 表 10-1

项目	pH	BOD_5	COD	SS	氨氮	总氮	总磷
数值	6～9	≤200mg/L	≤400mg/L	≤200mg/L	≤40mg/L	≤60mg/L	≤5mg/L

污水处理站出水水质 表 10-2

项目	pH	BOD_5	COD	SS	氨氮	浊度	总磷	游离余氯	总大肠菌群
数值	6～9	≤10mg/L	≤40mg/L	≤10mg/L	≤5mg/L	≤3NTU	≤0.5mg/L	管网末端 ≥0.2mg/L	≤3 个/L

2. 工艺流程及说明

本项目主要处理可生化性较好的生活污水、废水，景观环境用水对出水 BOD_5、COD、SS、氨氮、总磷提出了较高的要求。因此，传统生化处理较难达到，根据该项目的具体情况，采用具有脱氮功能的缺氧-好氧＋MBR 工艺作为污水的主体处理工艺，如图 10-8 所示，该工艺在大型污水处理厂污水深度处理中得到较广泛的应用，是一种成熟、可靠的运行工艺。

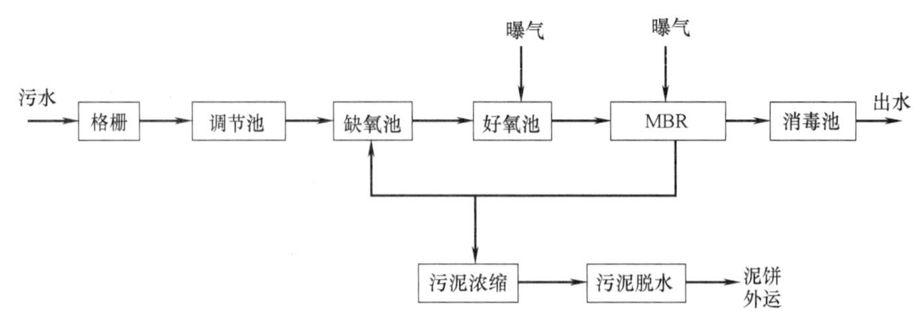

图 10-8 电厂生活污水处理工艺流程图

污水首先经 5mm 的细格栅，去除各种大的机械杂物，然后重力流入调节池，在调节池进行水质水量的均化。在调节池中安装穿孔曝气管，避免污水腐败，调节池的污水经过泵提升进入缺氧池，中间安装毛发过滤器，去除生活污水中的纤维状毛发，这些预处理措施能够保证膜生物反应器系统的正常运行。缺氧池主要功能是反硝化脱氮，达到提高可生化降解性和提高氨氮的目的。污水通过重力流从缺氧池进入好氧池。在好氧池中安装微孔曝气器，由鼓风机提供空气供氧，好氧段溶解氧浓度控制在 2～3mg/L。在好氧区完成碳化反应（有机物在好氧菌作用下分解为水和二氧化碳）和硝化反应（氨氮在自养菌的作用下被氧化为硝态氮），去除有机物和氨氮。

污水经过缺氧-好氧生物处理之后进入膜生物反应器，在膜区活性污泥作用下污水中有机物经过进一步降解，污水得到净化，出水达到要求指标。在膜区设置污泥回流泵，将污泥混合液回流到缺氧区，硝态氮在缺氧区反硝化生成氮气，达到生物脱氮目的。水通过产水泵的抽吸作用进入消毒水池，在消毒水池通过加药系统进行消毒保持管网末梢余氯，避免在中水输送过程中滋生细菌。部分剩余污泥通过污泥泵送至污泥浓缩池进行污泥浓缩、脱水处理。

经膜生物反应器（MBR）工艺处理后，出水达到设计的指标，可回用于洗车、冲厕、绿化等生活杂用。

3. 主要设备及构筑物

（1）格栅渠

格栅渠与调节池建为一体，格栅渠内放置机械格栅，有效尺寸为 5.0m×0.9m×5.0m，池体采用钢筋混凝土结构，渠内污水流速为 0.6～1m/s，数量 1 座。

格栅渠内设置旋转式机械格栅 1 台，主要通过机械格栅拦截去除生活污水中较大的悬浮物固体，保证后续设备和管路系统不被堵塞。自动旋转式机械格栅间歇运行，每隔 4h 时运行 0.5h（可调）。格栅宽 0.8m，安装角度为 75°，栅条间隙为 5mm，配套电机功率为 2.2kW。

（2）调节池

调节池主要功能是均化水质、水量。为防止格栅不能去除的颗粒物质在调节池中沉淀，在调节池安装一套曝气装置，曝气装置采用穿孔管曝气，防止污泥淤积、水质恶化。池内设污水提升泵以及液位计，污水经提升泵提升至后续处理装置，污水提升泵能根据液位自动启停。池体尺寸为 6.0m×6.0m×5.0m，有效容积为 120m^3，停留时间为 12h，池体为钢筋混凝土全地下式结构，数量为 1 座，池底穿孔管曝气，保证污水中溶解氧含量大于或等于 0.6mg/L。

调节池风机采用低噪声回转式鼓风机，数量为 2 台（1 用 1 备），风量为 2.5m^3/min，排出压力为 49kPa，电机功率为 3.7kW。采用耐腐蚀无堵塞潜污泵，数量 2 台（1 用 1 备），流量为 12m^3/h，扬程为 15m，电机功率为 3.7kW。

(3) 毛发收集器

毛发收集器能有效地拦截污水中夹杂着的毛发、泥沙等杂物，充分保证水处理设施的正常运行。规格为 ϕ600mm×900mm，进出口管道公称直径为 *DN*150，通过水量为 60m^3/h。

(4) 缺氧池

污水在缺氧池内进行反硝化反应，将硝态氨还原为气态氮（N_2），达到脱氨目的。数量为 1 座，地上式钢筋混凝土结构，池体尺寸为 4.0m×4.0m×4.5m，有效水深为 3.5m，有效容积为 56m^3，水力停留时间为 5.6h。缺氧池内设置高密度聚丙烯球状悬浮型填料，填料上生长有生物膜，填料直径为 ϕ100mm，比表面积为 5000m^2/m^3，相对密度为 0.92，装填比例为 1000 个/m^3，采用潜水搅拌机搅拌。缺氧池内设置潜水搅拌机 2 台（1 用 1 备），叶轮转速为 740r/min，叶轮直径为 260mm，电机功率为 1.5kW。

(5) 好氧池

好氧池数量为 1 座，地上式钢筋混凝土结构。池体尺寸为 7.0m×3.0m×5.0m，有效容积为 80m^3，停留时间为 8h，好氧池内设置醛化纤维圆形组合填料进行增强生物处理，该填料具有阻力小，布水、布气性能好，易长膜的优点，又有切割气泡作用。填料规格为 ϕ150mm×4500mm，比表面积为 2600m^2/m^3，单位重量为 600kg/m^3，装填比例为 560 个/m^3。好氧池内配备风机与调节池风机型号规格相同。

(6) 膜生物反应池

膜生物反应池分 1 格，池体为地上式钢筋混凝土结构，膜生物反应池主要功能是利用微生物降解污水中的有机物，同时进行硝化反应，为反硝化提供硝态氨。

池体尺寸为 3000mm×3000mm×5000mm，有效容积为 30m^3，停留时间为 3h，池底设置膜式曝气管，用于满足生化反应所需的空气量，溶解氧含量大于或等于 2mg/L。

膜组件是膜生物反应器的核心装置，其技术规格见表 10-3，利用膜的分离作用对生化反应池内的含泥污水进行过滤，实现泥水分离，由于膜的高过滤精度，出水水质优良，可直接回用。

膜组件的技术规格 **表 10-3**

项目	单支膜元件尺寸	有效膜面积	膜孔径	膜材质	膜通量	使用温度	pH
数值	534mm×450mm×1523mm	35m^2/个膜元件	0.2μm	PVDF 聚偏氟乙烯	11～15L/(m^2·h)	5～45℃	2～10

膜生物反应池内配备抽吸泵 3 台（2 用 1 备），流量为 3m^3/h，扬程为 15m，电机功率为 1.5kW，每台泵进水口配真空压力表 1 只，每 2 台泵配电磁流量计 1 只。膜生物反应池内配备混合液回流泵（兼作排泥泵）1 台，流量为 3m^3/h，扬程为 15m，电机功率为 2.2kW。膜生物反应池内配备膜生物反应器反冲洗泵 2 台（1 用 1 备），流量为 5m^3/h，扬程为 15m，电机功率为 1.5kW。

膜生物反应池风机采用罗茨鼓风机，数量为2台（1用1备），流量为$4m^3/min$，排出压力为49kPa，电机功率为5.5kW。

（7）消毒池

设置消毒池1座，主要功能是对污水进行消毒处理，减少大肠菌群等细菌数量，保证出水达标，同时保持部分余氯，起到持续杀菌作用。消毒池采用钢筋混凝土半地下式结构，池体尺寸为3.0m×2.0m×4.0m，有效容积为$20m^3$，停留时间为2.0h。

消毒池配备容积为1500 L的加药箱1个，计量泵采用隔膜式计量泵2台（1用1备），流量为50L/h，出口压力为1.0MPa，电动机功率为0.37kW。隔膜计量泵运行可靠、平稳、噪声小、安全无泄漏，计量泵的流量、压力满足系统出力要求，并留有一定的富余度。

（8）污泥处理

由于MBR工艺产生的污泥较少，直接用污泥泵打入工业废水处理系统中的污泥浓缩池，统一进行处理。

10.2 工业废水处理工艺案例分析

10.2.1 纺织废水处理工艺案例分析

纺织印染行业是工业污水排放大户，产生的污水主要来源于染色和印花工段，污水中主要含有纺织纤维上的污物、油脂、盐类以及加工过程中附加的各种浆料、染料、表面活性剂、助剂、酸碱等。印染废水主要特点是有机物浓度高、成分复杂、色度深且多变，pH变化大，水量水质变化大，可生化性差，属难处理工业废水。

漂染废水中含有染料、浆料、表面活性剂等助剂，该类废水水量大，浓度和色度均较低。浆染废水色度高、COD高，特别是近年根据国外市场开发出来的丝光蓝、丝光黑、特深蓝、特深黑等印染工艺，该类印染大量使用硫化染料、印染助剂硫化钠等，废水中含有大量的硫化物。

1. 工程概况

北方某印染厂，主要织物有麻、棉和化纤，使用染料有硫化染料、分散染料和直接染料；排放废水有退浆废水、煮炼废水、漂白废水、丝光废水、染整废水等，该厂日综合废水量为$5000m^3/d$，采用24h连续工作。经过现场勘察及水质监测分析，确定进水水质见表10-4。

某印染公司废水水质指标　　表10-4

项目	参数	项目	参数	项目	参数
BOD_5	≤700mg/L	SS	≤400mg/L	pH	6.0～9.0
COD	≤2000mg/L	氨氮	≤30mg/L	色度	≤400

排水污染物水质要求达到《纺织染整工业水污染物排放标准》GB 4287—2012直接排放标准，具体数值见表10-5。

废水污染物排放执行标准　　表10-5

项目	参数	项目	参数	项目	参数
BOD_5	≤20mg/L	SS	≤50mg/L	pH	6.0～9.0
COD	≤80mg/L	氨氮	≤10mg/L	色度	≤50

2. 工艺流程及说明

根据生产废水水质情况及处理后的水质要求，采用“水解酸化＋好氧＋物理化学”相结合的方法处理印染废水，印染废水处理工艺流程如图 10-9 所示。

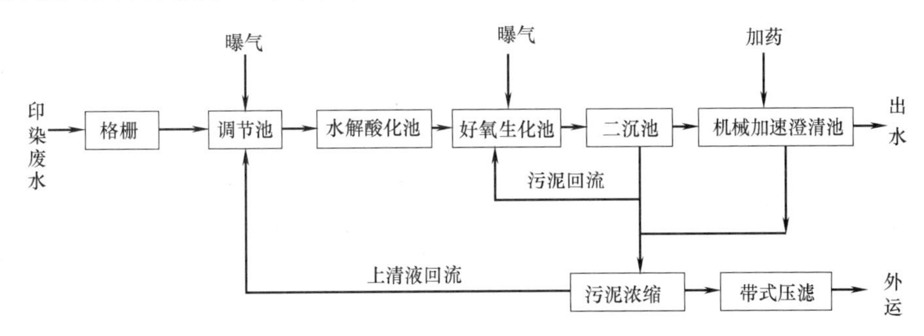

图 10-9 印染废水处理工艺流程

来自生产车间的染整废水，先经机械格栅或捞毛机拦截去除大块漂浮物及纤维物质后进入调节池。在调节池内实现水质、水量的均衡后，用泵将其提升至水解酸化池，通过厌氧菌的新陈代谢作用把大分子有机物转变为小分子，提高废水的可生化性。水解酸化池的出水自流至好氧生化池，通过好氧菌的生物好氧作用，使有机污染物进一步得到去除。好氧生化池出水中的大颗粒悬浮物通过二沉池进行泥水分离，二沉池出水再经机械加速澄清池，去除大部分悬浮物，出水达标后排放。

二沉池、机械加速澄清池产生的污泥，由污泥泵送入污泥浓缩池，浓缩后的污泥投加改性药剂以提高其透水性，改性后的污泥泵入压滤机进行压滤脱水，泥饼外运，压滤水和浓缩池上清液回流至调节池重新处理。

3. 主要构筑物及设计参数

(1) 格栅渠

主要功能是安置机械格栅，拦截大块杂物及纤维类物质，并保护后续传动设备。数量 1 座，格栅渠采用钢筋混凝土地下结构；设计进水流量为 $5000m^3/d$，有效尺寸为 4.0m×2.0m×2.0m。配套机械细格栅 2 台，栅距为 3mm，栅宽为 1000mm，采用机械清渣，刮渣设备电机功率为 1.1kW。

(2) 调节池

印染废水的水质水量变化很大，调节池可以调节水质水量作用。设计 1 座调节池，采用钢筋混凝土地下结构；有效尺寸为 20m×18m×5m，有效水深为 4.5m，有效容积为 $1800m^3$，水力停留时间为 7.8h。设 pH 在线监测装置 1 套。

调节池配套污水提升泵 2 台（1 用 1 备），污水提升泵流量为 $250m^3/h$，扬程为 15m，电机功率为 18.5kW。

采用穿孔曝气管对进入调节池的废水进行曝气搅拌，防止废水中的悬浮物沉淀，预曝气还可以改善印染废水的可生化性。穿孔曝气管公称直径为 *DN*40，气水比为 4∶1，材质为 UPVC，孔眼间距为 110mm，穿孔直径为 ϕ5mm，孔眼开于穿孔管底部与垂直中心线成 45°处。配套低噪声罗茨鼓风机 2 台（1 用 1 备），风量为 $24.18m^3/min$，风机出口压力为 39.2kPa，电机功率为 30kW。

（3）水解酸化池

印染废水处理中常将厌氧控制在水解酸化阶段，降解废水中部分污染物，一般 COD 去除率为 20%～40%，色度去除率在 40%～70%之间。水解酸化还可以提高印染废水的可生化性。

采用 1 座水解酸化池，钢筋混凝土半地下结构，地上高度为 5m，地下深度为 2.5m，有效尺寸为 25m×18m×7.5m；有效水深为 7m；总容积为 $3375m^3$，有效容积为 $3150m^3$，设计水力停留时间为 15h。

水解酸化池内均匀布置直径 ϕ150mm，高度 6m 的组合填料，填料有效体积为 $1800m^3$。配备潜水搅拌机 4 台，潜水搅拌机叶轮直径为 320mm，转速为 960r/min，电机功率为 4kW。

（4）好氧池

好氧池主要功能是通过废水和活性污泥的充分接触，微生物摄取废水中的营养物质，使污水得到净化。采用 1 座好氧池，分为 6 格，钢筋混凝土半地下结构，地上高度为 5m，地下深度为 2.5m，有效尺寸为 25m×30m×7.5m；有效水深为 7m；总容积为 $5625m^3$，有效容积为 $5250m^3$，考虑印染废水的可生化性较差，设计水力停留时间为 25h。

好氧池均匀布置 1500 套直径 ϕ260 盘式微孔曝气器，单支盘式微孔曝气器服务面积为 $0.5m^2$，工作通气量为 $0.5m^3/h$，淹没水深为 6m，充氧能力为 $0.3kgO_2/h$。好氧池配套 2 台（1 用 1 备）低噪声罗茨鼓风机，风量为 70 m^3/min，风压为 58.8kPa，电机功率为 75.0kW。

（5）二沉池

二沉池用于生化池出水活性污泥与净化后的泥水分离。采用辐流式二沉池 1 座，设计污泥回流比为 100%，设计表面水力负荷为 $0.80m^3/(m^2 \cdot h)$，沉淀池直径为 ϕ18m，深度为 4.7m，有效水深为 4.2m，配套污泥回流泵 2 台（1 用 1 备），流量为 210 m^3/h，扬程为 14m，电机功率为 15kW。采用 1 台周边传动刮泥机，周边线转速为 2m/min，电机功率为 0.75kW。

（6）机械加速澄清池

为了提高出水水质，设计机械加速澄清池 1 座，在机械加速澄清池内加入絮凝剂，实现二沉池出水彻底的泥水分离。设计表面水力负荷 $0.65m^3/(m^2 \cdot h)$，沉淀池直径为 ϕ20m，深度为 4.5m，有效水深为 4.0m，配套污泥泵 2 台（1 用 1 备），流量为 $50m^3/h$，扬程为 14m，电机功率为 10kW。采用 1 台周边传动刮泥机，周边线速度为 2m/min，电机功率为 0.75kW。

（7）污泥浓缩池

对二沉池和机械加速澄清池的污泥进行浓缩，减少污泥体积。设计 1 座钢筋混凝土辐流式污泥浓缩池，尺寸为 ϕ10m×5.0m，总容积为 $390m^3$。配备周边传动刮泥机 1 台，刮泥机周边线转速为 2m/min，电机功率为 0.75kW。

（8）污泥脱水装置

污泥脱水装置采用对压型带式压滤机，该设备具有自动化程度高，连续生产，处理能力大，渣饼含水率低，能耗低，运行成本低的特点。选择 1 套带式压滤机，滤袋宽度为 1.5m，电机功率为 2.5kW，处理能力为 8～$16m^3/h$。配备污泥螺杆泵 1 台，流量为

$40m^3/h$，电机功率为 7.5kW，扬程为 12m。

4. 运行费用分析

（1）电费

按年运行 360d 计算。装机容量为 265kW，使用功率为 130kW，功率补偿因素为 0.95，电价按 0.6 元/kWh，按废水处理量 $5000m^3/d$ 计算，折合水处理费用为 0.45 元/m^3。

（2）药剂费

硫酸投加量为 120mg/L，市场价按 1000 元/t；烧碱投加量为 120mg/L，市场价 2500 元/t；聚合氯化铁投加量为 200mg/L，市场价为 3000 元/t；PAM 投加量为 20mg/L，市场价为 25000 元/t。折合水处理费用为 1.5 元/m^3。

（3）人工费

按配备操作人员 8 名，人均工资 4000 元/月计算，运行成本为 0.21 元/m^3。

（4）污水处理总费用

0.45＋1.5＋0.21＝2.16 元/m^3。

10.2.2 制浆造纸废水处理工艺案例分析

造纸废水指制浆造纸工艺过程中产生的废水，包括备料废水、制浆蒸煮黑液和红液、洗涤废水、漂白废水与纸机白水等。造纸废水成分复杂，可生化性差，属于较难处理的工业废水，是我国工业废水的主要来源之一。备料废水包括湿法备料所产生的废水和干法备料所产生的废水。在木材湿法备料过程中产生的废水主要含泥沙，在草类原料备料中是水封除尘产生废水，这两种废水均可通过澄清后循环使用。造纸黑液指用烧碱法和硫酸盐法直接蒸煮原料（麦草、木片等）而产生的废水。黑液中含有大量的碱木素、半纤维素和纤维素的分解物等有机物，还含有各种钠盐。造纸红液指用亚硫酸盐法直接蒸煮原料而产生的废水。这部分废水的 COD 高达 100g/L，BOD_5 也有 50g/L，必须通过碱回收和红液综合利用回收原料并减少污染。漂白是在多段过程中进行的，并使用不同的漂白剂，常用氯气、次氯酸盐、二氧化氯和烧碱等。漂白废水是造纸厂外排废水的主要来源，有毒、处理难度较大。造纸白水中含有大量悬浮固形物，如纤维、填料和涂料等，还有可溶解的有机污染物。为了减少污染、节约用水，造纸厂通常采取部分回用或全部回用。剩余的白水可通过纤维回收、过滤、气浮或沉淀等方法进行简单处理后再回收使用。但是，长期循环使用会导致各类污染物在系统中积累，对造纸系统和产品质量造成不良的影响。

1. 工程概况

河北省某纸业有限公司为新建企业，主要产品为再生纸，废水量约为 $5000m^3/d$。废水处理后要求达到《制浆造纸工业水污染物排放标准》GB 3544—2008 中规定的标准。废水进、出水水质指标见表 10-6。

造纸厂废水进、出水水质指标 **表 10-6**

项目	BOD_5	COD	SS	pH	色度	氨氮	总氮
进水	≤1000mg/L	≤3500mg/L	≤1500mg/L	6.0～9.0	≤400	≤50mg/L	≤70mg/L
出水	≤20mg/L	≤90mg/L	≤30mg/L	6.0～9.0	≤50	≤8mg/L	≤12mg/L

2. 工艺流程及说明

制浆造纸废水处理工艺流程如图 10-10 所示。

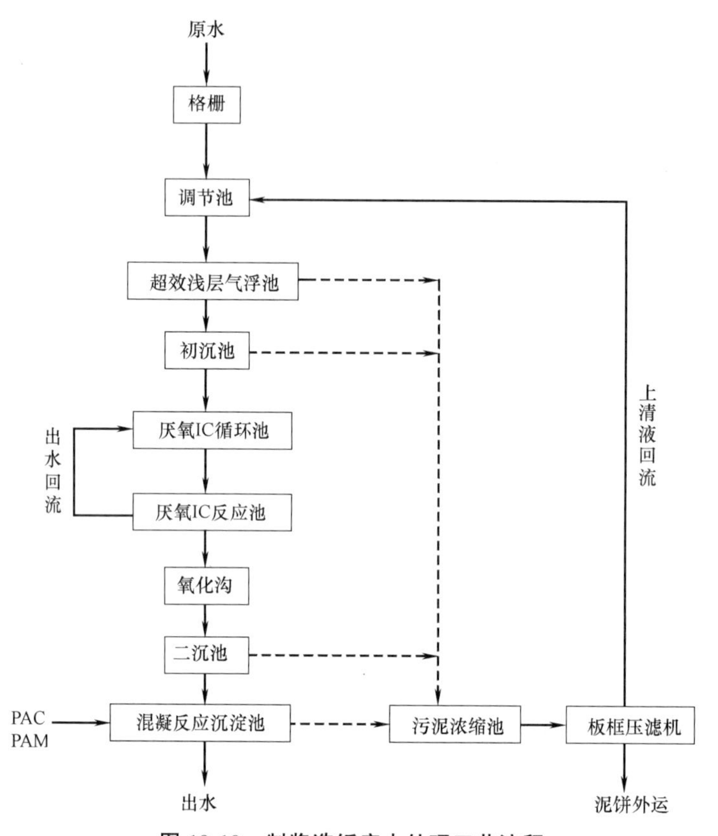

图 10-10　制浆造纸废水处理工艺流程

废水首先进入预处理系统。预处理系统包括格栅、调节池、超效浅层气浮池。格栅截留废水中较大的杂物与纤维，以降低其对后续工艺设备的破坏，调节池调节水质和水量。超效浅层气浮池可以有效去除一部分细小的悬浮物质。

预处理后出水进入两级 IC 厌氧系统，可继续去除 65％的 COD 和 68％的 BOD_5。出水进入改良的氧化沟系统，在曝气的作用下，经过好氧氧化作用消耗有机物，达到去除水中污染物的目的。废水在氧化沟中处理之后，进入二沉池，促使泥水的有效分离。二沉池出水进入混凝沉淀系统，对生化出水进行深度处理。

二沉池中的剩余污泥一部分经过污泥泵重新返回至氧化沟系统内，继续参与好氧生物反应，一部分排入污泥浓缩脱水车间，经脱水后外运处置。

3. 主要构筑物及参数

(1) 格栅渠

设计 1 座格栅渠，钢筋混凝土地下式结构。尺寸为 8.0m×0.7m×3.0m。设置 2 道格栅，安装于污水处理站调节池进口，格栅宽度为 0.6m，中格栅间隙为 10mm，细格栅间隙为 5mm，用以拦截较大杂物，保护后续处理设备运行安全。采用机械清渣，刮渣设备电机功率为 0.75kW。

(2) 调节池

采用 1 座地下式钢筋混凝土结构的调节池。尺寸为 10.0m×25.0m×6.0m，有效水深为 5.5m，容积为 $1500m^3$，有效容积为 $1375m^3$，污水水力停留时间为 6.6h。调节池中有不锈钢潜水搅拌机3台，叶轮直径为360mm，电机功率为2.9kW。污水提升泵3台（2用1备），流量为110m^3/h，扬程为11m，电机功率为5.5kW。

（3）超效浅层气浮系统

超效浅层气浮系统具有以下优点：净化池浅，池深仅为600～700mm，污水在气浮池中的停留时间短，仅需要3～5min，但表面有足够的浮渣储备空间，特别适用高浓度污水的处理。采用微气泡曝气系统，溶气效率高，处理能力强，表面负荷8～15m^3/(m^2·h)，占地面积小，也可不占地，架空、叠装或设置于建筑物上。其设有自动水位调节装置，流量适应范围大，刮起的浮渣含固率高。

设计水量为5000m^3/d，进水SS为1500g/L，圆盘总高度为1.1m，直径为ϕ8.5m，配套离心泵1台，主要作用是将气浮池废水泵入超效浅层气浮溶气罐，实现水气的良好混合。离心泵流量为80m^3/h，扬程为60m，功率为30.0kW。空气压缩机1台，主要作用是提供气浮系统需要的满足风量和风压要求的压缩空气，风量为0.6m^3/min，压力为1.0MPa，电机功率为5.5kW。高压储气罐1台，主要目的是存储压缩空气，同时起到稳定系统压力的作用，避免空气压缩机的频繁启动。其有效容积为1.0m^3，压力为1.0MPa。

（4）初沉池

采用1座辐流式初沉池，钢筋混凝土结构，沉淀池直径为28.0m，有效水深为3.5m，设计表面水力负荷为0.6m^3/(m^2·h)，有效容积为2154m^3。初沉池配套直径28m桥式周边传动刮泥机1台，周边线速度为3m/min，电机功率为0.75kW。污泥泵2台（1用1备），流量为40m^3/h，扬程为9m，电机功率为3.0kW。

（5）厌氧IC反应器内循环系统

厌氧IC反应器系统包含循环池和内循环厌氧反应器。循环池水力停留时间为1.0h，规格尺寸为5.0m×10.0m×4.5m，配备3台离心泵，2用1备，离心泵流量为160m^3/h，扬程为27m，电机功率为22kW（转速1450r/min），*DN*250进水流量计1台，*DN*300循环流量计1台。污泥泵1台，流量22m^3/h，电机功率为5.5kW，扬程为58m。

内循环厌氧反应器尺寸为ϕ8.0m×22.0m，反应器有效容积为1100m^3，设计容积负荷为16.0kg COD/(m^3·d)，产气量为2860m^3，配套进水泵3台（2用1备），流量为160m^3/h，扬程为27m，电机功率为22kW；*DN*250、*DN*300流量计各1台；配套pH计2台。

（6）氧化沟系统

采用改良型氧化沟1座。改良型氧化沟前面多了一个生物选择区，有利于微生物适应厌氧环境，占地面积小，集脱氮除磷为一体。设计污泥负荷为0.25kgCOD/(kgMLSS·d)，污泥浓度为3800mg/L，尺寸为86.0m×32.0m×5.0m，有效水深为0.5m，有效容积为12000m^3，采用4个廊道，钢筋混凝土结构。

设置3台伞形表面曝气机，叶轮直径为3300mm，电机额定功率为55kW，充氧量为115.5kgO_2/h，水流改变方向处共设置4台低速推流器，电机功率N=5kW，叶轮直径为2500mm，叶轮转速为56r/min。

（7）二沉池

采用1座辐流式二沉池，钢筋混凝土结构，完成活性污泥与净化后污水的分离，沉淀池直径为 33.0m，有效水深为 3.5m，表面水力负荷为 0.5m^3/(m^2 · h)，有效容积为 2992m^3。配备周边传动刮泥机 1 台，刮泥机外缘线速度为 1.5～3m/min，电机功率为 0.75kW。污泥回流泵 2 台（1 用 1 备），泵流量为 250m^3/h，扬程为 11m，电机功率为 15kW。剩余污泥泵 2 台（1 用 1 备），流量为 30m^3/h，扬程为 10m，电机功率为 3kW。

（8）混凝沉淀池

混凝沉淀池包括混合池、反应池和沉淀池，均采用钢筋混凝土结构，用于造纸废水生化尾水的深度处理，进一步降低出水污染物浓度，实现水质的有效净化。

混合池采用 2 组 2 格，单格混合池尺寸为 1.0m×1.0m×4.0m，有效水深为 3.5m，混合时间为 60s。反应池采用 2 组 6 格，单格反应池尺寸为 1.8m×2.4m×4.0m，有效水深为 3.5m，反应时间为 13.1min。沉淀池 2 座，沉淀池表面负荷为 0.8m^3/(m^2 · h)，沉淀池尺寸为 34m×16m×3.5m，沉淀池有效水深为 3.0m，沉淀池停留时间为 3.9h。

（9）污泥浓缩池

剩余污泥产量为 320m^3/d，含水率为 99%，采用 1 座钢筋混凝土结构重力浓缩池，浓缩池尺寸为 ϕ20.0m×3.5m，有效水深为 2.5m，有效容积为 942m^3，浓缩机电机功率为 3kW。

（10）污泥脱水机房

污泥脱水机房主要用于污泥脱水机及配套设备的存放，尺寸为 12.0m×22.0m×4.8m，主要设备为板框压滤机，压滤机滤室容积为 3m^3，过滤面积为 300m^2，过滤压力为 0.6MPa，电机功率 4kW。

4. 效益分析

此造纸废水采用此处理工艺的运行费用为 1.6 元/m^3，其中电费为 1.06 元/m^3，药剂费为 0.44 元/m^3，人工费为 0.1 元/m^3。系统稳定运行后，每年可削减化学需氧量 5.6t、氨氮 0.15t、悬浮物 4.6t、5d 生化需氧量为 2.2t，极大地减轻了水体污染，因此，也会促进区域经济和改善当地的生态环境。

10.2.3 电镀废水处理工艺案例分析

1. 电镀含氰废水处理工程实例

含氰废水处理工艺流程如图 10-11 所示。含氰废水中的氰离子（CN^-）能与镍、铜、铁过渡金属元素形成稳定的配位化合物，阻止了金属离子与氢氧根（OH^-）的结合，欲将其沉淀去除，必须先破坏其络合状态。目前，较为经济成熟的含氰废水处理工艺为碱性条件下氧化二级破氰法，常用的氧化剂为次氯酸钠，连续处理，机械搅拌，破氰的关键在于控制反应 pH 和氧化还原电位（ORP），为此采用两套 pH 控制仪和 ORP 控制仪准确控制加药量，多级反应，保证破氰充分进行。

一级处理 pH 控制在 11～12 之间，氧化还原电位控制在 300mV，停留时间为 20～30min，将氰化物氧化为氰酸盐，即局部氧化：

$$CN^- + ClO^- + H_2O \rightarrow CNCl + 2OH^-$$

$$CNCl + 2OH^- \rightarrow CNO^- + Cl^- + H_2O \tag{10-1}$$

次氯酸根与络合氰化物反应如下：

$$[Zn(CN)_4]^{2-} + 4ClO^- \rightarrow Zn^{2+} + 4Cl^- + 4CNO^- \tag{10-2}$$

$$2[Cu(CN)_3]^{2-} + 7ClO^- + H_2O \rightarrow 2Cu^{2+} + 6CNO^- + 7Cl^- + 2OH^- \tag{10-3}$$

二级处理加酸使 pH 在 7.5～8.0 之间，氧化还原电位控制在 650mV，停留时间约为 10min，将生成的氰酸盐进一步氧化成二氧化碳和氮，即完全氧化：

$$6CNO^- + ClO^- + H_2O \rightarrow 2CO_2\uparrow + N_2\uparrow + 7Cl^- + 2OH^- \quad (10\text{-}4)$$

氧化后的含氰废水排入综合废水调节池。

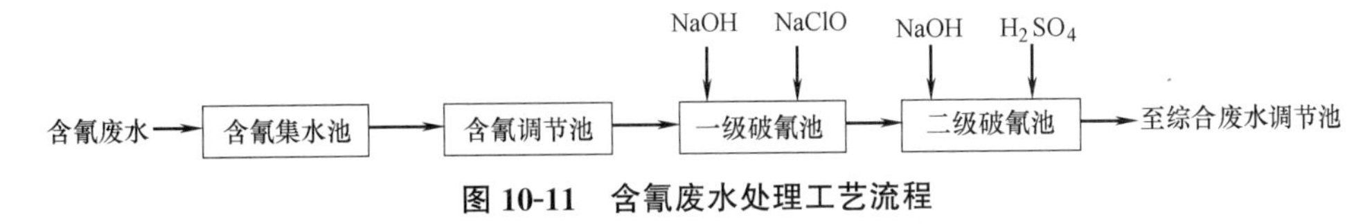

图 10-11　含氰废水处理工艺流程

2. 含铬废水

含铬废水中主要含有 $Cr_2O_7^{2-}$、CrO_4^{2-}、Cr^{3+} 等离子，在反应池中投加还原剂（药剂可选用焦亚硫酸钠、亚硫酸氢钠等还原剂），将 $Cr_2O_7^{2-}$、CrO_4^{2-} 中六价铬还原为 Cr^{3+}，然后采用化学沉淀从水中去除，如图 10-12 所示。反应机理为：

$$2Cr_2O_7^{2-} + 3S_2O_5^{2-} + 10H^+ \longrightarrow 4Cr^{3+} + 6SO_4^{2-} + 5H_2O \quad (10\text{-}5)$$

$$4H_2CrO_4 + 6NaHSO_3 + 3H_2SO_4 \rightarrow 2Cr_2(SO_4)_3 + 3Na_2SO_4 + 10H_2O \quad (10\text{-}6)$$

$$Cr^3 + 3OH^- = Cr(OH)_3\downarrow \quad (10\text{-}7)$$

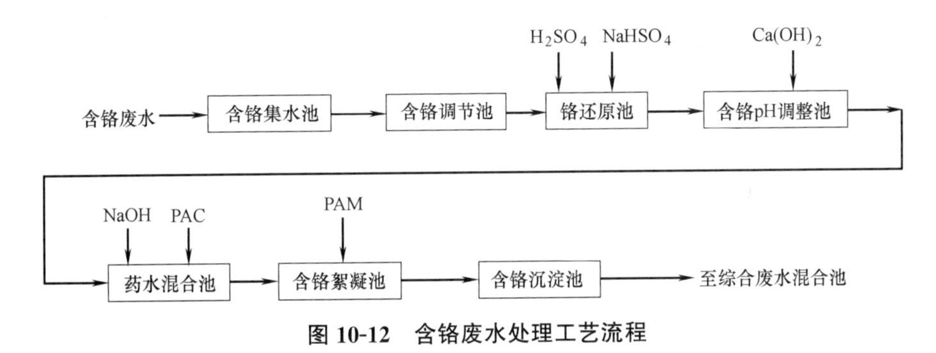

图 10-12　含铬废水处理工艺流程

采用连续操作，机械搅拌，还原反应的关键在于控制反应的 pH 和氧化还原电位（ORP），为此采用 1 套 pH 控制仪和 1 套 ORP 控制仪，由计量泵准确控制加药量保证六价铬还原为三价铬反应充分进行。反应 pH 控制在 2～3，氧化还原电位控制在 300mV 左右。还原后的含铬废水单独进行混凝沉淀，产水经滤池过滤后进入 CDMF 系统。将电镀前处理的酸洗废水，特别是酸洗槽的废酸加入含铬废水中，能起到既节约用酸量，又可引入 Fe^{2+} 作为还原剂和混凝剂的一举两得的效果。

3. 含镍废水

含镍废水是镀镍过程中产生的酸性废水，因为镍为一类污染物，故根据环境影响评价及环保要求，需在车间处理达标后方可排放或排入综合处理系统，含镍废水经含镍集水池收集后至含镍预处理系统，调节 pH 至 10～11 后，投加混凝剂/絮凝剂，絮凝后进入沉淀池，泥水分离后达标排放或进入综合处理系统，污泥自流入污泥井，经污泥泵泵入污泥浓缩池浓缩后压滤脱水，滤液流至调节池，含镍废水处理工艺流程如图 10-13 所示。

4. 除油废水

电镀工件如果表面有残留的蜡垢油污会影响整个电镀的镀层结构，电镀企业除油除蜡工

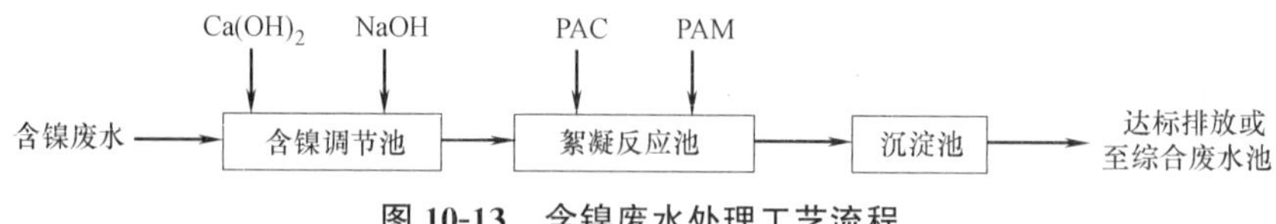

图 10-13 含镍废水处理工艺流程

艺包括化学除油、电解除油以及超声波除油 3 种方式，但除油除蜡溶液的基本成分大致相同，均为碱、磷酸盐以及表面活性剂等，因此，废水中石油类物质、COD 和磷酸盐含量较高，需单独收集后进入调节池，然后经提升泵打入强氧化池进行氧化后自流进入综合废水处理，如经处理后废水仍无法达到排放标准，则引入生化处理系统进行处理，如图 10-14 所示。

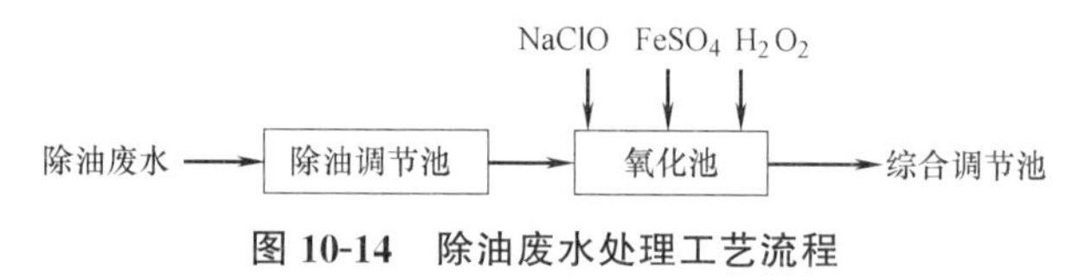

图 10-14 除油废水处理工艺流程

5. 综合废水

综合废水处理工艺流程（图 10-15）说明如下：含氰废水、含铬废水、含油废水、含镍废水、酸碱废水和其他金属电镀废水等经预处理后进入综合废水调节池调节水质和水量，然后用泵泵入反应池。在化合反应池 1 中投加石灰，加碱调节废水的 pH 在合适的范围内，使各种重金属离子生成氢氧化物沉淀。在混合反应池 2 中投加 PAC 和 PAM，在混合反应池 3 中投加 Na_2S 和 PAM，采用连续操作，机械搅拌，协助捕捉重金属沉淀，并去除水中的有机物质。反应后的废水进入辐流式沉淀池，经沉淀后的出水达到排放标准。对于沉淀池出水可能含有的悬浮颗粒，为确保出水达标，沉淀池后设滤池，过滤出水中的悬浮物。滤池出水进入废水脱盐净化（CDMF）系统，出水回用。沉淀池排泥将采用程序控制、污泥泵排泥。污泥排入污泥浓缩池进行浓缩以减轻污泥后续处理的负荷。污泥脱水系统采用板框式污泥脱水机。污泥浓缩池上清液、板框压滤机滤液和滤池反冲回流水因含有较多的悬浮物，这部分水先经回流水池，再由泵送至综合废水调节池。

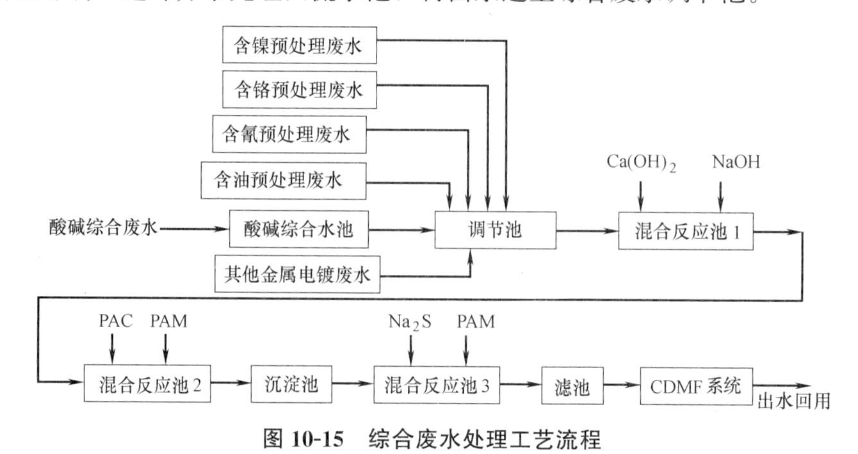

图 10-15 综合废水处理工艺流程

10.2.4 焦化废水处理工艺案例分析

焦化废水又可称为酚氰废水，是一种含有众多有毒有害物质的工业污水，如不经处理便进行排放，可对周围环境造成严重破坏，对人的健康和生命安全也造成一定的威胁。

某年产约120万t焦炭的新建焦化厂排放的污水水质不达标，酚氰超标严重，同时BOD和COD的排放量也远远超过国家标准，酚氰等有毒物质容易污染周遭地下水、土壤、大气，而BOD、COD的超标还可能会引起水体的富营养化，破坏生态健康，严重影响附近居民的生活健康以及生活质量。焦化厂排出污水主要来源于生产所排放的焦化废水以及工人生活排放的生活污水，污水产生量为4800m^3/d。该厂污水指标如下：

污水水质：COD 600mg/L，BOD_5 240mg/L，SS 350mg/L，NH_4-N 150mg/L；

挥发酚：190mg/L；

氰化物：100mg/L；

油类：50mg/L。

依据焦化废水国家二级处理标准，拟采用“预处理＋一级气浮＋A/O处理＋二级气浮＋深度处理”的方法，如图10-16所示。预处理包括格栅，隔油池除油以及调节池的水质水量调节，一级气浮则是为了降低进入生化段水中的酚氰含量，提高生化段去除效率，也减轻其负荷；生化段采用传统的A/O工艺；随后用深度处理对水质进一步提升，与传统的混凝—沉淀—消毒的深度处理有所不同的是，采用混凝气浮取代混凝沉淀，提高对各项物质，尤其是酚氰氮的去除率。

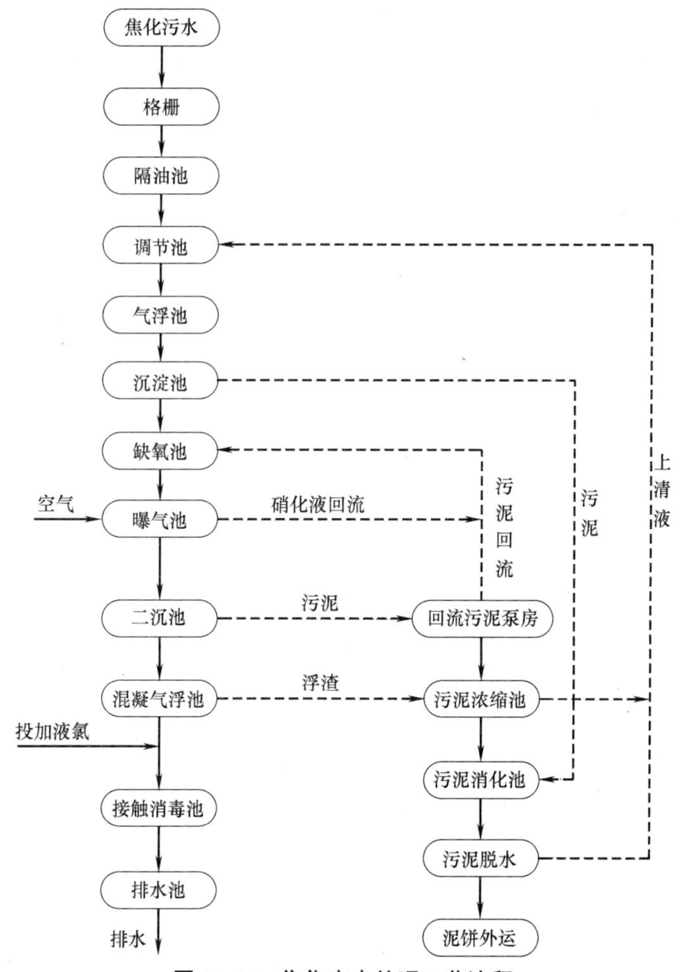

图10-16 焦化废水处理工艺流程

污水首先流经粗格栅，主要用来截留较大漂浮物，主要是工人的生活排水中的垃圾，后经细格栅拦截较小的杂物。之后进入隔油池，去除绝大部分焦化废水中的油类物质，使其符合后续生化进水对油类指标的要求。接下来，污水进入调节池，对水量和水质两方面进行处理。再利用提升泵使污水进入气浮池，气浮池除了可降低悬浮物外，还可以通过加入一定的絮凝剂去除相应的杂质，如投放一定量的硫酸铁使水中的氰絮凝，絮凝体由气泡带到水面，并通过刮渣机除去。从而大幅提高出水的各项指标。

在上述预处理工艺后，污水进入初沉池开始进行生化处理，污水生化采用反硝化-硝化工艺。

最后为满足回用水质要求，还需进行深度处理，因为考虑该厂的水质不达标部分以酚氰为主，故对比混凝沉淀和混凝气浮的工艺特点后选择以混凝气浮后接消毒的方法进行深度处理。出水最后还可送往城市中水池进行统一处理。

另外，对于水处理生成的污泥部分，将沉淀池分离的污泥排入污泥浓缩池，浓缩后的污泥经污泥脱水机脱水将送回调节池。

该工艺流程区别于一般处理工序的地方是“A/O 生化法＋二级气浮”，气浮工艺结构简单，运行稳定，电气可控性强，可达到的自动化程度高，可以很好地调整以适合不同的工艺以及不同的负荷，且构造简单，易于建设。同样，针对该厂的出水，A/O 生化法可以很好地满足处理后的出水要求，而且其结构和运行也简单方便。以这两种工艺为主要处理工序，不但简单易行，而且效果优良稳定。

10.2.5 燃煤电厂废水处理工艺案例分析

燃煤电厂废水主要有含煤废水、含油废水、酸碱再生废水、反冲洗废水、锅炉排污水、冲渣废水、冲灰废水、循环排污水、湿式脱硫废水等，如图 10-17 所示。

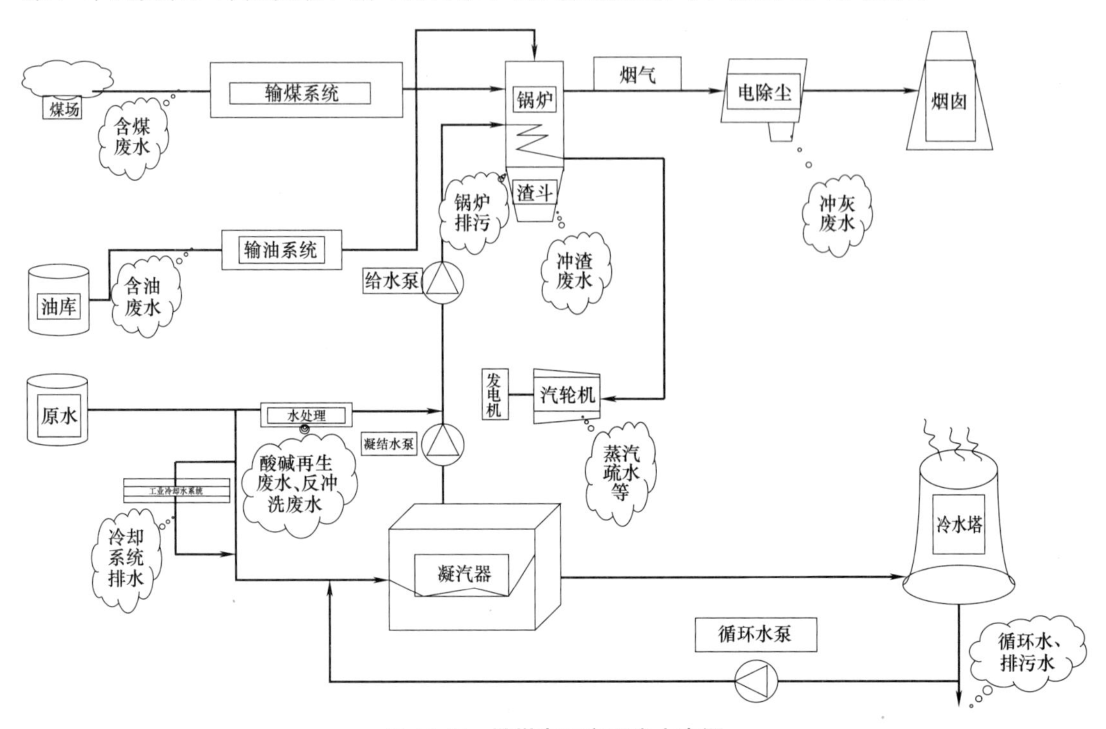

图 10-17 燃煤电厂主要废水来源

某火电厂的废水回收系统（图 10-18）设有主厂房排水池、生活污水池、工业回收水池，这三个水池的水出来先经过溢流式蓄水池存放，再通过液下泵排放到澄清池、与此同时还设有一个储矾箱，每次按照废水的量计量应该用多少矾，出来的矾添加到废水中，这样废水再到澄清池，经过无阀滤池之后直接进入冷水塔。

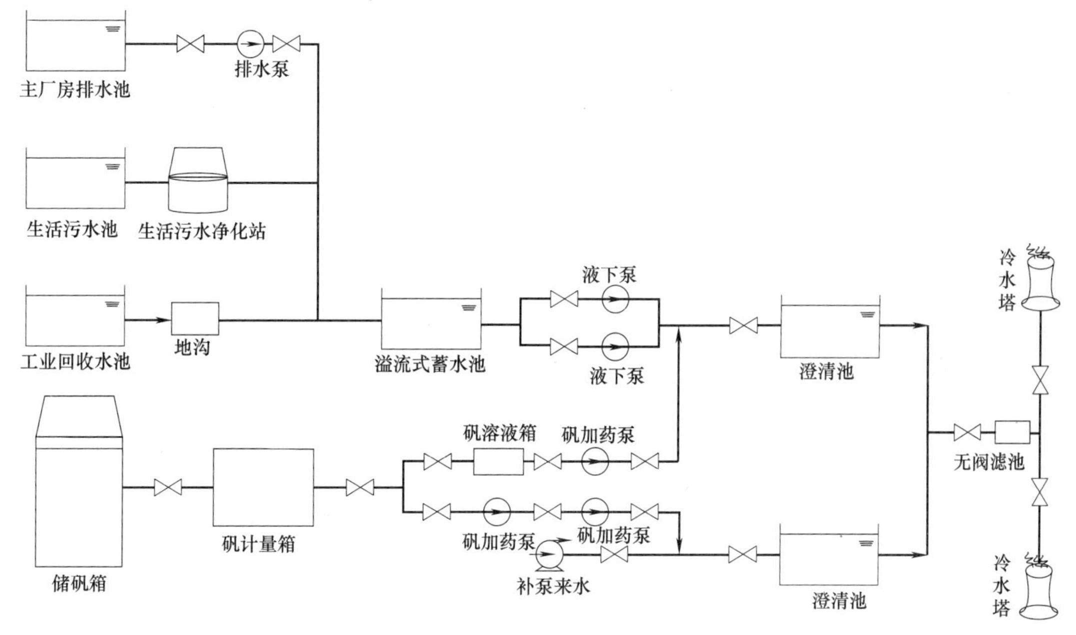

图 10-18　火电厂废水回收系统流程图

图 10-19 所示为某火电厂废水集中处理流程图。废水进入调节池，调节水量和水质，减少处理系统进水水质、水量的波动，废水处理池中的污水移入初始沉淀池以除去一些悬浮固体、油和化学需氧量，油罐流出物流入生物接触氧化池，浮游生物和污泥进入污泥池，然后进入冲灰处理系统。二氧化氯被添加到中间池中以除去大部分菌类和一些有机物质。经上述处理废水可以达到排放标准，循环补水的要求可以得到满足，实现废水资源化利用，达到节约用水、减少废水排放、提高环境效益和经济效益的作用。在燃煤电厂中采用的废水回收的方法比较多，但并不是所有的方法都适用于每个电厂，要根据电厂规模的大小、机组大小衡量，选择合适的方法。

1. 含油废水处理工艺案例分析

燃煤电厂废水也包括含油废水，指电厂油系统中的废水，油系统包括储油的设备、输油的系统等。那么这些油系统产生的废水就包括储油设备的排污，还有冲洗这些储油设备的废水。燃煤电厂使用的燃料油分为重油和轻柴油。轻柴油一般用于启动的时候点火，重油一般用来助燃；经常用到的乳化重油是重油、水、乳化剂的合成物；通过向油中加入乳化剂降低水的表面张力，当温度或其他环境条件改变时，这种油特别容易破乳，这样就会使油和水发生分离，因此，在乳化重油长期储存的过程中，油箱内会不断有水产生并且会沉积在储油罐的底部。这些积水就需要经常排除，这样就形成了油罐的排污水。重油罐排污水中含有特别多的重油，污染性极强，一般在储油场地设置专门的含油废水收集、处理系统，将大部分油污清除后再将废水排入厂区公用的排水系统。

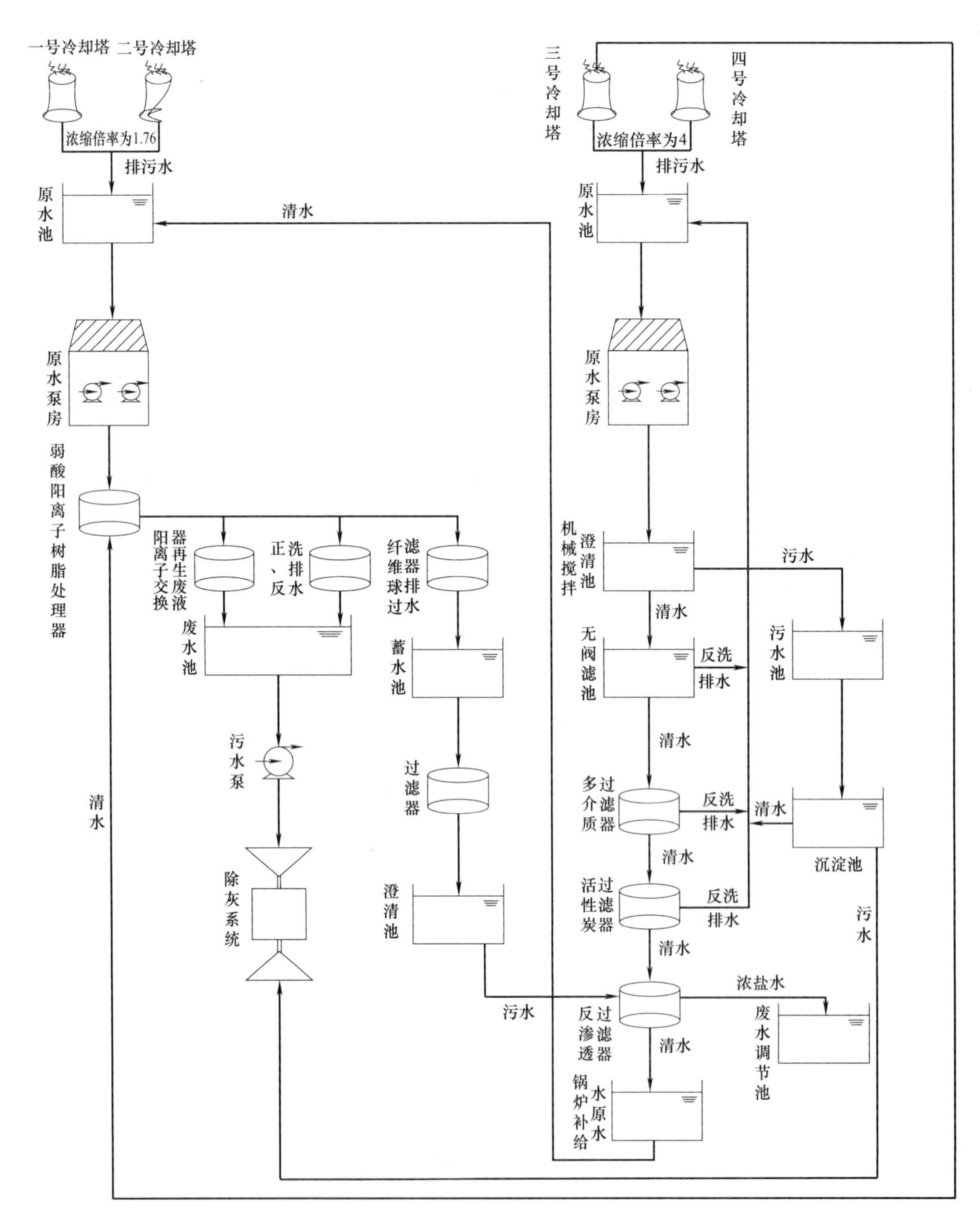

图 10-19 火电厂废水集中处理流程图

某电厂处理含油废水使用的是高效气浮法（图 10-20），其原理是设法使水中产生大量的微气泡，以形成水、气及被去除物质的三相混合体，在界面张力、气泡上升浮力和静水压力差等多种力的共同作用下，促进微细气泡黏附在被去除的微小油滴上后，因黏合体密度小于水而上浮到水面，从而使水中油粒被分离去除。表 10-7 是气浮改造前后水质及污染物去除率数据。

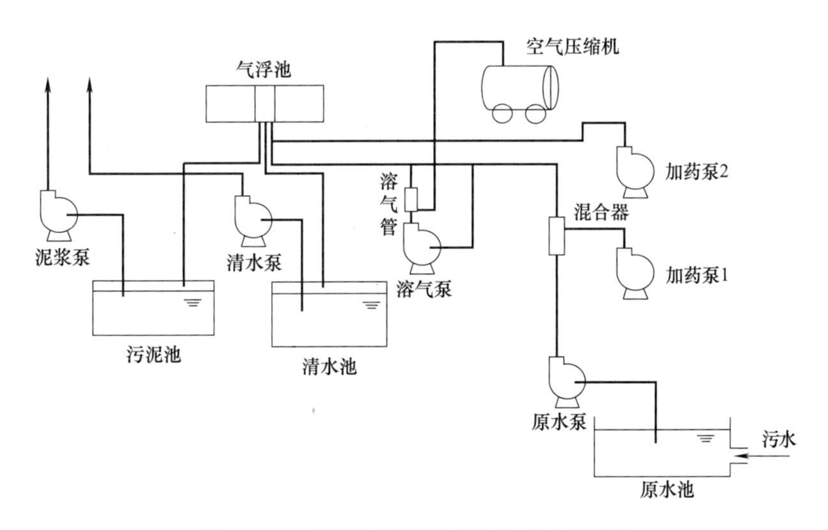

图 10-20　高效气浮法处理含油废水的流程

气浮改造前后水质及污染物去除率数据　　　　**表 10-7**

名称	石油类(mg/L)		挥发酚(mg/L)		COD(mg/L)	
	改造前	改造后	改造前	改造后	改造前	改造后
一浮进水水质	63	13.13	64	24.69	1198	504.98
一浮出水水质	48	12.65	51.96	29.08	951	481.09
二浮出水水质	43	12.29	68	31.01	1311	481.98
去除率	35%	37.6%	0	7.74%	0	2.72%

由表 10-7 的数据可以看出经过改造之后水质发生了明显的变化，一浮进水的水质还很不好，但是改造后水中的污染物明显减少了很多，各种污染物的去除率也提高了。表 10-8 是 pH 对除油效果的影响，随着 pH 的升高含油去除率越来越低，浮渣产量也越来越低。表 10-9 是溶气压力对除油效果的影响，可以看出溶气压力越大，含油去除率越高，浮渣产量也越多。气浮法是一种新型的水处理方法，不需要加入任何的药品，而且除完杂质之后浮渣也不多，废水对于设备的腐蚀也不强，设备的占地面积不大，并且该方法充分发挥了溶气气浮的优点，将溶气气浮与旋流和过滤技术有机地结合在一起，不使用化学药剂，减少浮渣产生量，提高过滤器的使用效率，提高工艺的适应性，大幅降低含油污水处理的

pH 对除油效果的影响　　　　**表 10-8**

pH	进水含油量(mg/L)	出水含油量(mg/L)	含油去除率(%)	浮渣产量[L/(m^3·原水)]
3.2	148	11	91.8	0.094
4.4	151	13	90.7	0.091
5.5	154	18	87.3	0.086
6.4	155	19	86.6	0.080
7.4	155	30	79.9	0.076
8.4	152	32	78.3	0.074

运行成本，同时具有清洁生产工艺特点，将成为新型含油污水处理技术。综合电厂的经济效益来看，这种方法可以广泛地运用。

溶气压力对除油效果的影响 **表 10-9**

溶气压力（MPa）	进水含油量（mg/L）	出水含油量（mg/L）	含油去除率（%）	浮渣产量［L/(m^3·原水)］
0.1	174	68	60.5	0.040
0.2	168	31	79.9	0.060
0.3	164	22	85.9	0.079
0.4	162	20	86.9	0.082

2. 煤炭废水处理工艺案例分析

北方某火电厂堆煤场的占地面积约为 10.5 万 m^2，其中堆煤区占地 7.2 万 m^2。该火电厂堆煤场废水来源于堆煤场喷淋水、输煤栈桥冲洗水、地面冲洗水和煤场雨水等，其中煤场雨水是废水的最主要部分。火电厂所产生的含煤废水统一排入厂内容积约为 2800m^3 的贮存池。按 1d 内将该池废水处理完，则该含煤废水处理能力为 100m^3/h。经过现场调查和废水取样分析，煤场废水的主要污染物为悬浮物（SS）和 COD，其中 COD 随 SS 的大小变化明显，沉淀后 SS 和 COD 均大幅降低，说明废水中 COD 主要来源于废水中煤粉的氧化过程，溶解性有机物较少。因此，煤场废水处理作用主要是降低废水中的悬浮物。根据现场取样检测及参照同类电厂含煤废水水质数据，含煤废水经过贮存池初步沉淀后水中 SS 小于 500mg/L。

含煤废水处理过后主要用于煤场喷淋和栈桥清洗，其水质应该达到《城市污水再生利用 城市杂用水水质》GB/T 18920—2020 的城市绿化用水的标准，经过综合考虑，该工程实例中含煤废水处理后的设计出水水质指标见表 10-10。

含煤废水处理后的设计出水水质指标 **表 10-10**

项目	pH	色度	溶解性总固体	BOD_5	NH_4-N	浊度	溶解氧	游离余氯	总大肠菌群
数值	6～9	≤15	≤1000mg/L	≤10mg/L	≤10mg/L	≤5NTU	≥2.0mg/L	管网末端水 ≥0.2mg/L	无

煤场废水处理工艺流程如图 10-21 所示。煤场含煤废水先通过沉淀池进行沉淀处理，水中大部分颗粒物得到去除，出水通过提升泵送往水力循环澄清池，同时从管道混合器中加入絮凝剂和助凝剂，使水中细小悬浮物凝聚变大，加快颗粒沉淀速率。水力循环澄清池上清液自流进入砂滤池，在砂滤池中填料截留吸附作用下去除水中剩余悬浮物及金属离子，砂滤池出水自流进入回用水池，通过回用水泵，将水送往煤场利用。水力循环澄清池产生污泥通过重力排泥管排往前段沉淀池进行沉淀处理，当沉淀池中废水处理完后，由铲车清理将泥煤送往煤场进行干化回用。

煤场废水处理主要设备及构筑物包括沉淀池、水力循环澄清池及加药系统、砂滤池、回用水池及回用水泵。

（1）沉淀池

沉淀池设 1 座，分成 2 格，钢筋混凝土地下式结构，进水口设 2 套手动闸阀，运行时交替使用。沉淀池末端设提升泵，将澄清后的含煤废水进行进一步处理。

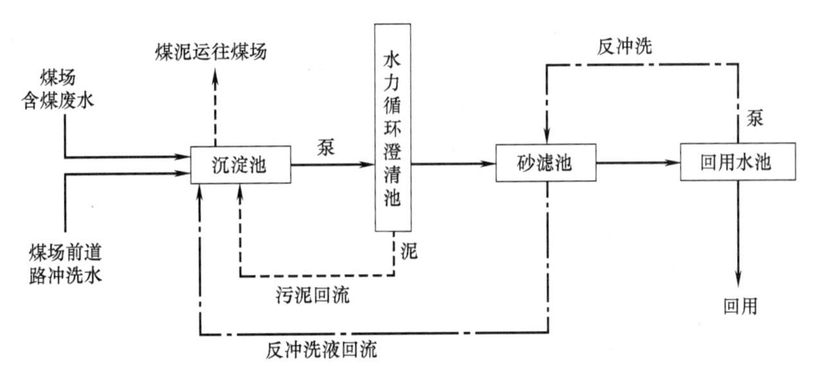

图 10-21　煤场废水处理工艺流程图

沉淀池外形尺寸为 24.0m×22.0m×3.5m，有效容积为 $1200m^3$，有效水深为 3.0m；进口闸门规格为 0.6m×0.6m。设污水提升泵 2 台，提升泵电机功率为 15kW，泵流量为 $130m^3/h$，扬程为 20m。污水泵的启停采用浮球液位控制，现场手动控制。

沉淀池出水管道上设置管式混合器 1 台，目的是将 PAC/PAM 与废水进行充分混合。混合器尺寸为 ϕ300mm×2000mm。

（2）水力循环澄清池及加药系统

水力循环澄清池作为煤场含煤废水处理核心工艺。设计 1 座钢筋混凝土水力循环澄清池，处理能力为 $100m^3/h$。水力循环澄清池进水悬浮物浓度宜小于 2000mg/L，短时间内允许达到 5000mg/L；水力循环澄清池尺寸为 ϕ8.4m×6.8m，其中清水区高度为 2m，超高为 0.3m，水力停留时间为 1.5h；

加药系统主要功能是配制贮存 PAC、PAM。箱体尺寸为 ϕ6.4m×5.5m，采用机械式搅拌机，功率为 0.75kW，转速为 104r/min，采用 2 套计量加药泵，流量为 500L/h，功率为 0.37kW。

（3）砂滤池

设置砂滤池的主要目的是过滤吸附去除水力循环澄清池出水中夹带的细小悬浮物。外形尺寸为 10.5m×2.5m×2.0m，过滤速度为 15.0m/h。

（4）回用水池及回用水泵

回用水池主要作用是贮存系统处理达标的出水，用于砂滤池反冲洗或煤场回用。外形尺寸为 20.0m×5.00m×3.50m，有效容积为 $300m^3$。回用水泵除将回用水池清水回用于煤场之外，还可以作为滤池反冲水泵。回用水泵流量为 $100m^3/h$，扬程为 30m，电机功率为 15kW，数量为 2 台。回用水泵启停可采用浮球液位控制，现场手动控制。

3. 脱硫废水处理工艺案例分析

烟气脱硫是目前世界上燃煤电厂主要的脱硫方式，是控制二氧化硫污染的主要技术手段。脱硫废水是烟气湿法脱硫过程中从吸收塔系统中排放的废水，主要来自于石膏脱水和清洗系统，水力旋流器的溢流水及皮带脱水机的滤液。脱硫废水中主要的污染物是悬浮物、过饱和的亚硫酸盐、硫酸盐以及重金属，其中很多是国家环保标准中要求控制的一类污染物。

案例一：某电厂脱硫废水处理的工艺流程如图 10-22 所示。脱硫废水含有石膏浆液废水、工艺冲洗废水、树脂再生废水、反渗透污水、循环冷却排污水。废水先经过 pH 中和箱利用氧化钙粉末进行调节，将 pH 调节至 9.5～11，此时部分重金属会形成氧化物沉淀下来，接着进入沉降箱用有机硫药瓶 TMT-15 进行沉降，这时铅、汞等重金属形成金属硫化物沉降，再进入絮凝箱里加入 $FeClSO_4$ 絮凝剂和助凝剂颗粒反应，这时就会形成氢氧化铁并将不易沉降的颗粒物粘连成大颗粒物，接着再进入澄清池当中，加入盐酸将 pH 调节至 6～9，上层清液就达到排放的标准可进行排放，另外澄清池底部的污泥接着进入污泥贮箱再到板框压滤机，滤液回到 pH 中和箱，下来的泥饼运走。

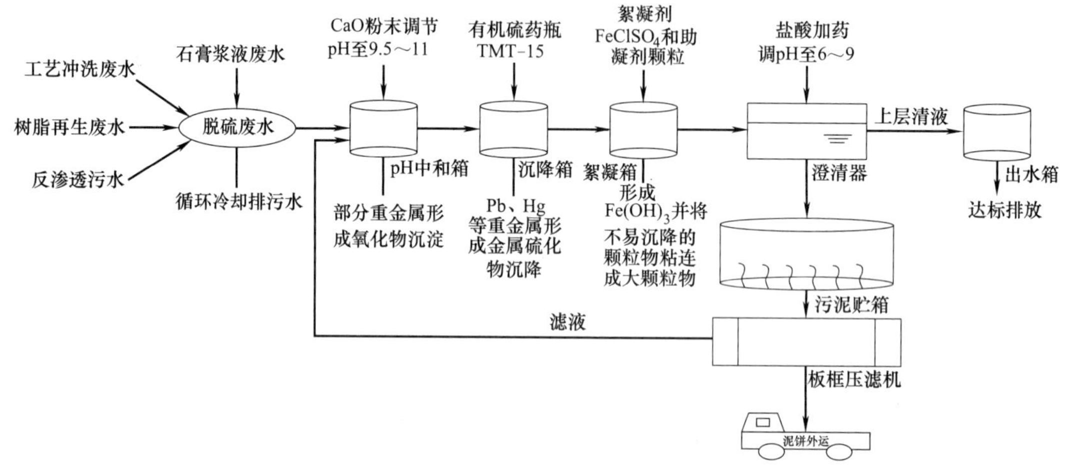

图 10-22 某电厂脱硫废水处理工艺流程图

案例二：某电厂使用烟道喷雾蒸发技术处理脱硫废水（图 10-23）。相对于其他脱硫废水处理技术来说，烟道喷雾蒸发技术的投资还有运行成本比较低，系统非常可靠，不需要额外的能量输入，系统所需要的能源全部来自空气预热器中抽取的少量热烟，对锅炉的热效率基本没有影响；系统设备占用的空间较小，投资运行的成本比其他技术要低的多。

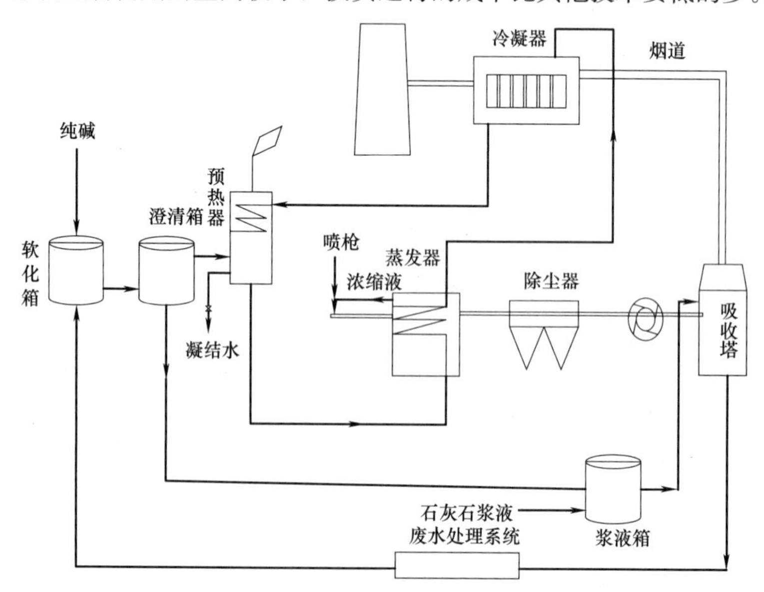

图 10-23 某电厂使用烟道喷雾蒸发技术处理脱硫废水工艺流程图

案例三：某电厂在脱硫废水处理方面一直尝试引用新型处理方法，反渗透预处理工艺以膜过滤为主，辅以杀菌工艺和沉淀工艺，目的是去除水中的悬浮物和微生物，使处理后的水质能够初步满足反渗透的进水要求。主体工艺通常采用两段反渗透系统，由于两段系统的进水为一段系统的浓水，需用专门的化学药剂对其进行处理，以确保两段系统的进水参数符合要求。同时在其进入两段系统前，可针对其水质情况，添加专业的阻垢剂和调节剂，确保系统稳定运行。产品水进入回用水池，系统中少量的浓水可用来冲渣，实现水处理系统的零排放。

该电厂脱硫废水零排放工艺流程如图 10-24 所示，排放的污水先经过混合池进行反应，再到沉淀池中进行沉淀，然后开始过滤，第一次过滤用的是全自动过滤器，然后再用活性炭过滤器过滤，过滤后的水排放到清水池中经过增压泵将水提升到精密过滤器中，然后再到全自动超滤装置，超滤后的水先存放到一个中间水罐中，加入阻垢剂和杀菌剂反应，在经过高压泵的作用下有一部分经过反冲洗泵再回到全自动超滤装置中，另外一部分则开始进入一段反渗透装置，接着再进入软化箱进行反应，然后再进入二段反渗透装置，处理后合格的水就可以回用，剩下的水也可用来冲渣。

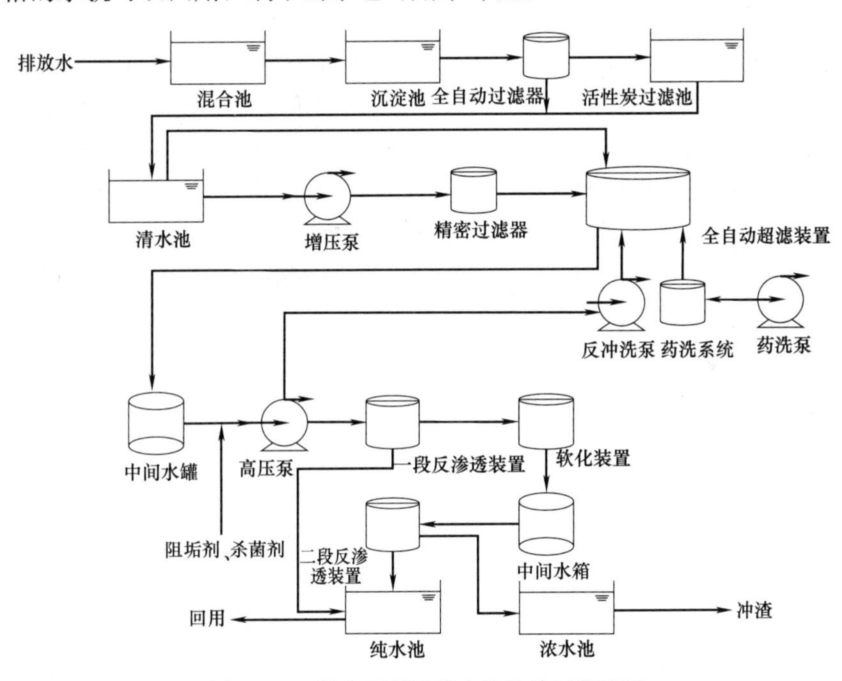

图 10-24　某电厂脱硫废水零排放工艺流程

案例四：某火电厂脱硫废水产生量为 $10m^3/h$，脱硫废水水质指标见表 10-11。

脱硫废水水质指标（除 pH 外单位：mg/L）　　表 10-11

项目	数值	项目	数值	项目	数值
氟化物	≤40	SS	≤12000	pH	5.0～6.0
COD	≤400	总铬	≤5	总镉	≤25
氨氮	≤15	Zn^{2+}	≤20	Cu^{2+}	≤20
总铅	≤5	总镍	≤6	总汞	≤4
Cl^-	15000～20000	SO_4^{2-}	4500	Ca^{2+}	1200
Mg^{2+}	4500	Fe^{3+}	600	K^+	140
Na^+	210	Sr^{2+}	20	NO_3^-	210

处理后出水要求达到《燃煤电厂石灰石-石膏湿法脱硫废水水质控制指标》DL/T 997—2020 排放标准。脱硫废水水质控制指标见表 10-12。

火电厂石灰石-石膏湿法脱硫废水水质控制指标（除 PH 外单位：mg/L）　表 10-12

项目	数值	项目	数值
总汞	0.05	SS	70
总锡	0.1	COD	150
总铬	1.5	氟化物	30
总砷	0.5	硫化物	1.0
总铅	1.0	pH	6～9
总镍	1.0	氨氮	25
总锌	2.0		

脱硫废水处理工艺流程如图 10-25 所示。将石灰浆配比成一定浓度通过加料管送入石灰浆制备箱。石灰浆根据废水的 pH、流量及石灰浆液浓度加入废水中。废水中的 pH 一般控制在 9.5±0.3，大部分重金属生成氢氧化物而沉淀，石灰乳中的钙离子与废水中的氟离子反应生成溶解度较小的氟化钙沉淀，与 As^{3+} 络合生成 $Ca_3(AsO_3)_2$ 等难溶物质。用 $Ca(OH)_2$ 作为中和剂，几乎可以使汞以外的重金属离子得到去除。中和池出水进入沉降池，在沉降池中加入有机硫（TMT-15），使其与水中剩余的 Pb^{2+}，Hg^{2+} 反应生成溶解度更小的金属硫化物而沉积下来。

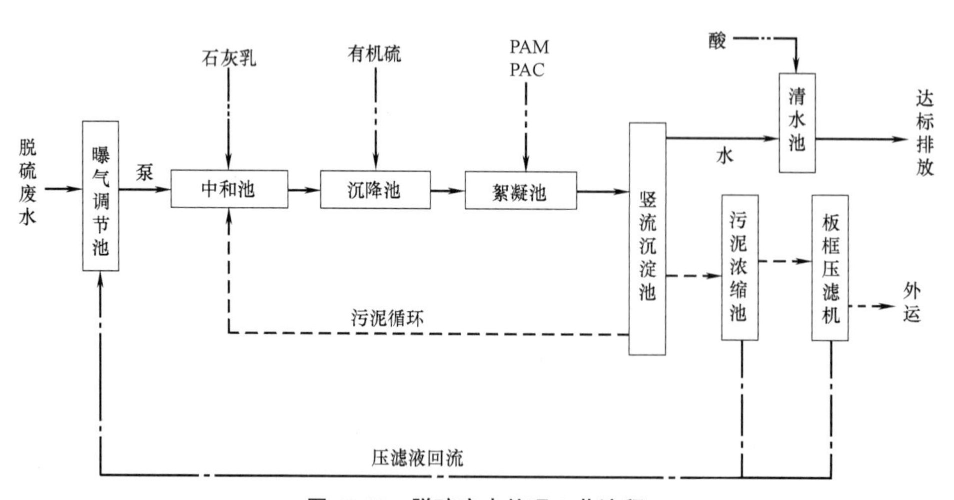

图 10-25　脱硫废水处理工艺流程

从废水中沉淀出来的氢氧化物和其他固形物、粒子都很细，分散在整个体系中难以沉降。在絮凝箱内加入 $FeClSO_4$，使水中的悬浮固体或胶体杂质凝聚成稍大的絮凝体，在絮凝箱出口处加入阳离子高分子聚合电解质作为助凝剂以降低颗粒的表面张力，强化颗粒的长大过程，进一步促进氢氧化物和硫化物的沉淀，使微细絮体慢慢变成更大、更易沉淀的絮状物，同时，也使脱硫废水中的悬浮物沉降下来。

废水自流进入沉淀池，絮凝体在沉淀池中与水分离。絮体因密度较大而沉积在底部，然后通过重力浓缩成污泥。大部分污泥经污泥输送泵输送到污泥脱水系统，小部分污泥作为接触污泥返回到中和池，提供沉淀所需的晶核。澄清池上部则为净水，净水通过澄清浓缩池周边的溢流口自流到出水箱，加盐酸将其 pH 调整到 6.0～9.0 后排放。

沉淀池底部的大部分污泥经浓缩后通过输送泵送到污泥脱水机，经压滤机脱水后，滤饼含固率为 45%左右，最后将滤饼运送到渣场贮存。污泥浓缩脱水的滤液进入回收池，由污水回收泵送往中和池内与新来的脱硫废水进入下一个处理循环。

（1）调节池：脱硫废水经收集后进入调节池，在调节池底部均匀布置曝气管，对脱硫废水进行充分曝气，鼓风曝气使未氧化的还原态硫化物进一步氧化，同时起到废水均化的作用。曝气调节池尺寸为 5m×4m×3.5m，有效容积为 $60m^3$，水力停留时间为 6h。池底均匀布置盘式曝气器，曝气器直径为 ϕ228mm，曝气量为 1.5～$4m^3/h$，服务面积为 0.3～$0.7m^2$/个，氧利用率为 25%～40%，充氧能力为 0.15～0.43kgO_2/h。罗茨鼓风机风量为 $8m^3/min$，2 台（1 用 1 备）。

（2）中和池：经过调节池的废水经过泵提升进入中和池，加入石灰的目的是使氧化后的硫酸盐转变为硫酸钙沉淀，调节废水的 pH，使废水的 pH 上升到 9.5 左右，使一些金属如铜、铁等形成氢氧化物沉淀。设有 pH 计，控制石灰的加入量。

中和池尺寸为 5m×2m×3.5m，有效容积为 $30m^3$，水力停留时间为 3h。中和池配备潜水搅拌机 2 台，电机功率为 2.2kW，叶轮直径为 320mm，转速为 740r/min。配套污泥泵 1 台，流量为 $2m^3/h$，扬程为 60m，电机功率为 1.5kW，转速为 960r/min。

（3）石灰消化池：主要目的是将生石灰粉消化制备为石灰石乳，数量 1 座，有效容积为 $20m^3$。石灰消化池配套搅拌机 1 台，电机功率为 2.2kW，叶轮直径为 320mm，转速为 740r/min，设计每天配药 1 次，加药量为 5%的消石灰，70～$80L/m^3$。石灰乳输送泵 2 台（1 用 1 备），石灰乳储存罐容积为 $2m^3$。

（4）沉降池：废水在中和沉淀处理后，进入沉降池，与加入的有机硫在搅拌机的作用下充分混合反应，使汞、铜、铅等重金属形成难溶的硫化物。沉降池尺寸为 5m×3m×3.5m，有效容积为 $40m^3$，水力停留时间为 4h。沉降池配备搅拌机 2 台，污泥泵 1 台，搅拌机和污泥泵型号规格同中和池。沉降池配备有机硫配药罐 1 个，溶液浓度为 15%，设计每天配药 1 次，加药量 $40mg/m^3$。

（5）絮凝池：废水经沉降处理后，进入絮凝池，在絮凝池中加入混凝剂和助凝剂进行絮凝反应，使废水中大量细小的悬浮物和溶解性无机离子形成大颗粒沉淀得到去除。絮凝池尺寸为 5m×3m×3.5m，有效容积为 $40m^3$，水力停留时间为 4h。配套搅拌机 2 台，污泥泵 1 台，聚合氯化铝、聚丙烯酰胺配药罐各 1 个。设计每天配药 2 次，聚合氯化铝溶液浓度为 40%，加药量为 40mg/L，聚丙烯酰胺加药量为 5mg/L，使用浓度为 0.1%。

（6）竖流式沉淀池：废水经絮凝反应进入竖流式沉淀池后，在重力沉降的作用下，絮体和水逐渐分离，使得上部形成澄清液，下部为污泥。污泥在静水压力的作用下浓缩后排出，一部分回流到中和池，提供沉淀物所需的晶核，有助于悬浮物的去除。另一部分进入污泥浓缩池，经板框压滤机后外运，滤液回流至中和池，竖流式沉淀池尺寸为 ϕ3m×6m，中心管直径为 0.7m，有效容积为 $30m^3$，水力停留时间为 3h。

（7）清水池：澄清液进入清水池，与加入的盐酸反应，使出水 pH 达到 6～9，达标排放。清水池尺寸为 2m×3m×3.5m，有效容积为 20m^3，水力停留时间为 2h。清水池设有潜水式搅拌机 1 台，使加入的盐酸快速起到中和作用，pH 计 1 台，控制出水 pH。

（8）污泥浓缩池：沉淀池池底污泥通过污泥泵提升至污泥浓缩池，对污泥进行重力浓缩，浓缩池尺寸为 3.5m×3.5m×6.0m，有效容积为 54m^3，污泥浓缩时间为 10h。

（9）板框压滤机：过滤速度快，耐高温及高压，密封性能好，滤饼洗涤均匀，含水率低。过滤压力为 0.4～0.6MPa，过滤周期不大于 4h；滤板外形尺寸为 420mm×420mm，滤板厚度为 50mm。

10.3 水处理装置观摩实验

1. 城市污水综合处理装置观摩实验

二维码 11

城市污水综合处理装置观摩实验

2. 火力发电厂补给水处理装置观摩实验

二维码 12

火力发电厂补给水处理装置观摩实验

参考文献

[1] 田禹，王树涛. 水污染控制工程［M］. 北京：化学工业出版社，2010.

[2] 戴友芝，肖利平，唐收印. 废水处理工程［M］. 北京：化学工业出版社，2017.

[3] 吴向阳，李潜，赵金如. 水污染控制工程及设备［M］. 北京：中国环境出版社，2015.

[4] 蒋克彬. 污水处理工艺与应用［M］. 北京：化学工业出版社，2014.

[5] 崔迎. 水污染控制技术［M］. 北京：化学工业出版社，2015.

[6] 李东升. 污水处理综合实训教程［M］. 北京：化学工业出版社，2009.

[7] 杨岳平，徐新华，刘传富. 废水处理工程及案例分析［M］. 北京：化学工业出版社，2003.

[8] 王春荣. 水污染控制工程课程设计及毕业设计［M］. 北京：化学工业出版社，2013.

[9] 彭党聪. 水污染控制工程课程实践教程［M］. 北京：化学工业出版社，2010.

[10] 徐新阳，于锋. 污水处理工程设计及毕业设计［M］. 北京：化学工业出版社，2003.

[11] 陈泽堂. 水污染控制工程实验［M］. 北京：化学工业出版社，2002.

[12] 沈晓南. 污水处理厂运行和管理问答［M］. 北京：化学工业出版社，2012.

[13] SIMON J，CLARIE J. 膜生物反应器——水和污水处理的原理和应用（原著第二版）［M］. 北京：科学出版社，2012.

[14] DAVIS M L，MASTON S J. 环境科学与工程原理（第二版）［M］. 王建龙，译. 北京：清华大学出版社，2008.

[15] GERARD K. Environmental engineering［M］. New York：Macgraw-Hill，1996.

[16] WESLEY W E. 工业水污染控制（第三版）［M］. 北京：化学工业出版社，2001.

[17] METCALF，EDDY. Wastewater Engineering Treatment and Reuse［M］. 北京：清华大学出版社，2003.

[18] ASANO T，BURTON F L，LEVERENZ H L，TSUCHIHASHI R，TCHOBANOGLOUS G. Water Reuse Issues，Technologies，and Applications，Ⅰ Ⅱ［M］. 北京：清华大学出版社，2008.

[19] 高廷耀，顾国维，周琪. 水污染控制工程下册（第四版）［M］. 北京：高等教育出版社，2017.

[20] 张志刚. 给水排水工程专业课程设计［M］. 北京：化学工业出版社，2004.

[21] 罗固源. 水污染物化控制原理与技术［M］. 北京：化学工业出版社，2003.

[22] 蒋展鹏. 环境工程学（第三版）［M］. 北京：高等教育出版社，2013.

[23] 王博涛. 水污染控制工程设计指导手册［M］. 北京：科学出版社，2017.

[24] 刘咏. 水污染控制工程课程设计案例与指导［M］. 成都：四川大学出版社，2016.

[25] 王春荣. 水污染控制工程课程设计及毕业设计［M］. 北京：化学工业出版社，2013.

[26] 崔玉川. 城市污水回用深度处理设施设计计算［M］. 北京：北京化学工业出版社，2016.

[27] 中华人民共和国住房和城乡建设部. 室外排水设计规范：GB 50014—2021［S］. 北京：中国计划出版社，2021.

[28] 中华人民共和国住房和城乡建设部. 拦污用栅条式格栅：CJ/T 509—2016［S］. 北京：中国建筑工业出版社，2016.

[29] 中华人民共和国国家市场监督管理总局，国家标准化管理委员会. 给水排水用格栅除污机通用技术条件：GB/T 37565—2019［S］. 北京：中国建筑工业出版社，2019.

[30] 中华人民共和国环境保护部. 生物滤池法污水处理工程技术规范：HJ 2014—2012［S］. 北京：中国环境科学出版社，2012.

[31] 中华人民共和国环境保护部. 水解酸化反应器污水处理工程技术规范：HJ 2047—2015［S］. 北

京：中国环境科学出版社，2015.
[32] 中华人民共和国环境保护部．污水气浮工程技术规范：HJ 2007—2010 [S]．北京：中国环境科学出版社，2010.
[33] 中华人民共和国环境保护部．升流式厌氧污泥床反应器污水处理工程技术规范：HJ 2013—2012 [S]．北京：中国环境科学出版社，2012.
[34] 中华人民共和国环境保护部．序批式活性污泥法污水处理工程技术规范：HJ 577—2010 [S]．北京：中国环境科学出版社，2010.
[35] 中华人民共和国环境保护部．生物接触氧化法污水处理工程技术规范：HJ 2009—2011 [S]．北京：中国环境科学出版社，2011.
[36] 中华人民共和国国家质量监督检验检疫总局，国家标准化管理委员会．膜生物反应器通用技术规范：GB/T 33898—2017 [S]．北京：中国标准出版社，2017.
[37] 中华人民共和国环境保护部．环境保护产品技术要求　膜生物反应器：HJ 2527—2012 [S]．北京：中国环境科学出版社，2012.
[38] 刘加强，郃传民，陈翠．某城镇污水处理厂二氧化氯消毒的设计计算 [J]．盐城工学院学报（自然科学版），2016，29（02）：45-48.
[39] 中华人民共和国住房和城乡建设部．污水自然处理工程技术规程：CJJ/T 54—2017 [S]．北京：中国建筑工业出版社，1994.
[40] 中国工程建设标准化协会．一体化生物转盘污水处理装置技术规程：CECS 375—2014 [S]．北京：中国计划出版社，2014.
[41] 中华人民共和国国家质量监督检验检疫总局，国家标准化管理委员会．二氧化氯消毒剂发生器安全与卫生标准：GB 28931—2012 [S]．北京：中国标准出版社，2013.
[42] 中华人民共和国环境保护部．厌氧-缺氧-好氧活性污泥法污水处理工程技术规范：HJ 576—2010 [S]．北京：中国环境科学出版社，2010.
[43] 余淦申，郭茂新等．工业废水处理及再生利用 [M]．北京：化学工业出版社，2013.
[44] 何圣冰．城市污水处理厂工程设计指导 [M]．北京：中国建筑工业出版社，2016.
[45] 杨晓惠，冯春宇，魏纳．固液分离原理与工业水处理装置 [M]．成都：电子科技大学出版社，2017.
[46] 张尊举，伦海波，张仁志．水污染控制案例教程 [M]．北京：化学工业出版社，2014.
[47] 王光裕．有机废水处理的基本设计与计算 [M]．北京：化学工业出版社，2016.
[48] 张统．污水处理工程方案设计 [M]．北京：中国建筑工业出版社，2017.
[49] 陈广飞．工业污水及渗滤液处理技术 [M]．北京：化学工业出版社，2015.
[50] 王阳．分散式污水处理技术与应用 [M]．北京：中国环境科学出版社，2016.
[51] 刘咏．水污染控制工程课程设计案例与指导 [M]．成都：四川大学出版社，2016.
[52] 中华人民共和国环境保护部，财政部．全国农村环境综合整治“十三五”规划 [EB/OL]．(2017-02-24) [2022-01-15]．http：//www.hbzhan.com/news/detail/115122.html.
[53] 国家发展改革委、住房城乡建设部．“十三五”全国城镇污水处理及再生利用设施建设规划（发改环资〔2016〕2849 号）[EB/OL]．(2016-12-31) [2022-03-10]．https：//www.ndrc.gov.cn/xxgk/zcfb/tz/201701/t20170122_962884.html? code=&state=123.
[54] 王腾飞．河北省农村生活污水处理技术优选体系的研究 [D]．石家庄：河北科技大学．2018.
[55] 谢雄，付军，张大为，徐慧．浅析农村分散污水处理的几种适用性技术 [J]．广东化工，2017，44（19）：125-127.
[56] 刘洪涛，韩长胜．一个农村生活污水处理的典型案例 [J]．水科学与工程技术，2016（01）：6-8.
[57] 罗轶．A^2/O 工艺在旅游度假村生活污水处理中的应用 [J]．环境与发展，2018，30（03）：

86-87.
[58] 黄伯平，李晓慧. 南京市江心洲农村污水分散处理技术及应用 [J]. 中国给水排水，2017，33 (06)：102-105.
[59] 中华人民共和国生态环境部. 2015 年环境统计年报 [EB/OL]. (2017-02-23) [2022-04-12]. http://www.mep.gov.cn/gzfw_13107/hjtj/hjtjnb/.
[60] 刘翠. 废纸制浆造纸废水处理工艺研究 [D]. 郑州：郑州大学，2016.
[61] 陈华东. 羊皮制革企业生产废水处理工程设计 [J]. 中国给水排水，2016，32 (16)：99-102.
[62] 刘丽，常亮，肖杰，欧阳白薇，肖学权. 畜禽养殖废水处理工程设计 [J]. 工业用水与废水，2017，48 (06)：74-77.
[63] 段云霞. 石岩. 城市黑臭水体治理实用技术及案例分析 [M]. 天津：天津大学出版社，2018.
[64] 中华人民共和国住房和城乡建设部. 农村生活污水处理工程技术标准：CJJ/T 163—2011 [M]. 北京：中国建筑工业出版社，2011.
[65] 中华人民共和国住房和城乡建设部. 镇（乡）村排水工程技术规程：CJJ/124—2008 [M]. 北京：中国建筑工业出版社，2008.
[66] 中华人民共和国环境保护部. 纺织染整工业废水治理工程技术规范：HJ 471—2020 [M]. 北京：中国环境科学出版社，2020.
[67] 北京市市政工程设计研究总院. 给水排水设计手册（第三版，第五册，城镇排水）[M]. 北京：中国建筑工业出版社，2017.
[68] 中华人民共和国环境保护部. 酿造工业废水治理工程技术规范：HJ 575—2010 [M]. 北京：中国环境科学出版社，2010.
[69] 中华人民共和国环境保护部. 屠宰与肉类加工废水治理工程技术规范：HJ 2004—2010 [M]. 北京：中国环境科学出版社，2010.
[70] 中华人民共和国环境保护部. 制革及毛皮加工废水治理工程技术规范：HJ 2003—2010 [M].. 北京：中国环境科学出版社，2010.
[71] 中华人民共和国国家环境保护标准. 制浆造纸废水治理工程技术规范：HJ 2011—2012 [M]. 北京：中国环境科学出版社，2012.
[72] 成官文. 水污染控制工程 [M]. 北京：化学工业出版社，2017.
[73] 李长波. 水污染控制工程 [M]. 北京：中国石化出版社，2016.